AFTER THE GREENING

The Browning of Australia

AFTER THE GREENING

The Browning of Australia

MARY E. WHITE

Kangaroo Press

ACKNOWLEDGMENTS

I acknowledge with deep gratitude the assistance and encouragement I have received from the very large number of people who, by their contribution, enabled me to complete this project.

I would not have known the wide brown land that was the end product of **After the Greening** if it had not been for my travels in the arid, semi-arid and out-of-the-way places. I could not have made those journeys without the assistance of Neville Alley, who took me to the deserts of South Australia and the Flinders Ranges; of John Casey, who organised an outback tour of western parts of N.S.W. and Queensland, and The Centre; of David Roots of Academic Tours whose trip took me out west to Camerons Corner, the Strzelecki and the Northern Flinders: and of the Western Australian SGAP with whom I went wildflower hunting on the sandplains and in the Stirling Ranges.

My thanks go to Jim Frazier who generously gave of his time to photograph the fossil specimens which form the major palaeobotanical part of this book - no one can photograph fossils better; and to Mantis Wildlife for use of photographs by Jim Frazier, Densey Clyne and Glen Carruthers. Others who kindly supplied photographs are: Neville Alley and Mark Benbow of South Australia; Dick Burns, Jeanette Closs, Bob Hill and N.J. Davidson of Tasmania; Brian Senior, J.C. Noble, Colin Simpson and Cliff Ollier of the A.C.T.; Mike Gray, Sydney; and Elizabeth George, Ian Tapley, Kevin Coate, J.S. Beard, Steve McLoughlin, K.W. Dixon, A. Lowrie, and the Kings Park Herbarium of Western Australia.

The South Australian Museum and the South Australian Department of Mines and Energy kindly lent specimens of silcrete fossils, and the Australian Geological Survey Organisation lent specimens from Melville Island, for photographing by Jim Frazier.

The **References** in this book are an indication of the help given to me by colleagues from whom I requested information and reprints of their more recent scientific publications. I thank everyone who answered my letters, spoke to me on the phone and in person, sent me reprints and generously gave of their time and advice. The hospitality and friendship offered in the course of the researching of the book is gratefully acknowledged.

Three friends in particular read my manuscript and corrected my information and generally kept me on track. To them — Neville Alley, Cliff Ollier and Dick Evans — a special thank you.

The year I have spent in the making of **After the Greening** has been one of the happiest and most rewarding of my life, and it has been the interaction with so many interesting and generous people which has helped to make it so.

Finally, my thanks go to the designer of this book, Linda Robertshaw, and the artists who made the maps and diagrams, Elizabeth Seymour and Penny Maxwell. Together they have turned my concept into reality.

Mary E. White
21 January 1994

Opposite: Bratchina Gorge in the Flinders Ranges. M.E.W.

White, M.E. (Mary E.).
After the greening.

Bibliography.
Includes index.
ISBN 0 86417 585 X.

1. Paleobotany - Australia. 2. Paleoecology - Australia. 3. Paleogeography - Australia. 4. Botany - Australia. I. Title.

561.1994

Cover design and photography by Jim Frazier.
Designed by Linda Robertshaw

First published in 1994 by Kangaroo Press Pty Ltd
3 Whitehall Road Kenthurst NSW 2156 Australia
P.O. Box 6125 Dural Delivery Centre NSW 2158
Printed in Singapore by Kyodo Printing Co. (S'pore) Pte Ltd

ISBN 0 86417 585 X

CONTENTS

PART TWO:

DRIFTING

PART THREE:

DRYING

FROM 2.4 MILLION YEARS AGO TO TODAY

PART FOUR:
UNBALANCING THE BIOTA
200 YEARS OF CHANGE

PREFACE

Voyages of Exploration and Discovery

The days of heroic voyages of discovery, when small ships set out to look for the Great South Land and found that there was no vast landmass to balance that which surrounds the North Pole, are long gone. Most of the major voyages of exploration have been accomplished and the globe is mapped. The 'romance' of journeys across uncharted seas, using the Westerlies to find new and faster routes to the Spice Islands is our fantasy. It was hardship and danger, not romance, for those early voyagers, making us realise how comfortable and safe is our travel and what a breed of men were those early explorers.

A pirate, an explorer and a writer of best-sellers — a strange mixture one might think — was the first Englishman to write a popular account of a landfall in 'New Holland'. The quest for Indonesian spices and the westerly winds brought the *Cygnet*, commanded by John Read, into King Sound (on the Kimberley coast of Western Australia) in 1688, with William Dampier, buccaneer, adventurer and soon-to-be-published author, on board. The ex-pirate was not impressed by what he saw of Australia or its inhabitants. To eyes hungry for the green of England after months at sea; stomach hungry for fruit or green food to counteract scurvy, and longing for fresh meat; and an urgent need for fresh water for the ship, the arid land offered nothing. But the sense of achievement for the adventurer must have compensated. Imagine the wonder in having arrived somewhere so new and remote that 'it is not yet determined whether it is an Island or a main Continent; but I am certain that it joyns neither to Asia, Africa, nor America'!

Dampier's description of the Australia that he had seen in *A New Voyage Round the World* did much to dampen any enthusiasm that Britain might have had to extend its empire down-under. But the buccaneer's book was avidly read as the greatest adventure story of the times and the writer received so much acclaim that Dampier was given command of HMS *Roebuck* in 1699 and repeated his voyage. This time his exploration of the Western Australian coast was south of the Dampier Peninsula and gave the name of his ship to Roebuck Bay. There, at the edge of the Great Sandy Desert his first unfavourable impressions of the land and its people were confirmed. His account of this visit is given in *A Voyage to New Holland in the Year 1699.*

William Dampier described the part of the Western Australian coast (in King Sound) which he visited in January 1688 graphically. 'This part of it that we saw is all low even Land, with sandy Banks against the Sea, only the points are rocky, and so are some of the Islands in this Bay. The Land is of a dry sandy soil, destitute of Water, except you make Wells: yet producing divers sorts of Trees: but the Woods are not thick, nor the Trees very big....There was pretty long Grass growing under the Trees; but it was very thin.' The red gum exuding from the trees he compared to 'Gum Dragon, or Dragon's Blood', a resin that was carried aboard ship.

The Aboriginals who lived in the area are recorded as tending their fish traps and apparently subsisting entirely on what these supplied, and a few shellfish, as there 'is neither Herb, Root, Pulse, nor any sort of Grain, for them to eat, that we saw: nor any sort of Bird, or Beast that they can catch, having no Instruments where withal to do so.' They attended their fish traps at every low tide whether that fell in the night hours or daytime. Dampier saw the inhabitants of that harsh country as 'the miserablest people in the world', comparing them unfavourably with 'The Hodmadods of Monomatapa, though a nasty people, yet for Wealth are Gentlemen to these; who have no Houses and Skin Garments, Sheep, Poultry....as the Hodmadods have.'

Small wonder that Britain did not immediately rush

to claim New Holland!

When James Cook made his first voyage to Australia in 1770 he wrote that the 'Eastern side is not that barren and miserable country that Dampier and others have described the Western side to be.' When settlement there started, the new arrivals could not wait to try to change the 'foreign' Australian environment into one as much like England as possible.

While the modern world is charted and mapped and heroic voyages of exploration and discovery are no longer there to be attempted, space exploration offers equivalent new challenges to the new Technological Man. But down-to-Earth 'voyages' of a different kind have emerged as a result of scientific discoveries, and exploring the prehistoric world and finding out how the Present has evolved out of the Past encompasses journeys of this kind.

To find out what went into the making of the dry and inhospitable land that William Dampier walked upon 300 years ago and what has been the history of Australia that results in its unique characteristics, come with me on a voyage of discovery back into geological time and through the stages that made our continent the way it is today.

LOCALITY MAP

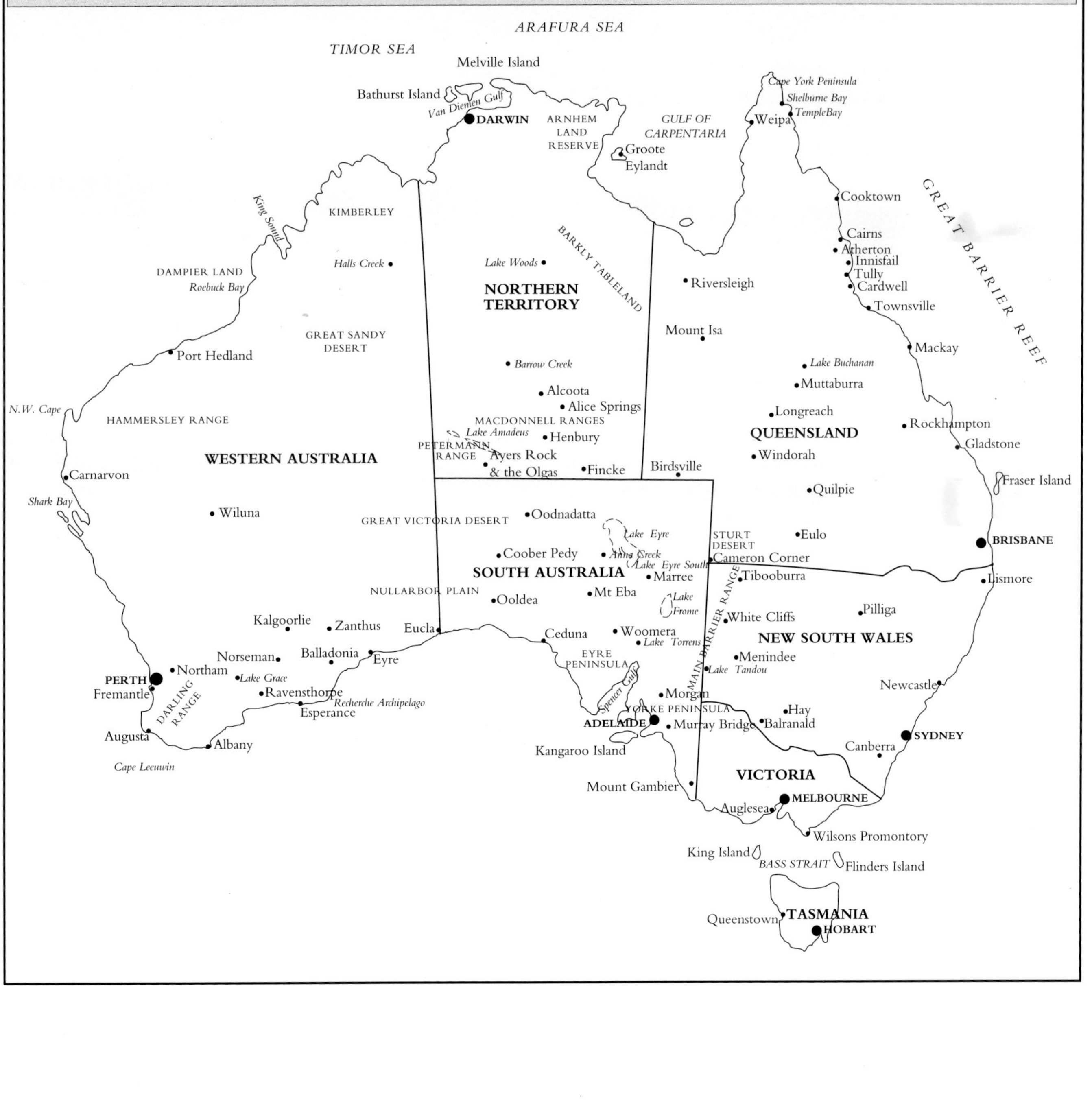

Wet tropical rainforest in northern Queensland, a remnant of the original Gondwanan forest that occupied much of the continent when it finally separated from Antarctica.

Jim Frazier

Spinifex rings, mulga and red wind-blown sand of the Red Centre.

M.E.W.

Ripple patterns in a sand dune east of William Creek.

N.F. Alley

INTRODUCTION

Rifting, Drifting and Drying

The making of the wide brown land

AUSTRALIA — The driest vegetated continent on Earth — a 'sunburnt country' in which more than two-thirds of the land is classed as arid, and half is desert. An island continent with a 'dead' heart that supports an amazing desert-adapted fauna and flora. A land of marsupials, fine-tuned to the vagaries of fickle climate — where floods and drought, food and famine alternate relentlessly, with lean years in between; and of eucalypt-dominated woodlands and forests; of mulga-lands and spinifex on glowing-red sands.

Australia's human population is mainly concentrated along its eastern coast and a narrow fringe inland, with scattered centres around its other margins. The outback is sparsely populated, and there the heartbreak of recurring drought is a fact of life for most of the primary producers — their plight little understood by the 88 per cent of the population that lives in cities.

This insular Australia, whose coasts are washed by three major oceans, is the youngest continent, yet it is composed of some of the oldest rocks known on Earth and has the longest fossil record of any land. In terms of geological time it has a short history as a separate entity. About 45 million years ago it severed its last connection with Antarctica, the keystone of Gondwana the great southern supercontinent, and started its independent journey northwards.

A salt crust rims an ephemeral stream in mulga land.

Jim Frazier

How was it long ago when this land lay in South-Polar regions? What was life like in such high latitudes, with a low-angle sun and mellow light part of the everyday scene and with the occasional beauty of the aurora australis? Was it the accustoming of its ancient faunas to long months of winter darkness that gave the ancestral marsupials of Australia-in-Gondwana eyes specialised for night vision and predestined the largely nocturnal habits of their modern descendants, or were they simply creatures of the night to escape the dinosaur-dominated dangers of days? What was the climate like in those distant times when Antarctica was vegetated and shared its forests of evolving flowering plants and ancient conifers with the lands which were to become Australia and New Zealand?

The processes by which a huge supercontinent fragments into separate continents, and the mechanisms by which lands move apart are on a scale almost beyond comprehension. That earth scientists can read the rocks and piece together the story of the separation of Australia from the lands which bordered it in Gondwana is our good fortune. We are privileged to be able to glimpse something of the amazing processes which have created our familiar modern land out of the very different world of the past. We are better able to understand the present when we know about the past, and we can be better equipped to plan for the future against the background of this knowledge.

The face of Australia has changed since it freed itself from other lands, and its landscapes and the Gondwanan cargo of plants and animals that were destined to travel with it altered progressively as the continent drifted northwards. The trials and tribulations for the Australian biota (its flora and fauna) on this odyssey through thirty degrees of latitude were the global changes in climate and the local environmental changes to which they had to adapt or perish as the distance from Antarctica increased through time.

From 45 million years ago the Southern Ocean steadily widened between Antarctica and the Island Continent Australia. A circum-Polar current was established, isolating Antarctica and leading to its eventual refrigeration and to the global ice age of the last 2.4 million years. How rapidly did world climates alter as lands were rearranged on the surface of the globe, and what were the consequences for the Australian flora and fauna?

The story of the genesis of our modern Australia and its evolution from the time when it started to emerge by the delineation of its margins within Gondwana to the present day encompasses 160 million years of geological time. It is a three-part saga of *rifting*, *drifting* and *drying*, like the gestation and birth, maturing and ageing of a living organism. The evolution of the independent continent through 45 million years unfolds as we establish the situation in time-slices down the ages. The journey has been from from a well-watered, green and forested land to an arid, sunburnt country — a progressive *browning*; from a Gondwanan biota to one that has evolved in isolation to be distinctly Australian and perfectly suited to the changing environment through time. We find that it is the ice age fluctuations of climate in the last 2.4 million years that have led to the central deserts and their marginal arid lands that now occupy two-thirds of the Australian continent.

The last 2.4 million years are of enormous significance as they represent the Age of Man as well as the time when the ice age had a profound effect on the physical landscapes, the climate and the biota of Australia. The Earth is now suffering ever more rapid and drastic environmental change as a result of human over-population that upsets the natural balance. If we are to be able to solve the environmental problems that face us today in Australia we need to understand what was the *status quo* of the land and its inhabitants before the arrival of humans in this continent. Then we must try to assess the impact of Aboriginal Australians on the time-worn land to which they came at least 60,000 years ago, before the last glacial stage of the ice age (and possibly as long as 140,000 years ago[20] — if the latest estimates based on the sudden great increase in fire frequency at that time are to be believed).

We need to understand the prehistoric story of this ancient land before we can appreciate that it is unique and that the problems that European settlement has created can only be solved with the wisdom gained from such an understanding. The question we have to ask, when we have established what the status of the continent was when humans first came to it, is: 'Are we turning the Island Continent into a desert island?'

The answer, at least up to this last decade, has unfortunately been in the affirmative. It seems that now the seriousness of the situation is at last beginning to be recognised and that some measures are being taken to reverse the process and to control the man-made element of environmental degradation.

Now, having introduced the theme of the book, it seems appropriate to introduce Australia — to view our time-worn continent from space and to see its basic features as cameras and remote sensing devices see them from satellites.

AN EYE-IN-THE-SKY VIEW OF AUSTRALIA [195, 196, 186, 367]

Australia today is under constant satellite surveillance. About ten non-military satellites are currently making regular passes across the continent, recording its features photographically and by many remote sensing devices from which a composite picture is compiled. For example, the Advanced Very High Resolution Radiometer (AVHRR) series of satellites of the National Oceanic and Atmospheric Administration (NOAA) produce images of the surface which reveal geological structures. The amount of heat reflected by the ground surface depends on the nature of the material that composes it and remote sensing equipment maps the different thermal signals given by different rocks and sediments.

Night-thermal imagery has been able to detect ancient drainage channels below the uniform cover of dune sands in the Canning Basin and other regions because they contain water and are cooler, allowing them to stand out in images. Remote sensing using radar can record heights on the terrain over which the satellite passes and topographical maps can be produced by piecing together all the digital information.

A topographical map of Australia made by remote sensing techniques shows many of the major features and regions which are discussed in this book. Though Australia is a flat land in that it has no really high mountains and few regions that approach mountain status, its surface is creased and wrinkled with varied low-relief topography overall. Only some of the basin areas appear smooth when such a map is side-illuminated to exaggerate minor differences in terrain heights. The ancient building blocks of the continent stand out as contrast to the basins.

A SATELLITE TOPOGRAPHICAL MAP OF AUSTRALIA

The relief of the continent, determined by radar, is exaggerated by illuminating the map from west to east. Map reproduced with kind permission of Director of the Australian Geological Survey Organisation (previously Bureau of Mineral Resources, Canberra).

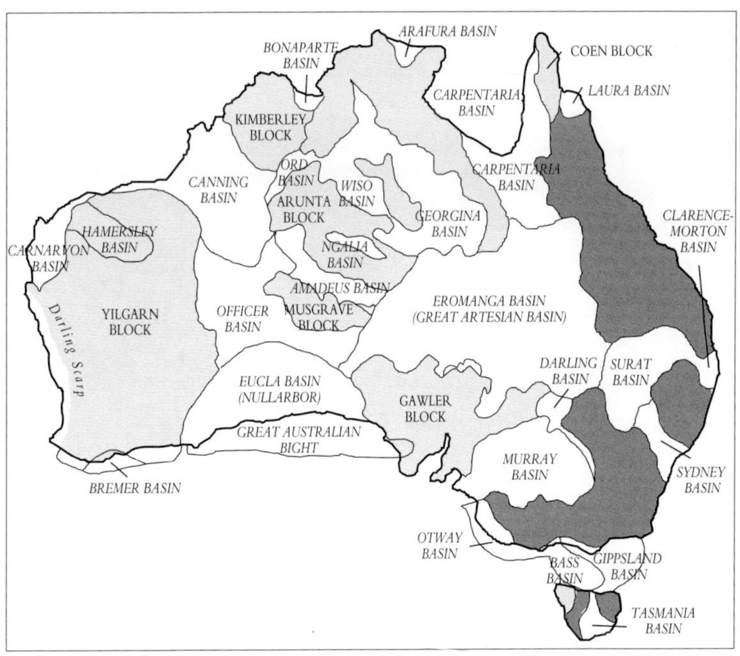

PART ONE

RIFTING

How Australia was Freed from Gondwana (From 160 to 45 Million Years Ago)

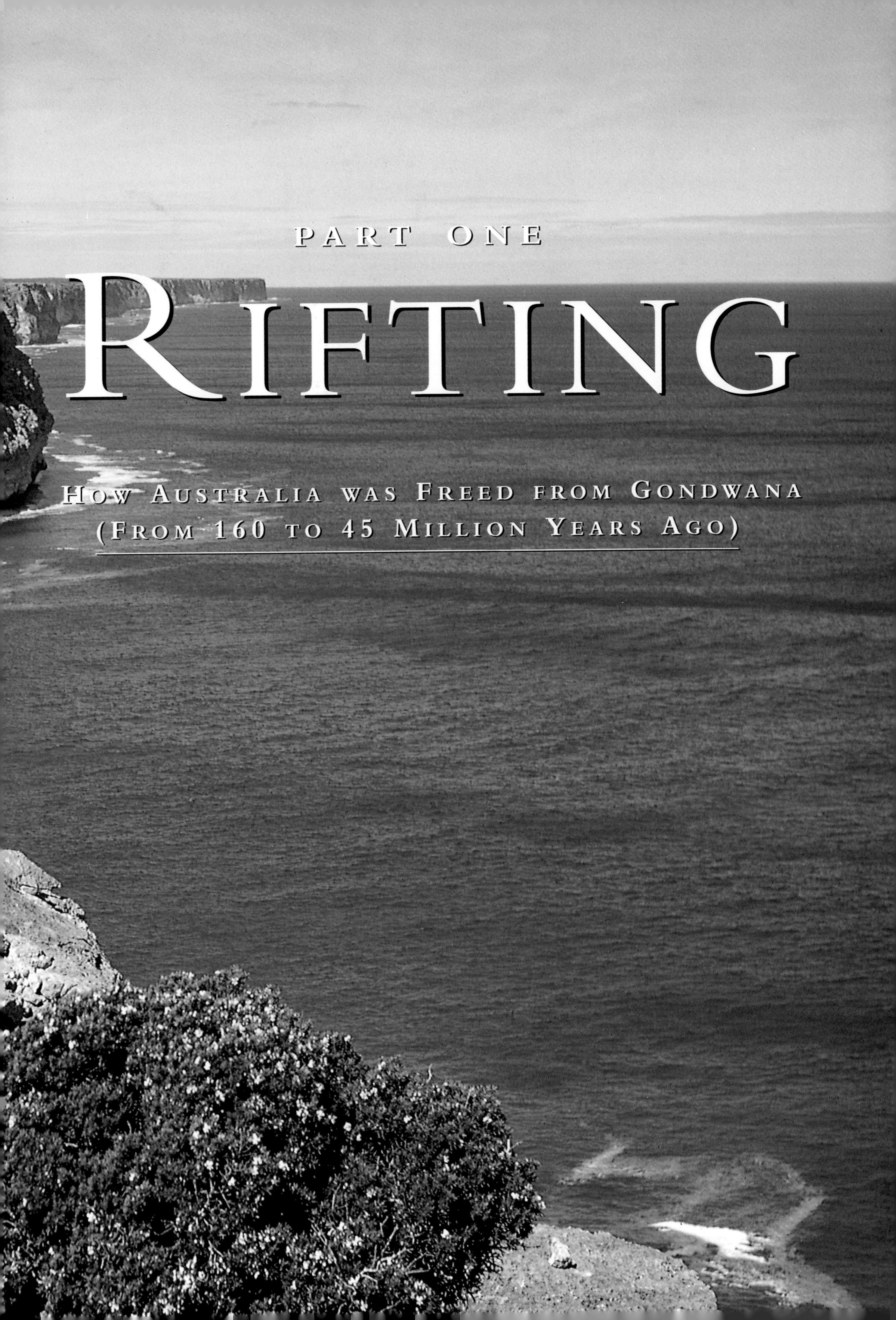

INTRODUCTION

THE MAKING AND BREAKING OF SUPERCONTINENTS

It is not easy to picture our world at the time of the great supercontinents. Consider Pangaea, the One Earth, and the one vast ocean, Panthalassa, during the time of dinosaurs when there was little to stop an extra-energetic individual from wending its way from the North Pole to the South Pole across land. Or imagine lesser journeys when a territory of a wide-ranging animal might include a bit of India, Antarctica and Australia all in its day's march! Is it any wonder then, now that it has been conclusively proved that supercontinents have existed in the past, that closely related plants and animals are found in lands that are now widely separated?

It was the shapes of modern continents that first alerted early geographers to the possibility that they had once been joined. Africa fits so neatly against South America when you remove the South Atlantic Ocean, and when other continents are brought together by eliminating the oceans which separate them an enormous jigsaw puzzle results with pieces whose edges fit.

It was the disjunct distribution patterns of modern plants that first brought scientific consideration to the idea of drifting continents. Sir Joseph Hooker was the botanist who accompanied James Ross on voyages of discovery in the Southern Ocean between

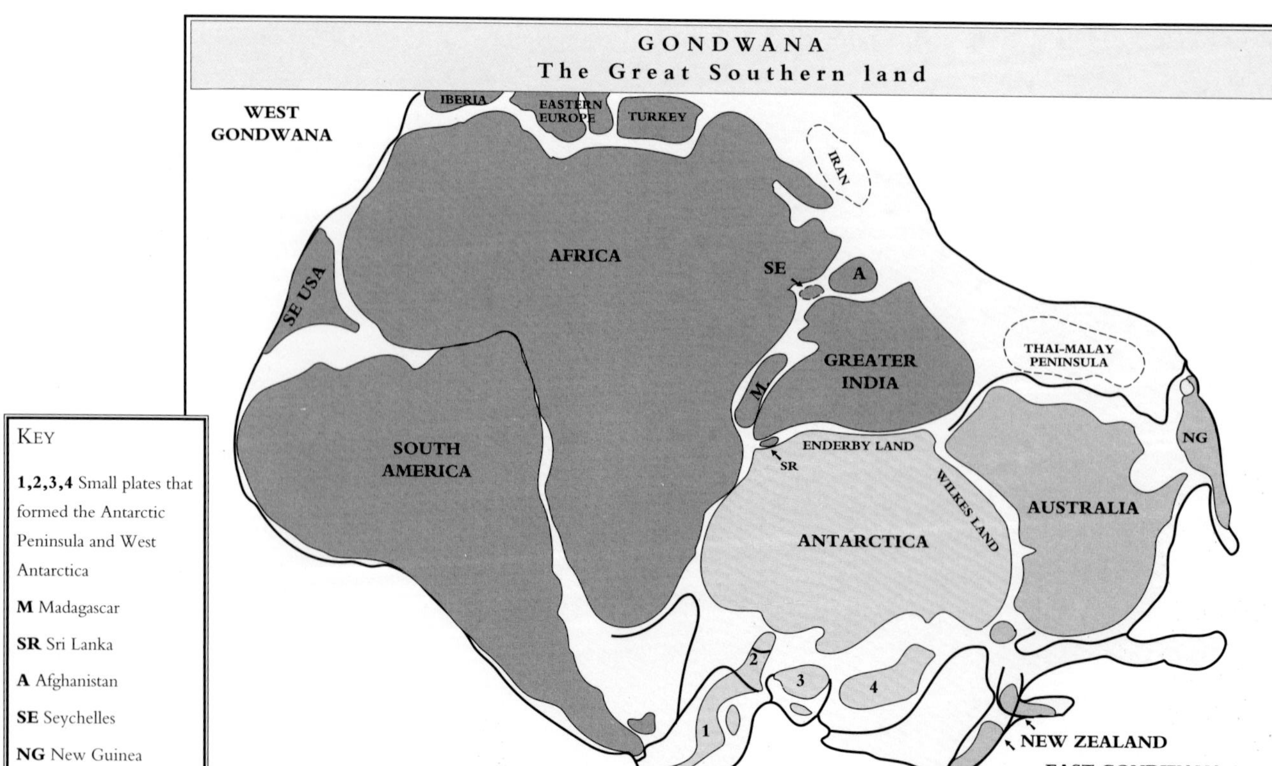

After de Wit *et al.* 371

1839 and 1843. He saw the forests of southern beech — *Nothofagus* — in southern South America, the sub-Antarctic islands, Tasmania and New Zealand; the distribution of southern conifers — the podocarps and araucarians; and the southern families of flowering plants, like the PROTEACEAE, with very closely related genera in the now widely separated lands. When he discussed the matter with Charles Darwin they agreed that such a distribution pattern implied a former land connection and that a vegetated Antarctica must have been involved.

In the years that followed, fossil floras were described in India where the Gondwana Series of rocks contained the *Glossopteris* flora associated with coal in a province where the native people were the Gondi. The name for the rock series with the fossils had been given for this reason. Gondwana means 'land of the Gonds'. Fossil floras in the southern lands where *Nothofagus* forests exist today were found to have the same *Glossopteris* flora in their coal-sequence rocks, indicating that India had had an ancient connection with southern lands even though it is now in the Northern Hemisphere. As more fossil faunas and floras and living plants and animals were described it became evident that a great number of disjunct distributions existed. It was also found that some of the geology on the edges of continents that were now far apart matched.

The Theory of Continental Drift was propounded to explain all the observed facts for which there was no other explanation. The name Gondwanaland was coined for the southern supercontinent, giving credit to the role of plant fossils in the formulation of the concept. (The accepted form for the name is now Gondwana.) The theory did not immediately gain universal acceptance because it was impossible to prove and the mechanisms by which bits of the Earth's crust might move in relation to each other could not be explained. It was only when technology had advanced to a point where the palaeomagnetism in rocks could be measured and other technologies enabled an understanding of the dynamics that acceptance came. Now the Plate Tectonic model, which explains the movement of continents, has replaced the Theory of Continental Drift.

The crust of the Earth consists of a number of plates which move. The continents ride on the plates of lithosphere (the rigid outer shell of the Earth). They are composed of continental crust made of SIAL, rich in silica and aluminium, which is lighter than the material which composes the plates and the oceanic crust. That material is known as SIMA because it is primarily silica and magnesium. Australia and India today share a plate — The Indian-Australian Plate — and have done so for the last 50 million years.

PLATES OF THE EARTH'S CRUST

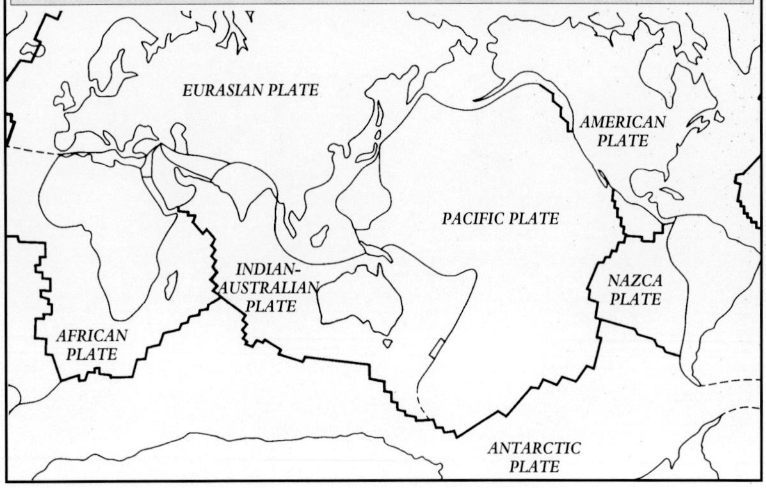

Continents move apart by a process that involves rifting, when fracturing of continental crust results in a rift valley developing between them, followed by production of sea-floor in the rift. Sea-floor spreading results from the production of basalt along mid-ocean ridges (which are like linear volcanoes). On-going eruption of basalt widens the sea-floor and moves the continents further apart. Sea-floor is recycled, being subducted in the deep-sea trenches in oceans. Oceans have oceanic crust beneath them, but seas can encroach on the land and form epicontinental seas when the sealevel is high or the land is lowered.

The zones where plates meet are active zones of the Earth's crust. Volcanic activity and earthquakes characterise them. For instance the 'ring of fire' that is the boundary between the Pacific Plate and the plates which surround the Pacific Basin is a highly active zone where sea-floor is being subducted and plate edges are riding over each other. New Zealand, on the outer edge of the Indian-Australian Plate, has active volcanoes, suffers earthquakes, and has on-going mountain-building on its Alpine Fault.

Where a continent is situated at the edge of the plate on which it rides, like India on the northern end of the plate it shares with Australia, collision with continental crust on

Above: Nothofagus cunninghamii, *the myrtle beech, with filmy ferns in cool temperate rainforest in Hellyer Gorge, Tasmania*

Dick Burns

Right: The Himalayas as seen here in Nepal, are the result of the collision of India with Asia. The lighter continental crust that was the leading edge part of the Greater India landmass when it seperated from Gondwana was crumpled up to form the mountains.

M.E.W.

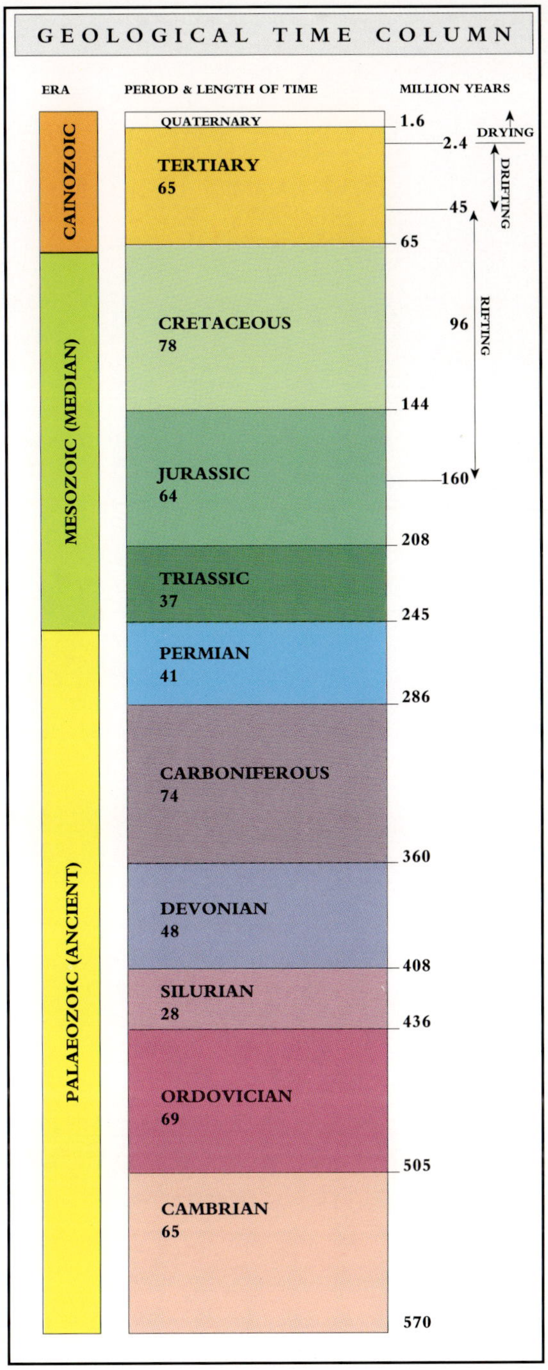

Deciduous southern beech, Nothofagus gunnii, *showing its autumn colour at Twisted Lakes, Tasmania with a dolerite massif in the background.*

DICK BURNS

Glossopteris indica *from the Lower Gondwana coal sequence in India, one of the fossil plants which occurs in Permian strata in lands which were once united in Gondwana.*

JIM FRAZIER

another plate results in mountain building. The continental crust is crumpled up, being lighter than the plate material on which it rides and therefore unable to be subducted. In the case of India, collision with Asia produced the Himalayas, and some of the plate material is subducted under the mountains.

Australia is fortunate in that it sits in the middle of its bit of plate and thus is not subject to the problems of active zones. However, the stresses and tensions within it are considerable as a result of the interaction that is going on with surrounding plates, particularly to the north where the plate is colliding with Southeast Asia.

The Plate Tectonic model for the movement[34] of lands on the surface of the globe applies to at least the last 1200 million[35] years of geological time, and it has been suggested that there have been three 400-million-year cycles of supercontinent building and fragmentation during this period. The arrangement of continents and oceans on the surface of the Earth before that time cannot be known with certainty. The modern world is still a dynamic one in which continents are moving in relation to each other, and in the case of Australia the northwards drift continues at between 6 and 7 centimetres a year, which represents several metres in an average lifetime.

THE 400–MILLION–YEAR CYCLES OF SUPERCONTINENT BUILDING AND FRAGMENTATION; AND OF ICEHOUSE TO GREENHOUSE CLIMATIC REGIME ALTERNATION[34]

A theory that a 400–million–year supercycle has operated, at least during the last 1200 million years of Earth history, has been proposed to explain the dynamics of supercontinent building and fracture. The driving force in each cycle is the differential heat distribution that results from the insulating effect of a very large area of land — Pangaea — that traps the heat of the convection currents in the molten zone of the Earth below its large area of crust. The land rises as the temperature increases below it and the Earth is said to be in a geocratic state — land-dominated. Sealevel in the one ocean — Panthalassa — is low in relation to the land.

While lands are united in Pangaea until the heat builds up beneath them, and while early spreading activity is slow, icehouse conditions prevail. When the heat builds up sufficiently the land starts to stretch and crack, ocean floor is generated between blocks and they move apart. During active sea-floor spreading carbon dioxide is progressively released from the Earth's mantle, resulting in greenhouse climatic regimes. (The mantle is the zone between the fiery core of the Earth and the crust.) A progressive loss of heat results from the sagging and rifting that occurs as the sea-floor spreads, and rapid opening of oceans further depletes the heat store. Thus spreading gradually slows and oceans begin to close by subduction, bringing the lands back together again in another Pangaea.

The cycle can be read in the rocks of the Earth. The sequence of sedimentary rocks reflects the stage of heat loss because sagging of the crust and formation of basin structures follows loss of heat, and sedimentary sequences are laid down in basins. The formation of volcanic rifts results in emplacement of volcanic rock in cycles that also correspond to phases in the 400–million–year supercycle.

Global sealevels are high during phases of rapid plate movement, affecting climate. Ocean currents are affected by distribution of land on the surface of the globe and ocean and air circulation patterns are influenced by the temperature differential between the Equator and the Poles. Glacial events of long duration are within icehouse states, but glacial episodes with glacial/interglacial fluctuations like that of the last 2.4 million years fall within greenhouse states. The next global icehouse state with glaciation of long duration should begin about 80 million years from now in tune with the 400–million–year supercycle which began 320 million years ago with the Late Carboniferous to Early Permian Ice Age.

A supercycle consists of 150 million years for the aggregated supercontinent (the time taken to build up sufficient heat to restart the cycle); and 250 million years for the rift oceans (the time from initial generation to final consumption).

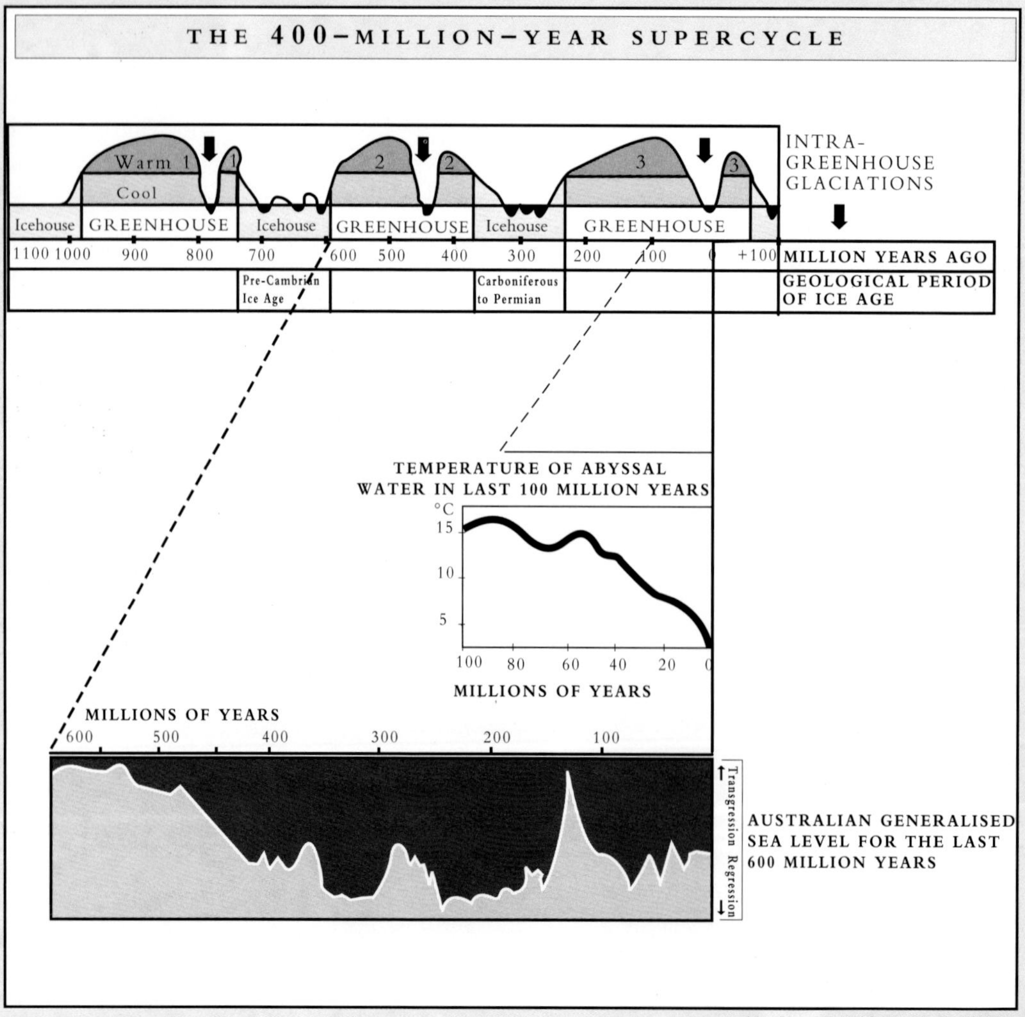

After Veevers 34. Sealevel after BMR 1

The acceptance of the concepts of Plate Tectonics in the 1960s resulted in a revolution in the Earth and Natural Sciences. Since then, accumulation of knowledge and improvements in technology have meant a continuing refinement of data and better understanding of the situations which existed through geological time. There are few areas of Earth Science in which more rapid advances are being made than in the reconstructions that show how the face of the globe has progressively altered. Within the last decade there has been much revision of ideas. The re-assembly of supercontinents is constantly fine-tuned to keep up with the inflow of improved data particularly from programmes of deep-sea drilling, seismic investigation of ocean floor geology and interpretation of palaeomagnetic information.

Even with the increased amount of information available, interpretation of the data by different earth scientists varies. The precise fit of the continents when they were united into supercontinents is still the subject of controversy. (In addition there are still some eminent scientists who do not accept the Plate Tectonic model but prefer the concept of an expanding Earth.) The effects of the rifting processes on the landforms of continental margins adjacent to rifts are hotly debated.

In order to give a coherent and reasonably simple account of this dynamic subject it is necessary to make a choice from the models proposed by the various experts and to tell the story that seems to be the most likely one, and the one most generally accepted at the time of writing. I have therefore decided to follow, in the main, Professor John Veevers and his team at Macquarie University (who are accepted as leading experts on Gondwana break-up) for the sequence of events in Australia's rifting; to follow one of Australia's leading geomorphologists, Professor Cliff Ollier, in matters related to landform evolution; and I take responsibility for the blend that results and the adjustments that have to be made to accommodate these two somewhat disparate views and the conclusions reached by other experts.

Australia the Island Continent was part of Gondwana, the great southern supercontinent, for most of its geological history. Its separation from other lands forms the first part of the saga of the evolution of our modern land. Our story starts in the Jurassic Period, in the Mesozoic Era — the median age of life when dinosaurs roamed the ancient forests worldwide and the southern and northern supercontinents were joined into one super-supercontinent, Pangaea the One-Earth.

EARTH MONSTER

If you prefer Science Fiction to scientific facts about the 400-million-year cycle of an ever-changing Earth, then fantasise with me...

Imagine the Earth Monster...

Not *the benign Earth-Mother Gaia, the largest living organism who maintains balances in all her systems, animate and inanimate, so that the biosphere remains life-friendly...*

...But the satanic, primaeval, cosmic one that was born of the Big Bang and lurks within. A fiery, molten-core creature with a thin air- and water-cooled skin: ocean crust of Panthalassa its sea-cooled underbelly and the itchy skin beneath the continental crust of its Pangaea carapace. Like a crab or a spider, or any number of invertebrates that outgrow their skins, which stretch and crack and move against their newly emerging skin below, so this satanic magma-creature finds that its heavy Pangaea carapace makes its back too hot to bear. It stirs, mobilising its sluggish circulation to stretch and crack its continental shell, driving the bits apart and bringing between them cooling seaways. And then, feeling its back too exposed, it gathers the fragments together again, reforming its carapace. A creature ever-changing from a geocratic Pangaea-monster to a thalassic one with lumpy skin of scattered continents and oceans. A breathing creature, expelling carbon dioxide as it drives its heat-holding carapace apart. A heat-powered engine, if you like, in a state of perpetual motion. The monster ails — febrile times alternate with chills and ague, and it has life-long problems with erupting volcanoes which, pimple-like, give warning of its overheated blood...

Lucky for us that this Big Bang monster is trapped and safely held in a cage of lithosphere and that he operates on a geological time scale. His earth-shattering movements are imperceptible in man-time, and only earthquakes and volcanoes tell us of his pent-up frustrations...

CRADLE MOUNTAIN

A legacy of Australia-Antarctica Rifting

During the Jurassic, continental extension between Australia and Antarctica resulted in the emplacement of an estimated 5000 cubic kilometres of molten rock beneath Tasmania.[46] Detachment faults meeting the asthenosphere in the region caused areas of crustal weakness. Molten magma was able to rise and when it cooled below the land surface it became dolerite. The strata which used to cover the dolerite have eroded and today many of Tasmania's dramatic landscape features, including Cradle Mountain, are created by the volcanic rock that resulted from supercontinent fragmentation.

Cradle Mountain is a remnant of the dolerite which was emplaced underneath Tasmania during stretching processes in its region. M.E.W.

CHAPTER ONE

STRETCHING

RIFTING BEGINS — BY CONTINENTAL EXTENSION IN THE JURASSIC AND EARLY CRETACEOUS[34-43]

(160 to 96 million years ago)

The Jurassic Period from 208 to 144 million years ago was a time of special significance for Australia as it was then that the rifting processes which were to determine its outline and many of its important topographical and geological features commenced.

Australia had been on the outer edge of Gondwana until Jurassic times, but early in the Period a wide strip of land[45] was elevated along the proto-Pacific coastline[32], comprising continental crust that would later move away and become mainly submerged. The submarine Lord Howe and Norfolk Rises, the Queensland Plateau, the islands of New Caledonia and New Zealand, and the Campbell Plateau are remnants of this eastern Gondwana province that formed the outer edge of the supercontinent and an extension of Australia. The present eastern margin of the continent was far from the sea as a result of the addition of this strip of land. It is currently fashionable to refer to this landmass as the New Zealand Subcontinent.[135] (In my *Nature of Hidden Worlds* I used 'Ancestral New Zealand Landmass'[45] for this territory.)

In the most up-to-date reassembly of Gondwana[42] the junction between Australia and Antarctica is positioned along Antarctica's Wilkes Land margin, and India fits with its Coromandel Coast lying adjacent to Enderby Land. The Indian landmass in Gondwana was considerably larger than the peninsular India of today. Greater India as it then was is reconstructed to include the crust that is folded up in the Himalayas and a portion that was subducted under the mountains when the continent collided with Asia.[22]

The use of seismic methods to determine the geology under the oceans has resulted in increasingly precise information on the stages of separation of Australia from other parts of

Fronds of a cycadophyte Otozamites *and very narrow stems of a conifer, closely covered in small scale leaves, in a Cretaceous fossil specimen from the Northern Territory.*

JIM FRAZIER

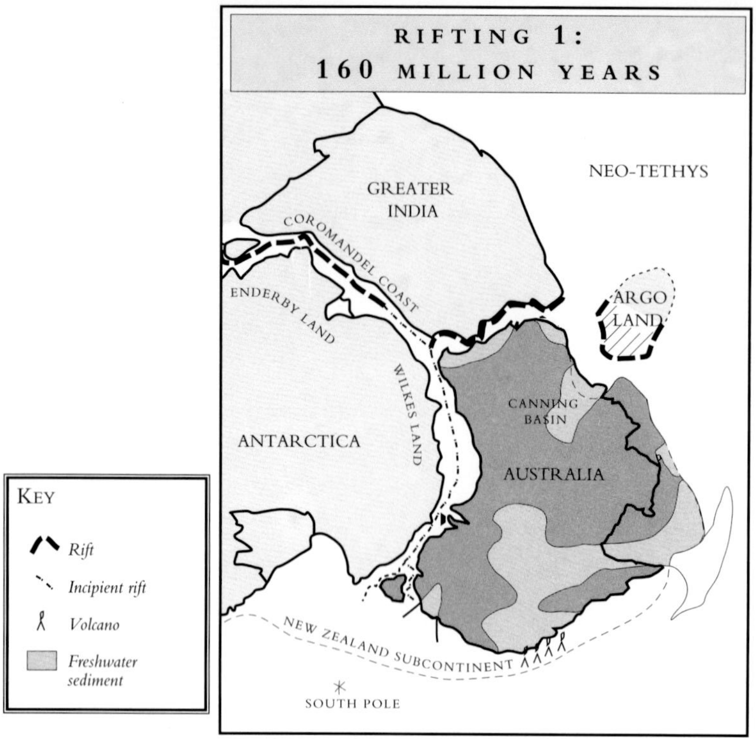

After Veevers *et al*. 42

Gondwana and has led to the revised positioning of Australia and India against Antarctica. Previous reconstructions which assembled the present-day continents into supercontinents had been done by fitting together the continental shelves of modern continents, based on the assumption that rifting between lands had been a largely vertical process where faults (cracks in the Earth's crust) caused a direct fracture zone between lands. With modern technology it has been possible to locate the position at which continental crust meets sea-floor volcanic rock — the *continent-ocean boundary*.[40] When the boundary lines between Australia and Antarctica and the sea-floor of the Southern Ocean are mapped,[10, 11] it becomes evident that there was stretching of the continental crust between the two continents before the oceanic crust started to develop between them. The stretching began about 160 million years ago.

It has been estimated that 360 kilometres of stretched crust lay in the centre of a 600 kilometre-wide rift system before there was any production of basaltic sea-floor.[37] Thinning of the crust from its 40 kilometres depth to about 10 kilometres thick in the centre of the rift was accomplished before sea-floor spreading started to form the true floor of the Southern Ocean. (Ocean floor is different from continental crust. Sea-floor spreading results from the emplacement of basalt along mid-ocean ridges.) Continental crust thinning was also occurring in the New Zealand Subcontinent during the Middle and Late Cretaceous and was accompanied by the formation of fault-bounded basins as tension increased.

Rifting that involves stretching prior to fracture and the upwelling of magma to form sea-floor is described as a passive margin continental extension model, and it was occurring in many places while supercontinent fragmentation was taking place. A wide rift system had developed by crustal stretching between Africa and South America prior to formation of the South Atlantic basaltic floor, starting during the Middle Jurassic about 175 million years ago. In other cases separation of continents could have been by movement along transcurrent or strike-slip faults, where lands slip past each other without continental extension, or by fracture without prior stretching.

In the first stage of rifting between Australia/India and Antarctica between 160 million years and 132 million years ago (from the Middle Jurassic to the very early Cretaceous), there was stretching and thinning of crust between Antarctica and Australia on a north-northeasterly azimuth. Movement between India and the Kron Prins Olav coast of Enderby Land was mainly slip along a transcurrent fault. About a fifth of the continental crust extension between Antarctica and Australia is believed to have taken place in this first stage of rifting, and the movement was occurring at the very slow rate of 6 millimetres per year. Sagging followed extension and sedimentation commenced in the rift zone.

During this first rifting phase a continental fragment of uncertain origin, known as Argoland, was already moving away from the Canning Basin region of the Western Australian coast by a process of active sea-floor spreading. The resulting sea-floor formed the Argo Abyssal Plain. The land moving away is believed to have formed islands that perhaps maintained some island chain connection between Australia and Asia before they attached to that continent to become Sundaland.

The Canning Basin was entered by the sea as the ocean between it and Argoland opened up, and the fault-bounded troughs that were associated with rifting seaward of the Basin

THE STAGES OF RIFTING

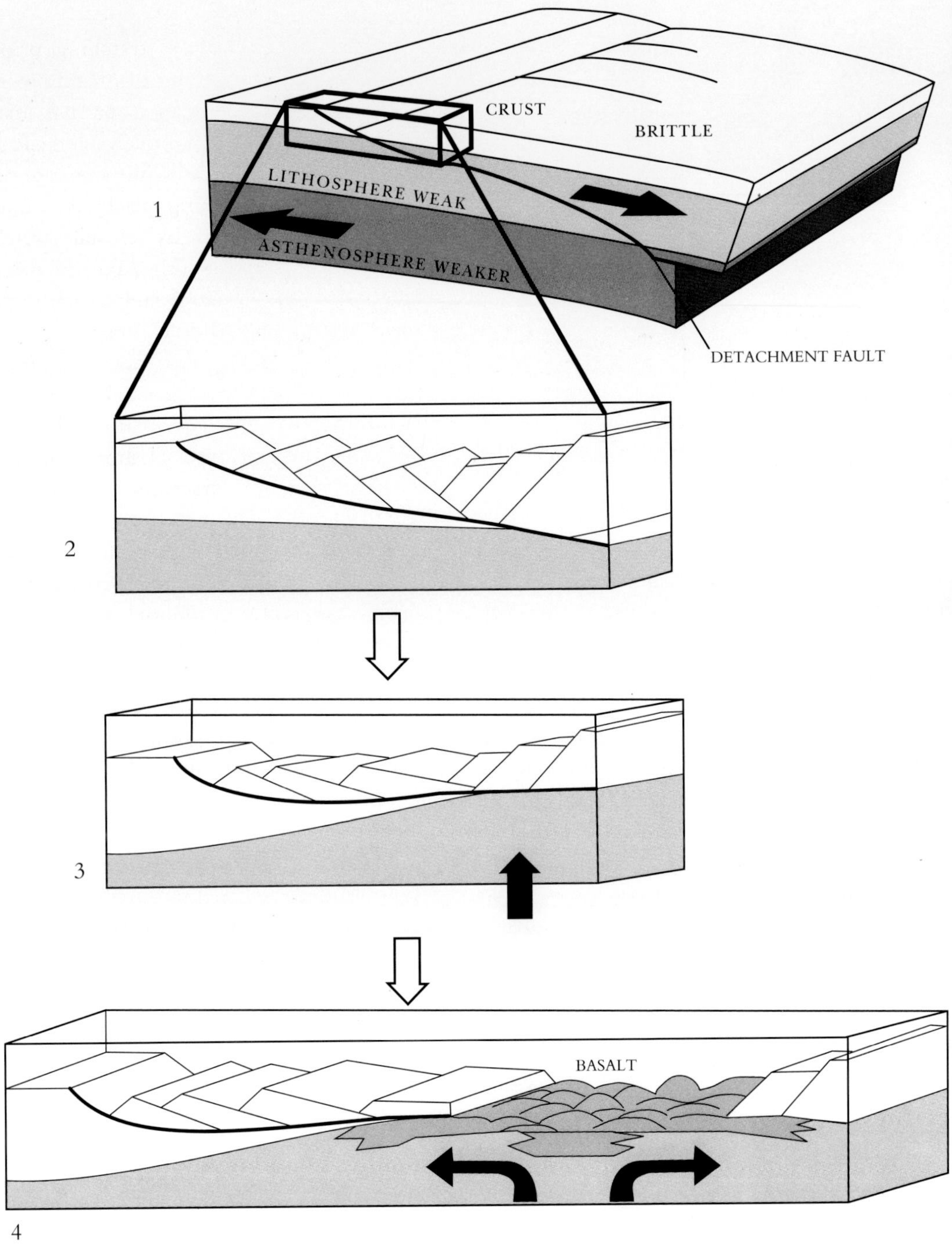

1. A master detachment fault, which extends into the asthenosphere, enables a block of crust to be extended during the initial stretching.

2. Extension of cooler, brittle crust is accomodated by a conjugate series of faults. The collapse and partial rotation of the fault blocks results in formation of half grabens as the crust thins progressively.

3. As a result of thinning of the crust, there is an isostatic adjustment to restore equilibrium which causes the thinned crust to be bowed up. Hot lithosphere comes closer to the surface where maximum thinning and bowing-up has occurred.

4. The upper lithosphere becomes molten with changes in pressure and as the crustal blocks part, basalt emerges to coat the rift-zone floor.

Block diagram prepared by Dr P. R. Evans

AUSTRALIA'S CRETACEOUS BASINS

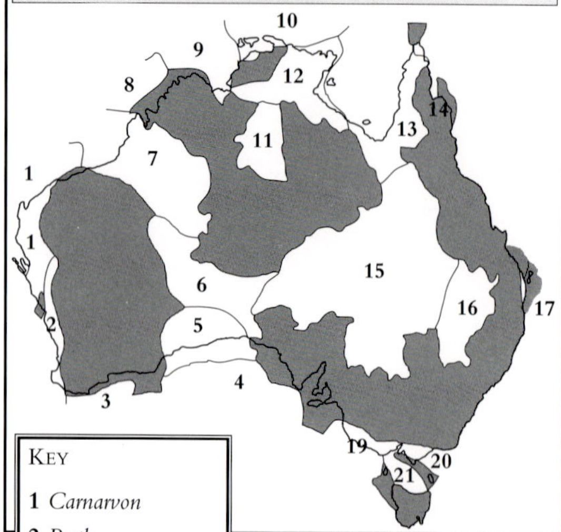

Key

1 *Carnarvon*
2 *Perth*
3 *Bremer*
4 *Great Australian Bight*
5 *Eucla = Bight Basin*
6 *Officer*
7 *Canning*
8 *Browse*
9 *Bonaparte*
10 *Money Shoal and Bathurst Terrace*
11 *Wiso*
12 *Northern Territory*
13 *Carpentaria*
14 *Laura*
15 *Eromanga*
16 *Surat*
17 *Maryborough*
19 *Otway*
20 *Gippsland*
21 *Bass*

Above right: An artist's impression of the vegetation of the Early Cretaceous, with Muttaburrasaurus *dinosaurs, another small dinosaur, and pterosaurs (about 120 million years ago). This was still the age of conifers, cycads, ginkgos and ferns, and flowering plants were rare.*

Art work by Dorothy Dunphy

accumulated plant material which later became the source of oil and gas on the Northwest Shelf.

From this time on rifting of Australia's margins was to proceed in an anti-clockwise direction, so that the oldest rifted margin is in the northwest and the youngest in the northeast off New Guinea. This infers a clockwise rotation of Australia away from the other parts of Gondwana.

In a second stage of rifting between 132 and 96 million years ago the remaining four-fifths of continental extension took place between Antarctica and Australia, and India started a north-westward rotation away from Australia. The subsiding of India's east coast basins from 132 million years ago when its movement began is matched by subsidence in the Perth Basin. (The sea is known to have invaded the rift and entered the basins because marine sediments of this age occur in them.)

Both rifting stages — from 160 to 132 and from 132 to 96 million years ago — were part of a regime of sedimentation and activity that preceded the formation of any basaltic sea-floor in the rift systems between Australia and Antarctica.

An over-simplified explanation of how crustal extension takes place[10], and its resulting alteration of landscapes by forming a major depression which has a significant effect on regional drainage patterns, and thus on sedimentation, follows.[11]

To understand and to visualise the processes involved and the structure and nature of the rift valley that was formed requires a knowledge of some basic facts about the structure of the Earth. When material as rigid and strong as that which forms the Earth's crust is subjected to the unimaginably large forces involved in plate movements, the lack of uniformity in the material involved makes any stretching process very complex.

The rigid outer layers of the Earth which comprise the lithosphere are composed of continental and oceanic crust and the portions of lithospheric mantle which form the plates in which the crustal material is embedded. Below the lithosphere lies the asthenosphere, the part-molten outermost region of the mantle which surrounds the dense core of the planet.

Continental crust, whose thickness varies from 10 to 50 kilometres, is divided into three layers: the upper is cool and brittle, the middle is warmer and more ductile, and the lower is hot and plastic. Between the base of the lower layer of the crust and the lithospheric mantle which forms the plate beneath it is the 'Moho' (or Mohorovicic Disconformity) which is clearly seen when seismic methods are used to determine Earth structure. The material below the Moho in the lithospheric mantle has very different properties and is much hotter, and its plasticity increases with depth. Convection currents within the mantle below the lithosphere (which come closest to the surface in the asthenosphere) are responsible for the movement of the plates and the portions of continental and oceanic crust which they support.

Because continental crust varies in nature from very strong and brittle to comparatively weak and plastic at depth, it does not extend

FEATURES OF A RIFT VALLEY

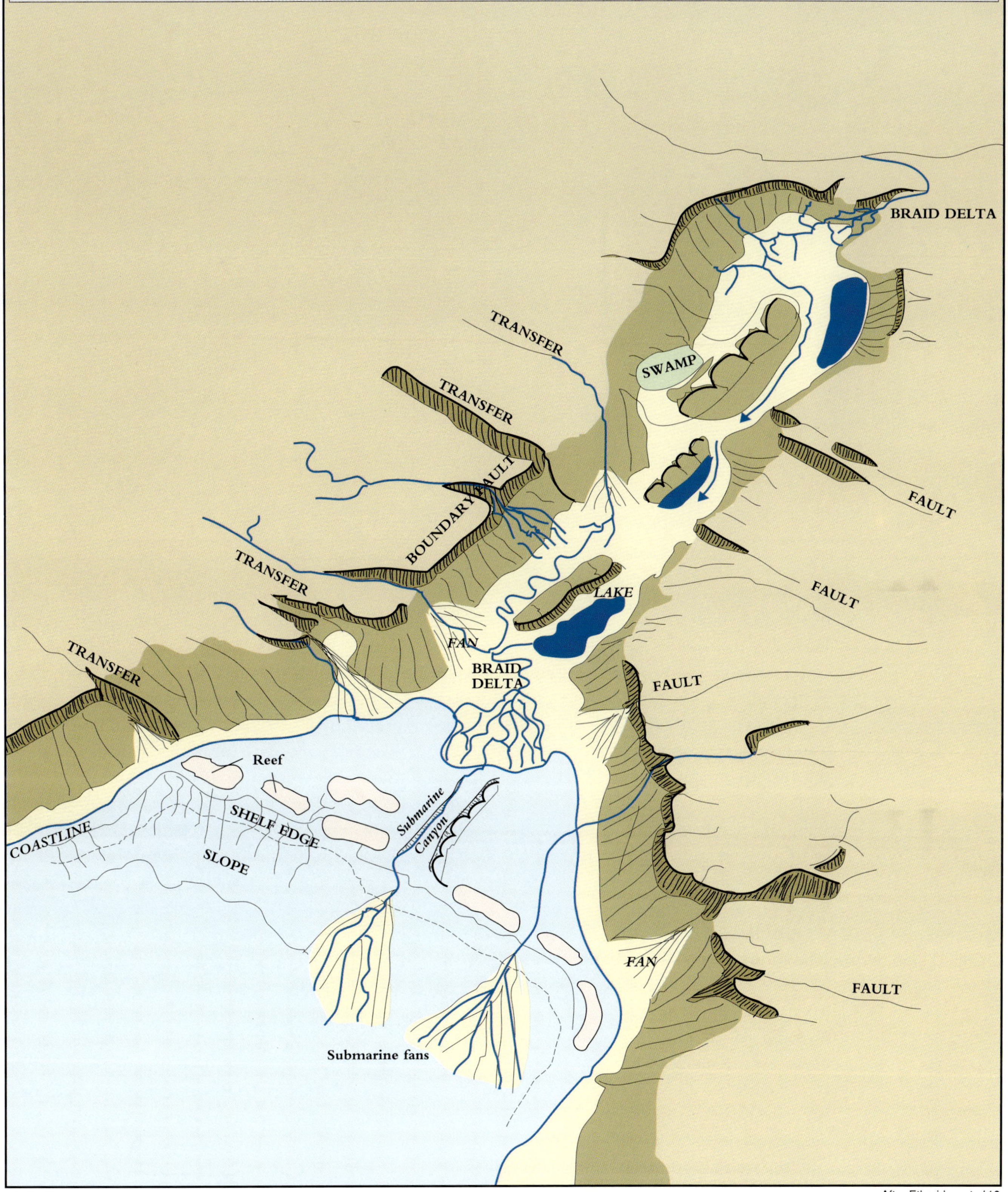

After Etheridge *et al* 10

evenly when it is stretched. (Think of chewing gum which extends freely when fresh, and how it behaves differently when it has dried and has a surface crust. In the latter case the dry surface cracks into blocks of crust which are pulled out on the surface of the more ductile fresher material towards the point at which the thinning is sufficient for the extended material to break. Bearing this model in mind makes the processes of continental crust stretching more easily understood.)

When a rift is opening up by continental extension, the block of crust at the head of the rift (comprising surface cool and brittle,

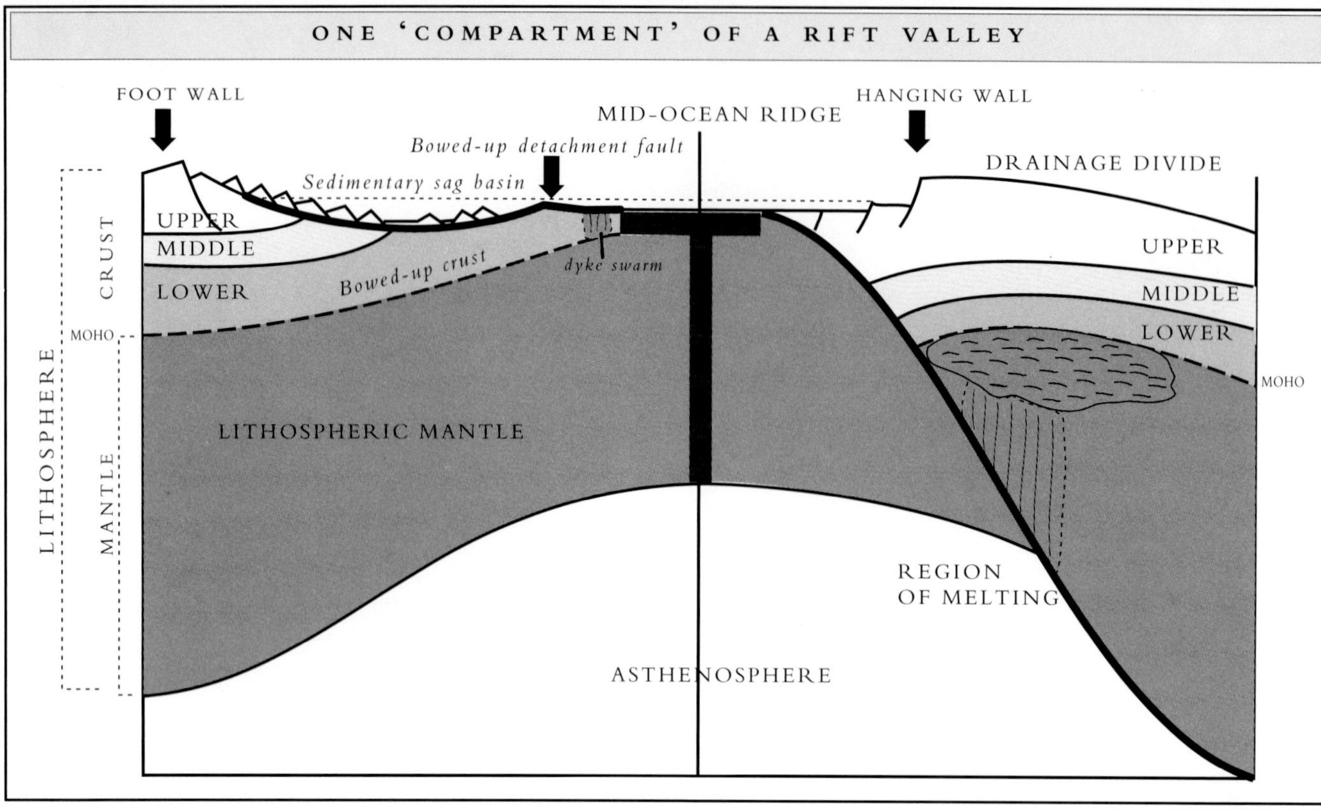

After Etheridge *et al.* 11

middle warm and more ductile, and lower plastic material) shears off along a low-angle 'detachment fault' which runs from the surface and down through the lithospheric mantle. Where the fault intersects the asthenosphere the weakened area that it produces results in molten magma being able to rise up through the lithosphere to produce a mass of volcanic rock below the continental crust, bowing the crust up and forming a drainage divide. The edge of this divide is progressively down-faulted towards the forming rift and becomes the hanging wall of that section of rift.

On the other side of the rift, the side from which the detachment fault originated, the brittle upper crust breaks into blocks which tilt against cracks (faults) which result from its inability to stretch. These blocks are pulled along on the surface of the more ductile lower layers of crust. (The V-shaped structures that result from tilting of the small blocks between faults are known as half grabens.)

With the thinning of crust during stretching, sagging follows and a sedimentary basin (with its quota of half graben secondary areas of sedimentation within it) develops and widens. As the process continues, the thinned crust beneath the basin becomes progressively bowed up as a result of isostatic adjustment which brings the system back into gravitational equilibrium. (Parts of the Earth's crust go up or down on the plastic sub-crustal zone in response to their changing weight, maintaining balance.)

Towards the middle sector of the rift the weakness of the thinned crust becomes critical and swarms of dykes (vertical faults up which molten basalt rises) occur. A point is reached in the rift centre where the thinning of the bowed-up crust and the increased heat result in a major fracture. Basalt then starts to emerge from this linear volcano — the mid-ocean ridge — which builds up ultimately into an under-sea mountain range. The basalt that emerges here forms oceanic crust and sea-floor spreading results in the continental crust on either side of the mid-ocean ridge being pushed further apart.

The processes involved in stretching of continental crust[10] on detachment faults in the manner described induce a basic asymmetry[11] in the resulting rift landscapes with the two sides having contrasting topography, regimes of sedimentation and subsidence history. In effect the result is one of upper and lower crust conjugate pairs. The upper crust margins form the uplifted hanging wall of the detachment, while lower crust margins comprise the footwall to the detachment, commonly overlain by highly extended remnants of upper crust, which form half grabens.

The extension events in the formation of a rift are spasmodic and constrained by the geology of the region where lines of weakness, described as 'transfer faults', allow a block of crust to slip up to that line and then interrupt the process. When stretching and thinning of crust resumes on the detachment faults between that transfer fault and the next, another pair of contrasting landforms is produced. Thus up the side of a passive margin rift valley system there is an alternation of cliffs and slopes and areas of deep and shallow subsidence. A rifting zone, controlled by transfer faults, becomes a series of compartments.

Because extension events between Australia and Antarctica occurred over a very long period of time, erosion was filling the subsiding rift area with sediment. Accumulation of great piles of sediment resulted as the area subsided. These formed suites of rocks whose derivation from continental sediments which incorporated plant material made them likely source rocks for oil and gas deposits. In order to prospect for commercial fields along Australia's southern margin, it has been necessary for geologists to understand the complex structures that were formed when separation of Australia from Antarctica was occurring by continental extension in the Jurassic and Early Cretaceous. The vast rift system with its fault-bounded compartments and troughs has potential traps for hydrocarbons.

In order to visualise what a passive margin rift system like that which had developed between Australia and Antarctica during the phase of continental extension would have looked like, three geologists of the Bureau of Mineral Resources in Canberra attempted a reconstruction of part of such a rift system at the stage when the sea was just starting to enter the rift zone.[10] With their permission their reconstruction has been used for a redrawn version of the rift zone.

The 360 kilometre wide strip of stretched crust, bounded by altered continental edges on either side, resulted in a zone up to 600 kilometres wide where conditions were very different from those in the pre-rift eastern Gondwana. Throughout the Jurassic phase of continental extension — between 160 and 144 million years ago — the rift zone was a terrestrial environment. Rivers flowed on the floor of the rift and deep lakes, swamps and marshes were features of the landscape.

Australia was rotated compared to its present orientation with its north-south axis then its east-west axis now, and it lay between latitudes 70° and 40° South. In those high latitudes the winters in the southern part of the continent were times of little sunlight, but the world was believed to be warm, without polar ice. (Recent research has suggested that there was some glaciation in the Allan Hills region of Antarctica[76] in Early to Middle Jurassic times[134], so world climates may not have been uniformly warm throughout the whole Period. An alternative explanation might be that mountain peaks existed in the region and were sites for glaciers and snowfields.) Antarctica lay in similar latitudes and supported a flora similar to that of India, Australia, New Zealand and South America.

This was the age of conifers, cycads and ferns. Though similar forests of these ancient plant groups occurred worldwide, extending into Arctic regions in the Northern Hemisphere, it is now evident that the genera comprising the southern floras were distinct from those in Northern Hemisphere lands[12] — the Jurassic flora of Gondwana was in fact Gondwanan[21] (like the Permian and Triassic floras that preceded it) and not a part of a cosmopolitan flora as had previously been thought. The southern conifers of the families ARAUCARIACEAE and PODOCARPACEAE[44], the pentoxylalean and bennettitalean cycadophytes,

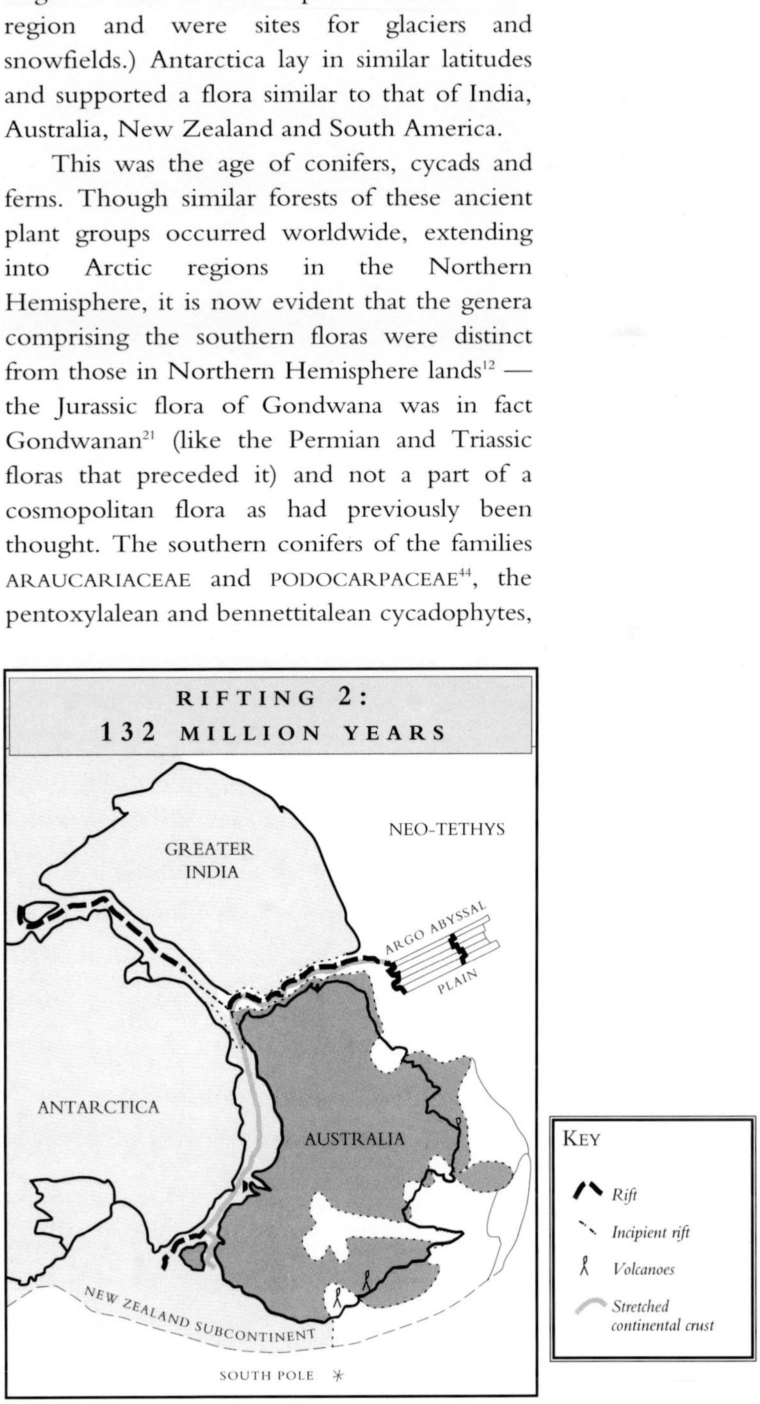

After Veevers *et al.* 42

MOUND SPRINGS [132, 55, 188]

The Mound Springs National Park southwest of Lake Eyre has a moon landscape feel about it. A glary white crust and pebble scree cover the ground. Occasional saltbush is the only vegetation and flat-topped 'mesas', which are in fact extinct mound-springs, rise out of the flatness. Active mound-springs, ringed with reedy rushes, surrounded by a baked crust of lime and gypsum, some with a thin blue river cutting its way through the crust, occur among the extinct mounds. [132, 55, 188]

At 'the Bubbler', a mound-spring where mud bubbles up and the water flows away through salt crust into the saltbush plain, the stream is filled with mats of algae, green and blue-green, thriving in the warm water. The Blanche Cup mound-spring has a reed-ringed pool on top of a considerable mound and *Frankenia* bushes grow as dense cushions on its flanks. There is evidence in the form of fans of pebbles and gravel that this spring was much more active in the Pleistocene. Now only a trickle runs down one side of the mound after rain.

The abundance of mound-springs in the region results from the fact that this is the edge of the Great Artesian Basin. Here the underground artesian water that has flowed across the Basin for about 2 million years since it fell as rain on the Great Divide comes up against the geological structures against which the basin edge abuts. The water rises and emerges at the surface carrying mud and bringing salts with it. The mud is the product of weathering of the Mesozoic rocks within the Basin through which the water has travelled.

The extinct 'mesas' result from the lowering of the ground around the plug of mud and salts (calcrete, gypcrete) that had accumulated during the springs' active phase. Hamilton Hill and Beresford Hill attain a height of about 40 metres as a result of the deflation of the surrounding landscape which has occurred during ice age dry and windy times and they probably date back to Pliocene times. Mounds older than that which must have existed throughout the lifetime of the artesian flow across the Basin have disappeared, eroded away.

The extinct mounds have shallow, circular silt-filled depressions on top, the sites of the original vents. The massive limestone of which they are made is yellow-grey or pinkish-brown or pale grey, and is fine-grained. Fossils in it comprise gastropod shells, reed casts and moulds and what appear to be other plant fragments. Presumably in non-arid phases during the life of the mound-springs *Casuarina* and other plants grew near the springs in a vegetated landscape.

From the top of the Blanche Cup mound, with its reed-fringed pool in the foreground, Hamilton Hill can be seen across the salt, sand and pebble desert plain. Live and extinct mound-springs in one amazing landscape.

M.E.W.

A great concentration of springs at Dalhousie Springs is responsible for discharging more than 90 per cent of the total water from springs in South Australia. The local geology and position of faults causes this concentration.

Some active springs are not precipitating carbonates and their mounds are unconsolidated mud. Fred Springs is an example of a mud mound. At Eulo in western Queensland such mud mounds without significant deposition of calcrete and with no surrounding crust on the ground also occur. At Eulo the water is welling up against a granite body which was an island in the Mesozoic Eromanga Sea when the artesian water-bearing sediments were being laid down, and the mound-springs are not a basin-edge phenomenon.

Springs also occur at many places in the Basin where faults interrupt the flow of water in an aquifer, making deep artesian water available at the surface.

Mound-springs have been an important water resource for animal life in a dry continent. In times of drought they can be so trampled by the animals which have no other water that the mound is broken and the water dispersed to evaporate uselessly. Many have suffered in this way where sheep and cattle have done the damage. In the past giant *Diprotodon* have died at sites of old mound-springs, as evidenced by their bones, found near Bundeena Station in the Eulo district of western Queensland.

Aboriginals followed the line of springs through arid lands on their travels, and many mound-springs have an abundance of stone artefacts scattered around them. Early European explorers, like John Macdougall Stuart (who made the first crossing from Adelaide to Darwin in the 1860s), E.J. Eyre and others, used the springs as stepping stones through a waterless land. Later the Ghan Railway and the Overland Telegraph both followed a route based on the availability of water from artesian springs along the Great Artesian Basin's edge.

While Elizabeth Springs once had 'enough water running to drive a flour mill' according to early settlers it now does not flow, and the volume of outflow on all other springs is similarly depleted since Basin waters were exploited by sinking thousands of wells. Bearing in mind that the main recharge occurs on the eastern edge of the Basin on the Great Divide, and predominantly in the better-watered northern sector, it has to be questioned whether recharge can really ever balance usage, such is the time lag. The drying up or very much reduced flow of mound-springs must surely be a serious symptom of depletion of the resource.

Low-growing cushion-shaped Frankenia *bushes grow in the salty, lime- and gypsum-rich sand on the slopes of the Blanche Cup mound-spring. These plants produce delicate pink flowers and their small hard leaves have rolled edges and are covered with salt crystals. The plants actually excrete salt, and the crystals on their surfaces have the added advantage of reflecting heat in this desert environment.* M.E.W.

A mound-spring trampled and destroyed by cattle, some of which died when the water dried up. Near Eulo. BRIAN SENIOR

A blue ribbon of water runs through the salt and chemical crust that surrounds The Bubbler, an active mound-spring whose pool is a mud cauldron in which large bubbles form as the water comes up under pressure. The water of the drain that cuts down through the crust is heavily infested with algal mats, brilliant green and blue-green, among reed clumps and fringing samphires. The small dunes around the crusted area support saltbush and nitre bushes. M.E.W.

and the tree-ferns that were major components of the Gondwanan vegetation are distinct from the northern forms that look superficially similar but are merely closely related. Under conditions of generous rainfall and warmth forests were widespread. (Although the southern and northern supercontinents were joined to form Pangaea, aridity in middle and low latitudes in Permian and Triassic times had formed a barrier to the mixing of floras. The more generally humid Jurassic probably had no comparable barrier to the mixing of flora, and the differences resulted from provincialism over the vast expanse of Pangaea.)

The eastern half of today's Australia (which was the southern part of the continent during the time when continental stretching was taking place) was a very different realm in the Jurassic. Rivers flowed into the gigantic depression that occupied that part of the continent, carrying sediment which would form rocks of great significance to modern Australia. These rocks are the main aquifers of the Great Artesian Basin. (The significance of artesian water in a dry continent today is enormous.) The whole region was a network of lakes, swamps and streams through which large rivers meandered, spreading sandsheets. Plant material accumulating in the swamps formed deposits of coal.

Early in the Cretaceous when India separated from Australia's Western Australian margin and the sea entered the proto-Indian Ocean rift, the Darling Scarp became a feature of the landscape. This was partly the result of some elevation of the continental margin but largely due to the down-warping of the strip of continent to its west. The Scarp follows the edge of the ancient Yilgarn Block, one of the original building blocks of the continent.

At the same time, marine invasion of the rifting zone between Australia and Antarctica began. Early Cretaceous marine sediments were laid down in the Albany Sub-basin, offshore from the Bremer Basin, and in the Eucla Basin. (The Albany Sub-basin is known only from seismic investigations. It contains a Jurassic to Cainozoic sequence of sediments similar to that in the Eyre Sub-basin of the Eucla Basin. The evidence for the earliest invasion of the rift zone between Antarctica and Australia by the sea will be in this region in the form of the oldest marine sediments in the sequence. Seismic investigation cannot readily distinguish between marine and non-marine rock. Only drilling, such as has been done in the Eyre Sub-basin in the course of exploration for oil and gas, will confirm the entry of the sea in this region and the time of inception of the proto-Southern Ocean.)

A shift of the South Pole 10° closer to the edge of the New Zealand Subcontinent (bringing a point at the New South Wales/Queensland border on the present-day Australian coast to about 83° South) by 132 million years ago moved Australia into very high latitudes. Further movement in relation to the South Pole continued until the Pole lay close to Tasmania by 120 million years ago. The cooling that marks the start of the Cretaceous and extends into the Early Cretaceous no doubt stems partly from this higher latitude situation which came at a time when global climate was changing in response to altered distribution of land and sea. The months of winter darkness that are a feature of regions so close to the Pole created added problems for all living things. The golden age of continuous warmth and humidity that had applied during Jurassic times in latitudes almost as high was coming to an end.

At the same time as a seaway was starting to separate Australia and Antarctica, the Carpentaria Basin on the opposite side of the continent was entered by an arm of sea which penetrated into the middle of the Eromanga Basin, starting the formation of a vast epicontinental sea. Global sealevels were rising with the increased plate-tectonic activity that was breaking up the supercontinents world-wide. Much of the world's land would be covered by shallow epicontinental seas at some time during the Cretaceous Period. Australia was such a flat continent with its interior basins interconnected, that it was flooded during the Early Cretaceous. Northern Hemisphere lands had their major flooding later in the Period.

On Australia's Pacific Ocean margin the New Zealand Subcontinent was partly submerged during the times of high sealevel and was eroding rapidly so that even as sealevels fell less of it was above water.

Sealevel in the Australian epicontinental sea (the Eromanga Sea) peaked between 115 and 110 million years ago, when all the continental basins were inundated and land was effectively reduced to four island blocks. Drainage patterns tended to become radial on these blocks during the time when their rivers made their way to the nearest seaway. The

NEW ZEALAND—THE LARGELY SUBMARINE SUBCONTINENT

The present-day New Zealand landmass is only a minor portion of a very large area of submerged continental crust[135]. Had it not been for activity associated with the development and evolution of the modern Australian/Pacific Plate boundary during the last 25 million years, the whole Subcontinent might have become submarine[372].

The Campbell Plateau, Chatham Rise and Bounty Trough lie to the south and east of South Island, forming a submerged area three times larger than subaerial New Zealand. The Lord Howe Rise and Norfolk Ridge extend to the northwest from the Challenger Plateau that lies to the west of North and South Islands.

The breakup of Eastern Gondwana and sea-floor spreading to create the Tasman Sea were the cause of the foundering of the bulk of the continental crust that comprises the New Zealand Subcontinent. Stretching and thinning of continental crust of the Campbell Plateau and Lord Howe Rise occurred during the pre-rifting phase. Continental crust under the Campbell Plateau and the Lord Howe Rise is known to be only 20 to 25 kilometres thick, in contrast to the 30-35 kilometres thickness below central South Island and the 35-40 kilometres thickness under Wellington.

The area of the Campbell Plateau, Bounty Trough and Chatham Rise was increased by 25 per cent by this continental extension and thinning which took place during the Mid to Late Cretaceous (from 105 to 85 million years ago). Foundering of the thinned continental crust followed and by 35 million years ago the phase of broad regional subsidence was completed.

During the phase of crustal thinning, fault-bounded areas subsided to form basins. The Great South Basin south of South Island resulted from intra-continental rifting between South Island and the Campbell Plateau as a result of crustal tension.

The Taranaki Basin, important for its coal and gas reserves, was one of the other areas of subsidence that resulted on the Subcontinent as crust was extended and thinned.

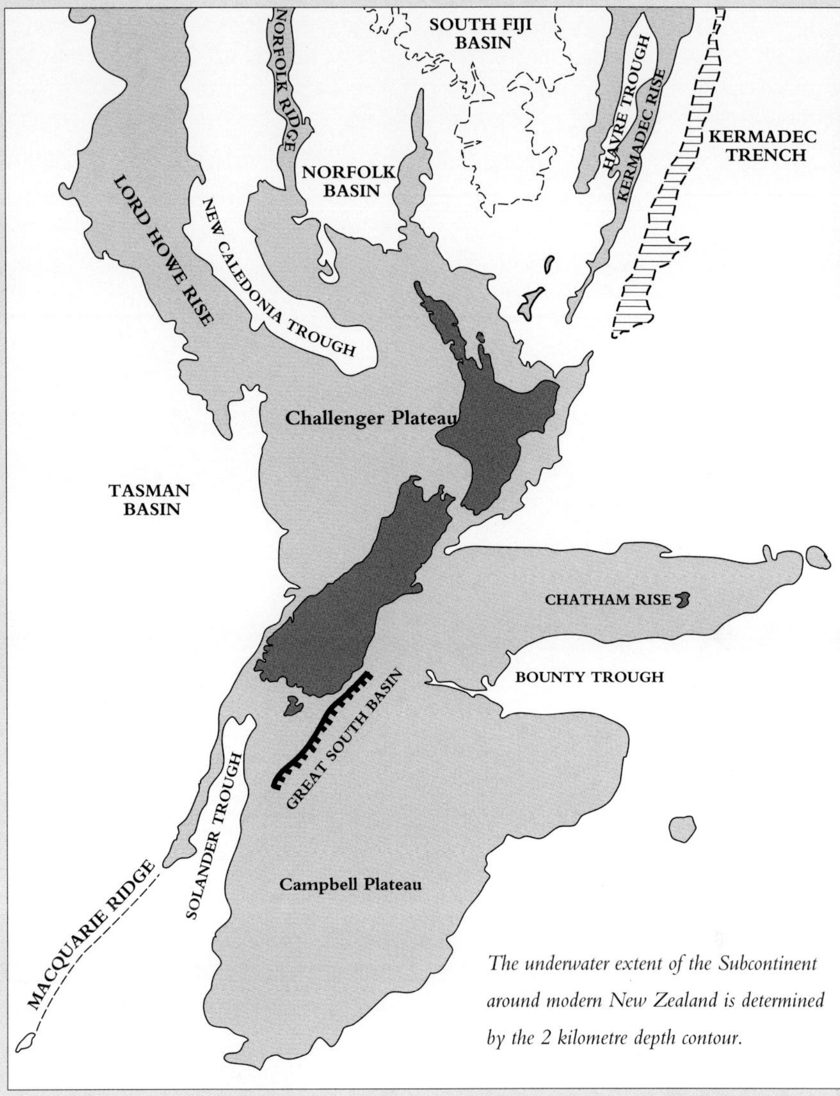

The underwater extent of the Subcontinent around modern New Zealand is determined by the 2 kilometre depth contour.

After Kamp 135 and Griffiths 372

ancient drainage patterns in the vast area of the continent that was flooded during the Early Cretaceous were largely obliterated. Pre-Cretaceous drainages on parts that remained as land can often still be traced on aerial photographs or satellite pictures.

After the time of maximum impact of high sealevels between 115 and 110 million years ago in Australia, epicontinental seas retreated and by the end of the Cretaceous global sealevel had stabilised at slightly lower than the steady Jurassic level.

There appears to be evidence of winter ice in the Eromanga Sea[13] during the Early Cretaceous[14]. Dropstones which are presumed to have been ice-rafted are found in some of the sedimentary deposits. Isolated rounded stones up to small boulder size have been found in fine-grained strata in the eastern Northern Territory and northwestern Queensland. The sediments containing the stones were deposited on the shallow marine marginal shelf in the Carpentaria Basin. (Ice is not the only carrier of dropstones. During

THE GREAT ARTESIAN BASIN

The name 'Great Artesian Basin'[188] is given to that part of the Eromanga Basin[373] which contains the Jurassic and Cretaceous aquifers and the stored water resources which are of such great importance in the dry Australian continent.

The Great Artesian Basin underlies about one fifth of the Australian continent, an area of 1.7 million square kilometres. Parts of New South Wales, Queensland, South Australia and the Northern Territory have the benefit of the water which is stored under pressure deep below areas which are mainly arid and lacking permanent surface water.

The reason why the water is under pressure relates to the saucer-shape of the basin and its elevated rim which is the recharge zone. Just as a water tank in the roof of a high building or a water-tower in a flat landscape ensures water pressure at taps, so the topping up of the artesian system at a greater elevation than the storage aquifers ensures a pressurised system.

It might be argued that as a result of the discovery of artesian water in 1880, which enabled arid lands to be opened up for grazing, the degradation of these areas has been such that it would have been better for the health of the continent if this enormous water resource had never been discovered. Much of the marginal land bordering the central deserts has been turned into desert as the result of overgrazing and unsuitable management practices and is now beyond repair.

Hamilton Hill, an extinct mound-spring, rises out of the plain like a mesa in the Mound Spring National Park. The calcrete and gypcrete of which it is composed was brought to the surface by the artesian water which emerged as a spring when it rose against the Basin rim. Chemical weathering of the aquifer rocks along the water's path across the Basin provided the lime, gypsum and other salts that built the structure which is now exposed as a hill. The hard and resistant nature of the material which forms the hill enabled it to resist the erosion which lowered the landscape around it during dry and windy Pleistocene times.

M.E.W.

The water is stored in sandstone layers which were deposited in the Basin during the Jurassic and Cretaceous Periods. In the Jurassic the area was not unlike the Amazon Basin with rivers and lakes spreading and accumulating vast amounts of sand. During the Cretaceous the Eromanga Sea invaded the area and marine sandstones were laid down over the Jurassic sediments. When the epicontinental sea retreated, freshwater lakes and rivers added more layers to the sedimentary sequence that filled the Basin.

Some of the sedimentary layers in the Basin are aquifers, carrying water in their porous structure; others are not permeable and cap the aquifer layers. Thus water-bearing layers are intersected in wells at different depths, from mainly about 500 metres but in some cases to 2000 metres down. The water is under pressure and flows from the wells. Until fairly recently there was little thought of conserving the water and bores simply ran along drains through the arid land and became reed- and bush-fringed and blue-green and green algae-infested. About 90 per cent of the free-running water was wasted by evaporation and seepage, and little thought was given to the fact that such underground water is a finite resource. Today about 3000 bores are still free-flowing, but 20,000 are not running to waste and are equipped with windmills or engine pumps. In the interests of conservation of the water resource a more responsible attitude is now evident and, increasingly, the level of water in troughs and dams is controlled with ball-cock devices to minimise waste.

The quality of the water in different-depth aquifers varies, some being more palatable than others, and some being sulphurous and unpleasant-smelling. Temperature of water varies from 30° to 100°C in wells and from 20° to 45°C in springs. Recharge of the basin occurs mainly on the eastern side along the Great Divide. Here the sedimentary rock layers of the basin bend upwards, forming a receptive area of something like 15,000 square

THE GREAT ARTESIAN BASIN

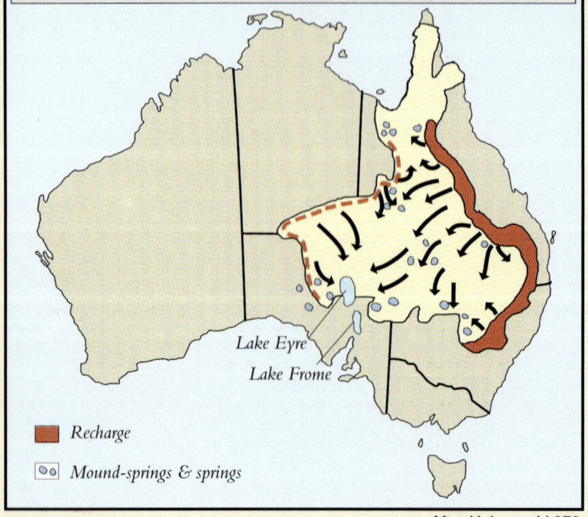

After Habermehl 373

kilometres in total. Rainfall in this zone penetrates to sufficient depth to enter the porous rock and add to the artesian store, particularly where the soils are yellow earths that saturate rapidly and shed runoff to red earths down-slope which absorb the water.

Even if only about 4 per cent of the average annual rainfall in this area is absorbed, the recharge of about 4 billion cubic metres probably almost balances annual usage. However, this depends on the 'health' of the reception area. Soil degradation and loss of plant cover in this recharge zone would eventually greatly decrease recharge. A further factor has to be taken into account. Water travels through the rock at only one metre a year. The patterns of groundwater flow have been mapped and over most of the Basin water flows from northeast to southwest, following the main structural and topographical gradient of the Eromanga Basin. Some water takes two million years to travel right across, as for instance, the water that reaches the Lake Eyre region from its start in the recharge zone east of Longreach. Near Windorah in the central deepest part of the Basin the water is about 1.4 million years old.

When the water reaches the far side of the Basin or when an aquifer layer comes in contact with a fault or a mass of granite or other geological body that blocks its way, the water may come to the surface as a seepage which forms a salt-pan area, or as a mound-spring or spring.

The same sediments which hold the water can be source rocks for hydrocarbons. Organic matter incorporated in them matures at depth to produce oil and gas. Accumulation of hydrocarbons (which are lighter than water) occurs where an impervious layer (cap rock) lies above a source rock. The artesian water co-exists with the hydrocarbons and many small oil fields occur in the Great Artesian Basin in southwestern Queensland and South Australia.

A salt lake has formed where saline seepage has been dammed by an embankment that used to carry the Ghan railway line alongside the Oodnadatta track. The seepage of artesian water and its heavy load of salts in this region results from the geological structure here where the Bulldog Shale of the Artesian Basin abuts against the Algebuckina Sandstone at the Basin rim. Samphires grow in the salt crust at the water's edge.

M.E.W.

An artesian bore at Yowah Opal Field. The hot water, from which steam can be seen rising, runs away along a drain through the landscape. The water here is foul-smelling and sulphurous. M.E.W.

floods very large boulders can be floated long distances among the roots of trees.) The sediments that contain the dropstones are of the same age as the Bulldog Shale, one of the widespread sedimentary rock strata laid down in the Eromanga Sea.

In parts of the Eromanga Basin that were the most southerly during Cretaceous times some of the Bulldog Shale contains pineapple-shaped crystal clusters known as glendonites[31]. (These are essentially carbonates which have replaced other minerals.) The presence of the particular kind of glendonite crystals in the Bulldog Shale is regarded by some geologists as evidence for near-glacial conditions in the areas in which they were formed. Such crystals are known to form only under very cold water conditions.

Elevation of parts of the Australian margin adjacent to the rift zone between Australia and Antarctica resulting in high mountains, and strongly seasonal climates on elevated areas in the high latitudes of the times, have been suggested as explanations for this postulated evidence of extreme cold. Rivers flowing from such elevated areas into the inland Eromanga Sea might have frozen periodically, and shallow embayments of the sea might also have done so. However, there is no firm evidence that any really high land existed adjacent to the rifting margins at this time, and carbonates are versatile and mobile and different pseudomorphs may form under a range of conditions in nature. Widespread Early Cretaceous glaciation is by no means proven, though high latitudes no doubt meant cold winters and generally cool temperate conditions over at least the southern part of Australia. Dark winters may well have seen ice in rivers, lakes and even shallow seas, enough to account for glendonites without invoking widespread glaciation.

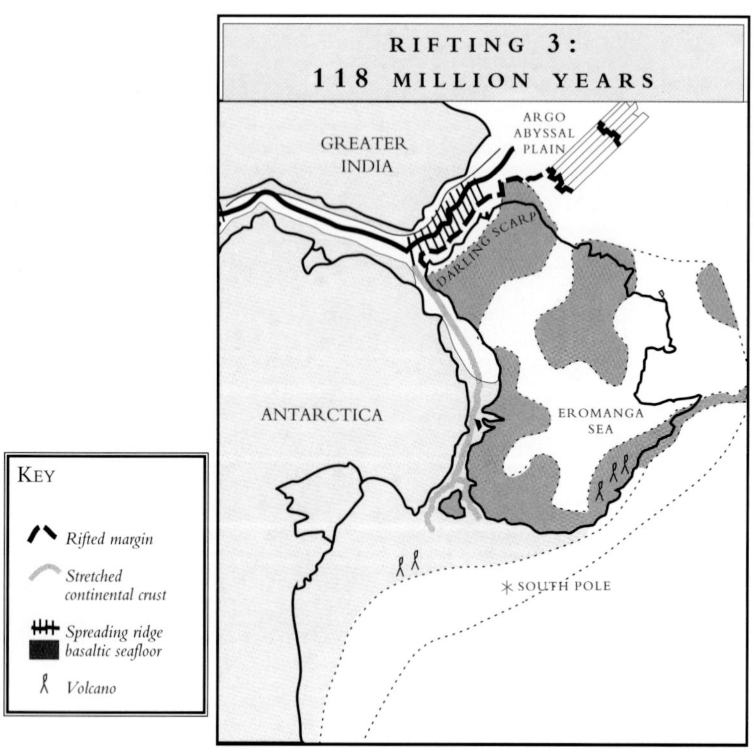

After Veevers *et al.* 42

The marine sediments laid down in the epicontinental sea and the terrestrial sedimentation that followed when the sea retreated and freshwater deposition returned to the Eromanga Basin formed sedimentary rocks that cap the Jurassic aquifer rocks of the Great Artesian Basin (the artesian water-bearing part of the Eromanga Basin). Water has presumably been travelling into and across the Artesian Basin, and has been stored under pressure there, from the time when the Great Divide formed in the Late Cretaceous. Such an underground water reservoir is obviously of great importance to Australia today. In addition it is the deep weathering of the Cretaceous rocks that were laid down in the ancient sea that has resulted in Australia's opal deposits.

India was moving away from Australia at a rate of a few (3-5) centimetres a year during the Indian Ocean sea-floor spreading stage between 132 and 96 million years ago, while continental extension between Australia and Antarctica continued at a slower rate of about 6 millimetres a year. The seaway in the developing rift between Antarctica and Australia was connected to the epicontinental Eromanga Sea via its Eucla Basin arm.

Rifting by crustal extension, which had proceeded from the Western Australian end and through the Great Australian Bight, changed direction when it formed the Otway and Torquay Basins and progressed further to create the Bass Basin. The Gippsland Basin[123] formed as a result of incomplete rifting of an arm of the Tasman Sea some 15 million years later.

Crustal stretching around Tasmania resulted in a change in Tasmania's position relative to mainland Australia. It came to lie in its present situation in relation to the Victorian coast, having moved southwards about 130 kilometres. (Between 160 million years ago when crustal extension started and 96 million years ago when the stretching phase was over and sea-floor spreading began, Tasmania constituted a third crustal plate, operating independently of Antarctica and Australia. The South Tasman Rise behaved like a fourth plate. It was by no means inevitable that Tasmania would go northwards with Australia in the final separations.)

The Early Cretaceous was a time of maximum global change, prompting evolutionary changes in flora and fauna. Lands were moving apart, and a network of rift zones partitioned the continental crust all over the world. These rift zones were very significant contributors to the evolution of the flowering plants. They provided diverse microclimates, new niches, areas of reduced competition and rapid change — environmental opportunities which became the driving force of a burst of evolutionary activity such as the diversification of angiosperms represented.

The rift zone between Africa and South America on the western side of Gondwana appears to have been the place where evolution of angiosperms began more than 130 million years ago, and the early flowering plants spread globally from there while lands were in close proximity. It was some time before any flowering plants reached the Australian region of eastern Gondwana, according to pollen evidence.

Not only rift zones provided opportunity and impetus for newly evolving angiosperms. After the continental flooding that affected all parts of the world to some extent, sealevel dropped. As the epicontinental seas retreated a wide belt of deltas, brackish lagoons and estuaries became established around the margins of the inland seas. A further suite of newly emerging habitats appeared — marshes, swamps, sea-bed sands, disturbed areas, zones where competition was reduced, and places that were unsuited to the ancient-style vegetation. Like weeds that invade disturbed areas on roadsides or cleared patches in forests, the angiosperms that arose to fill all the new niches so generously provided for them took advantage of the new opportunities. The very rapid evolution of all the families of flowering plants with plant types for all regions and seasons is a reflection of the wide range of choices that they had.

In southeastern Victoria the Otway and Gippsland Basins'[26] fossil flora and fauna provide a surprisingly complete picture of the life and times there between about 116 and 106 million years ago[27]. A great thickness of sediments accumulated in the rift zone of which these Basins were a part and they provide information on the history of the region. Rivers flowed from volcanic highlands on what Professor Ollier calls the Tasman Divide on the adjacent Lord Howe Rise (part of the New Zealand Subcontinent). These rivers brought volcanic sedimentary material into the rift basins. Braided streams flowed across the rift valley, spreading sand and gravels. The rift valley streams and rivers experienced seasonal flooding with meltwater from the snows on the volcanic mountains.

The climate is estimated to have been temperate and humid with seasonal extremes, and the three dark months of winter must have imposed a strain on the biota. In spite of this, more than 150 taxa of plants, invertebrates and vertebrates, including dinosaurs, have been recorded. The dominant dinosaurs were small- to medium-sized hipsilophodontids. They were agile creatures with large eyes adapted for night vision and it has been suggested that they might have been warm-blooded. Other vertebrate fossils include lungfish, fish, turtles, lepidosaurs, pterosaurs, freshwater plesiosaurs, theropods and birds. No crocodiles or mammals have been found. A rich invertebrate fauna has a preponderance of organisms that lived in the water — nymph stages of mayflies (Ephemeroptera) and flies (Diptera) being particularly abundant. Numbers of other insect groups are represented, including cockroaches, dragonflies, grasshoppers and beetles.

More than fifty different plant taxa have been recorded. No flowering plants have been conclusively identified. Ginkgoes and southern conifers dominated the forests. Wood of tree trunks shows marked annual rings, implying seasonality of climate, though dormancy due to months of darkness can account for times of no growth as well. The understorey contained pentoxylalean cycadophytes, tree-ferns, horsetails and mosses. Bracken-like ferns grew on the forest fringe, and the more open moorland was covered by clubmosses and peat mosses.

Most of the continent supported similar forests. In the New South Wales part of the Murray Basin there is evidence of catastrophic fires in the podocarp forests in the Early Cretaceous[84]. Large amounts of wood that has been converted to charcoal, with cell structure preserved, have been discovered in some strata.

Near Muttaburra in Queensland the *Muttaburrasaurus*, an ornithopod dinosaur, left its bones in the sediments in the Eromanga Sea. A crocodilian, an ankylosaurian dinosaur (*Minmi*), a sauropod dinosaur, a pterosaur and lungfish have also been found in Queensland.

In the opal fields near Lightning Ridge a diverse vertebrate fossil fauna occurs. Here the oldest Australian mammalian fossil — the opalised jaw of a monotreme, *Steropodon galmanii*, 110 million years old — was found. Turtle and crocodilian fragments, dinosaur bones, lungfish and bony fish remains are also part of the rich fauna. Some of the material has been converted to precious opal.

A Glendonite crystal from the Bulldog Shale in the Eromanga Basin. An indicator of cold water conditions in the Early Cretaceous.

JIM FRAZIER

The salt-scalded plain near the Oodnadatta track where the Bulldog Shale of the Great Artesian Basin meets the basin rim and the water in its aquifers comes to the surface.

M.E.W.

CHAPTER TWO

RIFTING BY SEA-FLOOR SPREADING

IN THE LATE CRETACEOUS AND EARLY TERTIARY

(96 to 50 million years ago)

At 96 million years ago there was a major change in the rifting processes round Australia[39]. India changed direction, by the abandonment of the spreading ridges on which the Indian Ocean had been widening up to this time[22]. These abandoned ridges are now the Lost Dutchman and Dirk Hartog undersea ridges and they trace India's early movement on a northwesterly azimuth. India's movement up to 96 million years ago was slow — a few centimetres a year. On its new due-north flight path its movement became rapid (between 15 and 20 centimetres a year) and continued apace until it collided with Asia between 55 and 50 million years ago. Relative movement between Australia and India ceased at 45 million years ago with the extinction of activity on the ridge that lies now in the Wharton Undersea Basin.

At the same time as India started its new phase of rapid movement, sea-floor spreading started in the rift zone between Australia and Antarctica. It was to proceed at a slow rate of only about 4 millimetres a year throughout the interval when India's northward movement was rapid[43]. During the rifting phase between 96 and 45 million years ago the seaway between Australia and Antarctica was to grow to a width of about 500 kilometres[37]. It widened by another 2600 kilometres at about 6 to 7 centimetres a year during the 45 million years of the island continent's rapid drifting phase.

The seaway between Australia and Antarctica (which was the proto-Southern Ocean while it was just a marine invasion of the sagging rift zone during continental extension) was becoming the Southern Ocean from 96 million years onwards as its true sea-floor was generated in the middle of the rift. At the start of sea-floor spreading the seaway was narrow and it only extended as far as the Otway Basin. The margins of the 360 kilometres of extension terrain remained as regions of rift floor

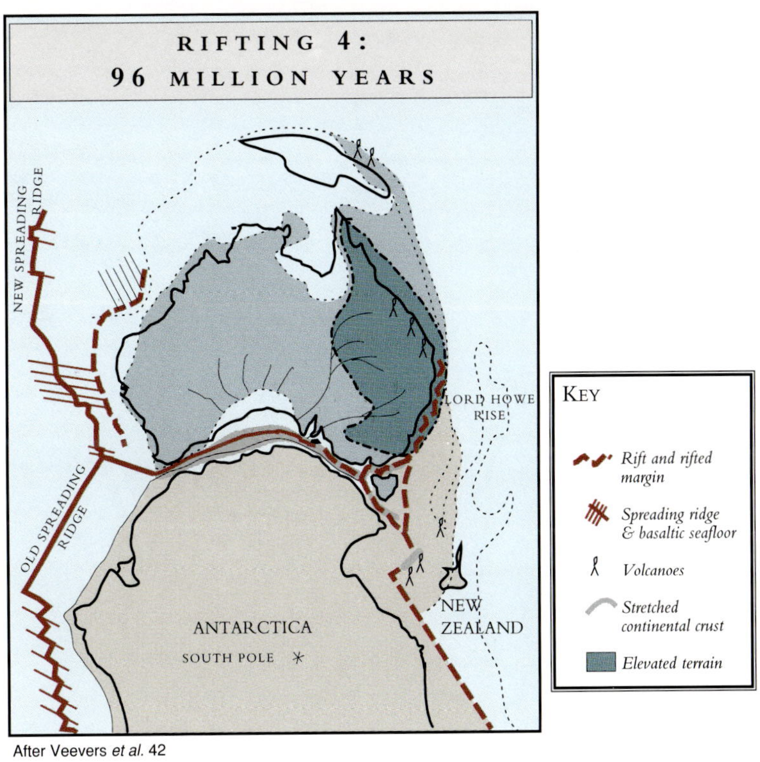

After Veevers *et al.* 42

GONDWANA DISPERSION AND CLIMATIC ZONES AT 80 MILLION YEARS

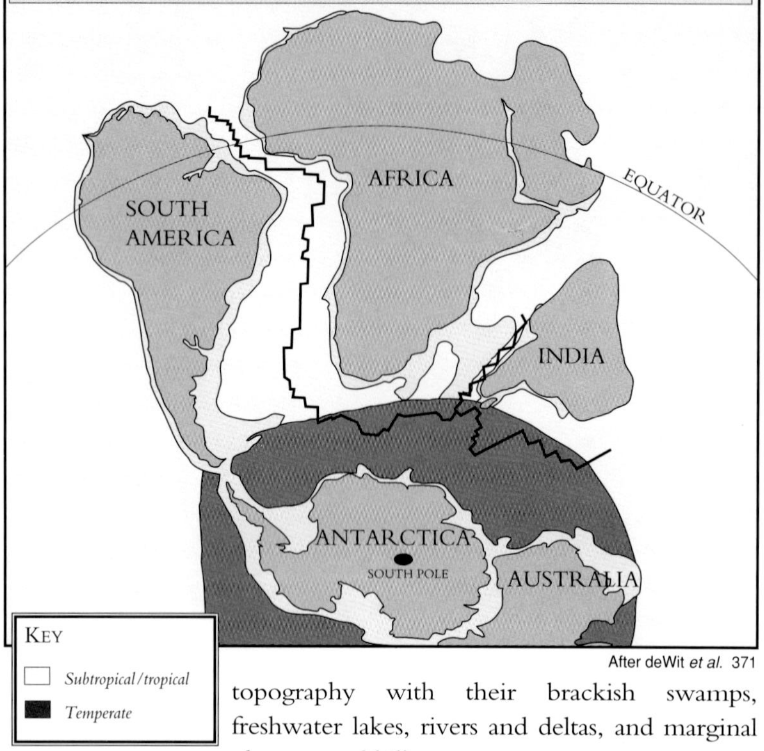

After deWit *et al.* 371

In the warm ice-free world of the Late Cretaceous there was little gradation in climate from the South Pole to the Equator.

topography with their brackish swamps, freshwater lakes, rivers and deltas, and marginal plateaux and hills.

The episodic nature of rifting and its speed in different zones was a globally orchestrated phenomenon. While one bit of crust was having a strong move, others were constrained. Movement between Antarctica and Tasmania/ South Tasman Rise continued to be slipping on a transcurrent fault during the slow spreading phase of the Southern Ocean. A land connection was maintained in that region by this trailing edge of the Australian continental crust.

Rifting between Australia and the Lord Howe Rise complex (the New Zealand Subcontinent) started also in the new phase of rifting activity that began at 96 million years ago. The Tasman Sea was opening from the South Tasman Rise northwards but the New Zealand sector was still in continental connection with Antarctica across the Campbell Plateau. Spreading of the Tasman Sea would cease during the 55 to 50 million year interval when India was colliding with Asia, this phase of seafloor spreading being one of the pulses of activity that were controlled by global tensions and relaxations. The Coral Sea Basin was starting to form between Australia and eastern New Guinea as rifting proceeded northwards.

Volcanoes had been active along what is now the Queensland coast (the edge of the continent there was then much further into the Pacific, incorporating the Queensland Plateau) and on the Lord Howe Rise during the earlier part of the Cretaceous[39]. From 96 million years ago, and perhaps related to the change in plate-tectonic regime, volcanic arc activity moved away from the Australian margin to end up some 3000 kilometres away on a line that passes through New Zealand today. The only volcanic activity on the Australian continent since has been due to the passage of the continent across 'hot spots' in the Earth's crust.

Most of the continent was well above sealevel by the Late Cretaceous and there was extensive and deep weathering. (Deep weathering was a global phenomenon in the hot and humid Late Cretaceous and again in the Early and Middle Eocene.)[81] Shallow sea still covered parts of the Perth and Carnarvon Basins and carbonate deposition was taking place along the western continental margin.

The eastern edge of the continent rose significantly from 96 million years ago to form a watershed — the Great Divide — in response to the Tasman Sea rifting[39]. On the seaward side of the Divide a strip of continental edge sagged, exaggerating the topography. Rivers on this side of the Divide began to cut gorges as they ran to the sea, and erosion of the edge of the watershed started the scarp retreat that has resulted in the dissected gorge and spur terrain that now characterises the 'ranges' of the Great Escarpment.

Drainage patterns altered on the continent and all the major rivers draining the highlands are believed by some authorities to have crossed the eastern half to enter the Ceduna Depression, a major depocentre (place where sediments

RIFTING 5: 84 MILLION YEARS

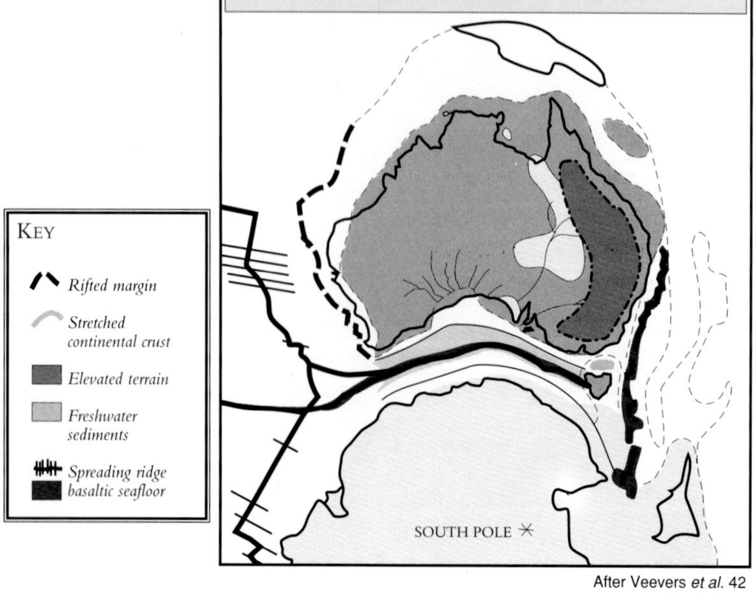

After Veevers *et al.* 42

were dumped) in the Great Australian Bight. There rapid subsidence was taking place as a result of the rifting and formation of the Southern Ocean. Other experts can find no evidence that major rivers from the northern sector of the Great Divide had this drainage pattern and, based on the distribution of Winton Formation rocks, believe that they drained towards the Gulf of Carpentaria. Rivers from the southern continental edge adjacent to the Bight flowed into the Ceduna depocentre.

By 84 million years ago, in Late Cretaceous times, a seaway about 100 kilometres wide separated Australia and Antarctica, but Tasmania was still in land connection with Antarctica. The Bass Basin was not connected to the sea — Bass Strait was a region of river floodplains, swamps and lakes — and the types of sediments that accumulated in the Basin reflect their river and lake deposition.

Drainage from the Eastern Highlands was still in a southwesterly direction with some of it reaching the Bight. A northward-flowing river system is believed to have extended across the Officer and Canning Basins in Western Australia to enter the Indian Ocean.

While the Tasman Sea was widening the New Caledonia Trough developed between the Lord Howe and Norfolk Rises on the sinking New Zealand Subcontinent. It is not certain whether there was island chain connection from Australia across the Chesterfield Rise to New Caledonia and perhaps via the Norfolk Rise towards New Zealand. We do not know, if there was such connection, for how long migration routes, important to the study of modern floras and faunas, might have persisted.

At about 80 million years ago, Madagascar, which had originally been attached to India's western margin on one side and the East African coast on the other, became a separate entity and it has maintained its island status since then. It has an interesting biota of Gondwanan origin and has primates that are related to African groups.

During the last 20 million years of the Cretaceous Period (between 85 and 65 million years ago)[36] sea-floor spreading in the oceans surrounding Australia continued. The transcurrent fault on which Tasmania was moving against Antarctica became a narrow seaway but there was crustal extension of the South Tasman Rise, maintaining links with the Antarctic continent.

Australia still lay in high latitudes during the

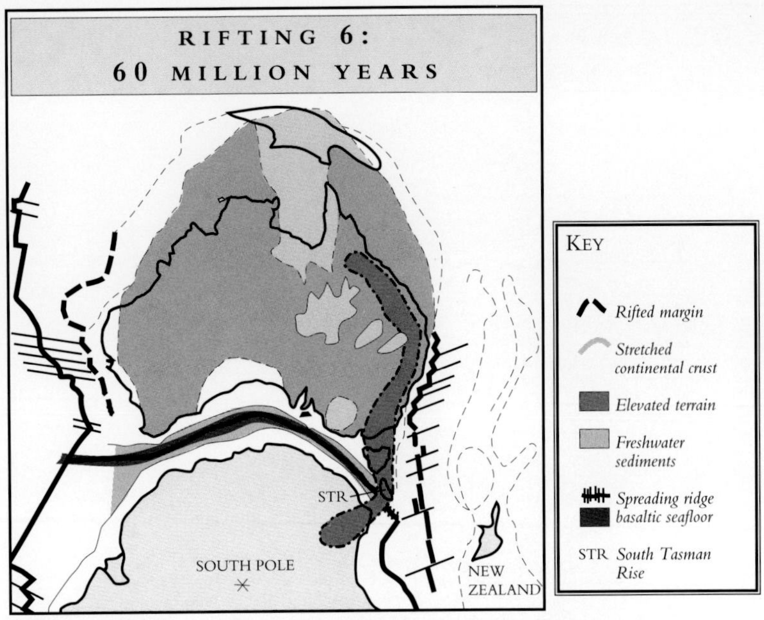

After Veevers *et al.* 42

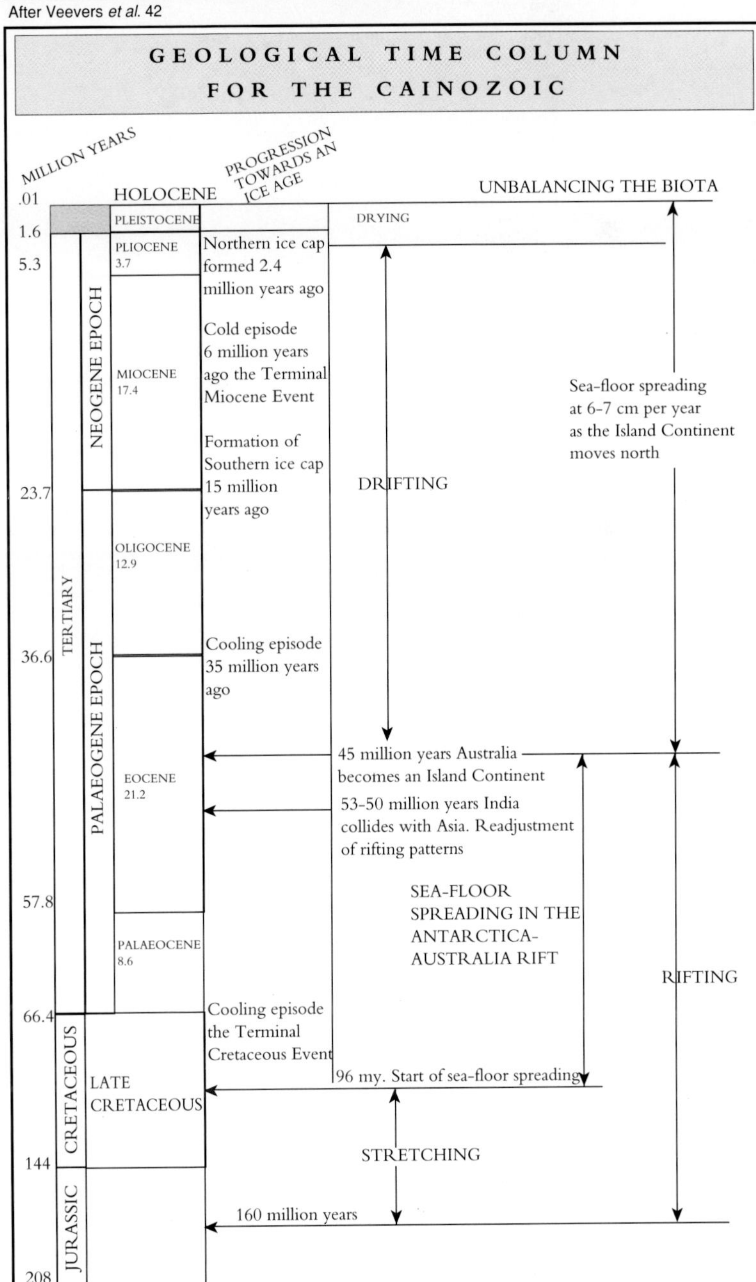

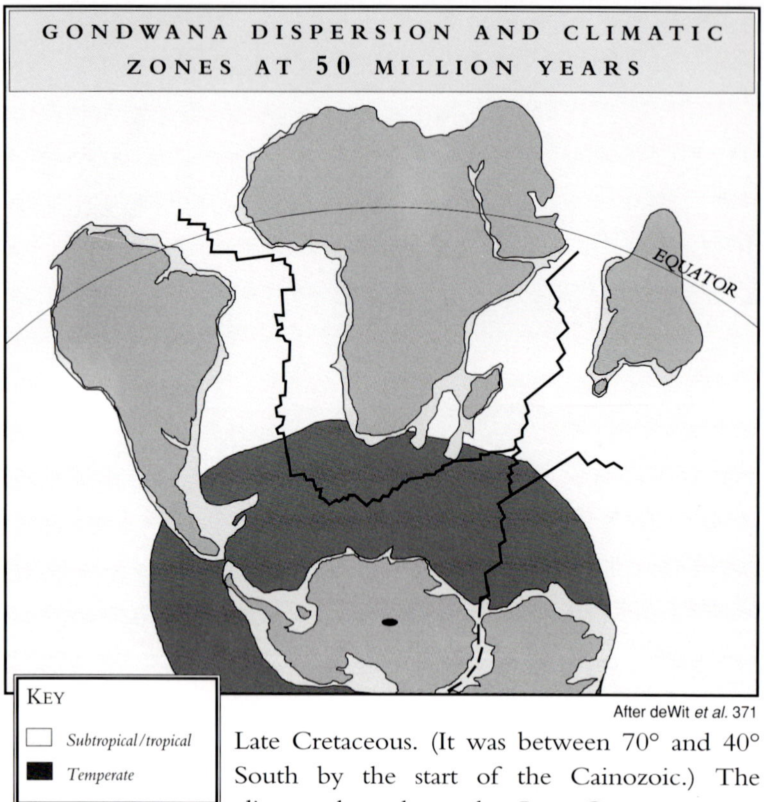

After deWit *et al.* 371

During the very warm 'golden age' Eocene, an ice-free world saw temperate climate in high latitudes and little gradation between the Equator and the Poles.

Late Cretaceous. (It was between 70° and 40° South by the start of the Cainozoic.) The climate throughout the Late Cretaceous was warm to hot and moist. Extensive river systems developed in the western half of the continent; drainages in the Eucla Basin region that had started to flow into the Ceduna depocentre after the inland sea retreated about 110 million years ago now connected to rivers flowing from the rising catchments further afield; major rivers entered the Indian Ocean along the Western Australian coast. In the eastern part of the continent, rivers spread thin sheets of sand over wide areas centred on the Gulf of Carpentaria, Lake Eyre and the Murray Basin.

The Wollombi Falls cascade down the rugged Great Escarpment, joining the flat plateau country (which rises gently to the Great Divide) to the coastal plain.

CLIFF OLLIER

The overall relief of the continent was becoming more like it is today. Deep chemical weathering was widespread, reflecting humid and hot conditions.

Much of the sediment carried by rivers found its way right off the continent to be deposited on the rifting margins. This left little in the way of sediment to form rocks of this age on the continent.

The end of the Cretaceous was marked by a sudden cooling event, which, coming on top of all the steady changes that had been occurring throughout the Period, added impetus to the change from ancient to modern-style flora and fauna. Dinosaurs and many other creatures became extinct in this Terminal Cretaceous Event. The cooler interval was brief and there was a return to warm to hot conditions in the Palaeocene. There is no evidence in Australia or New Zealand of massive extinction of flora during the Event.

In the Palaeocene epoch, the beginning of the Tertiary Period[17] between 65 and 58 million years ago, world oceans were warm.[18] The temperature of abyssal waters (those at the bottom of deep sea) around Australia was 12-15°C (in contrast to today's 2°C). Ocean circulation was sluggish, lacking temperature gradient from deep to shallow. Westerly wind systems were the major influence on the continent, but in a land as large as Australia, circulation in the interior was probably chaotic. Climate in the south was wetter than in the north. Conditions are believed to have been hot and perhaps seasonally arid in the northeast; temperate with high rainfall in the southeast.

In the Carnarvon Basin of Western Australia warm-temperate to subtropical conditions prevailed under the influence of an early Leeuwin Current.[24] In the Lake Eyre region the mean annual temperature is estimated to have been between 18-19°C with rainfall of about 1400 millimetres a year.

Slow sea-floor spreading at a few centimetres a year was continuing between Australia and Antarctica during the Palaeocene and Early Eocene. Between 55 and 50 million years ago, while India's northward rush was grinding to a halt as its path was blocked by Eurasia, patterns of rifting globally were again

The Great Divide and the Great Escarpment

The Great Escarpment, a line of 'ranges', runs east of the Great Divide watershed for almost the whole length of eastern Australia.[103, 105] It attains a height of several hundred metres and separates the coastal zone from the tablelands.

The Great Escarpment and the Great Divide originated about 80 million years ago when the New Zealand Subcontinent rifted from the eastern Australian margin in the rifting phase that initiated the Tasman Sea 96 million years ago. The continental edge was progressively upwarped, forming a watershed — the Great Divide — that is far inland (tens to hundreds of kilometres) from the new continental margin. The Great Escarpment has been formed by scarp retreat (the eroding-back of the edge of the upwarp).

The suggested order of events in the creation of the Divide and Escarpment are: erosion of the plains; initiation of major river patterns; widespread emplacement of basalt along the region; uplift and formation of the Divide and retreat of the Great Escarpment.

An undulating plain rises from Central Australia to the Great Divide[123], which is for the most part quite remarkably flat, and then descends towards the east coast where the Great Escarpment abruptly separates it from the coastal strip. The height of the Divide is variable along its length. From only 700 metres elevation near Goulburn, it rises to about 1000 metres in the north. To the south it rises steadily to Cooma and then rises abruptly to the 2230 metre section at Mt Kosciusko. Here major faulting with block uplift involving up to a kilometre of vertical movement occurred in Miocene times.

In some cases the present-day Divide seems to follow an ancient line which may represent the original palaeodivide. For example, the Gourock Range is a part of the Divide where tributaries of the Murrumbidgee River and the Upper Shoalhaven arise to flow on opposite sides of the watershed. They have the same gradient and no incision of their stream beds, features which are interpreted as showing great age. In other places the Divide has become a sharp ridge as a result of headward stream erosion on both sides. Plateaux like the Bulga and Comboyne Plateaux have been isolated by erosion east of the Divide in some regions.

West of the Divide, on the tablelands, relief is low and preservation of ancient landforms is common. On the eastern side of the Escarpment, in the coastal belt, the relief is great, geomorphic processes are rapid and few old landforms are preserved.

Elevation of the continental edge is probably a continuing phenomenon, though whether it was sporadic or steady over geological time is debatable. Whether it was rapid at first and the rate of uplift later tapered off is also uncertain. The episodes of volcanism that emplaced basalts along the Divide caused local uplifts, complicating the picture, and some authorities say that three major periods of volcanic activity[102] have coincided with three major phases of general uplift, rather than the uprising being an on-going process. The uplift of the Great Divide must be seen as two lines of uplift, with axes north-northwest from Brisbane northwards, and north-northeast south of Brisbane.[100]

Similar landforms which involve great divides and great escarpments have been produced in a similar manner on passive rift margins of other continents during supercontinent fragmentation.

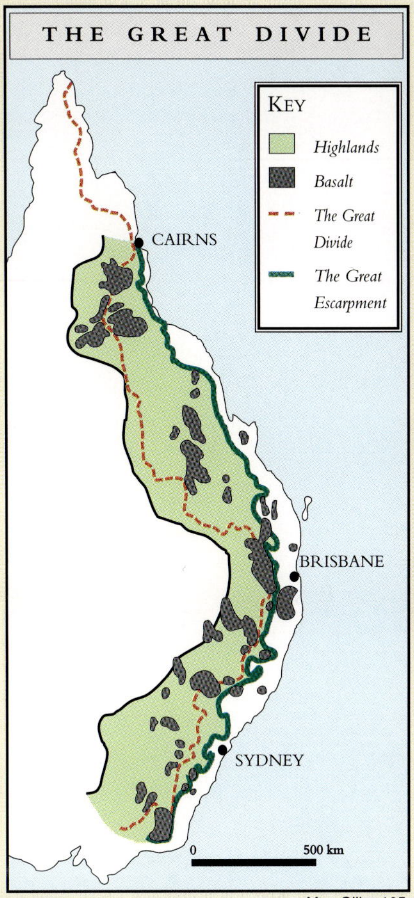

After Ollier 105

The Great Divide was created by upwarping of the edge of the continent and emplacement of basalt. The ranges of the Great Escarpment have been formed by erosion of the edge of the Divide.

Cliff Ollier

The Drakensberg in South Africa, bordering the Natal coastal plain; the Eastern Ghats in India; and the Guyana highlands in South America all formed when the continental margin was upwarped following rifting, and the 'ranges' were formed by scarp retreat.

Associated with the uprising of continental margins, has been the subsiding of the centre of the continents. In the case of Australia this has resulted in the 15 metres below sealevel 'sump' in the Lake Eyre Sub-basin.

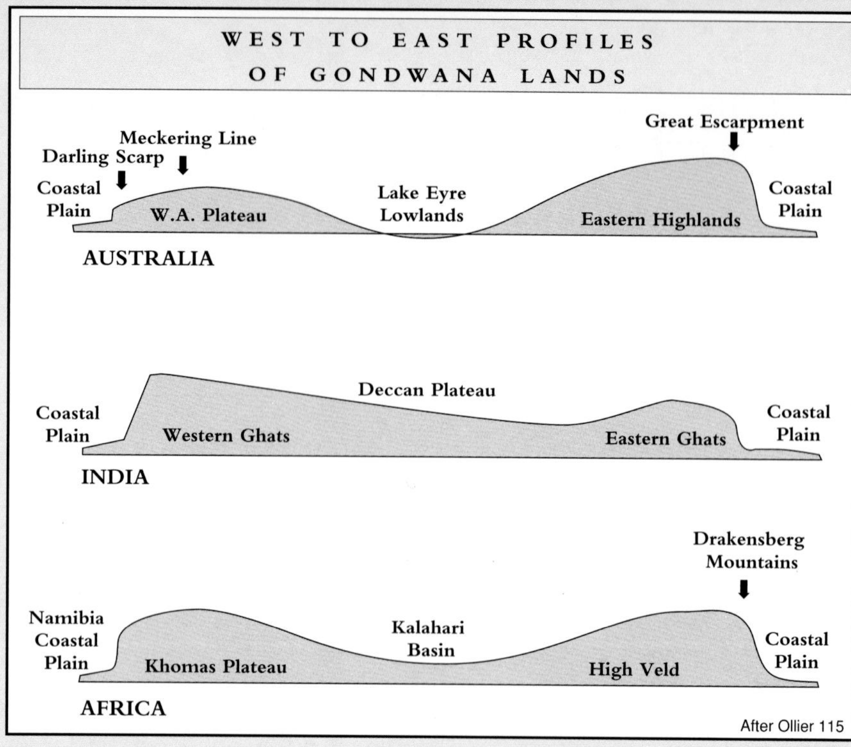

undergoing change. Spreading in the Tasman and Coral Sea Basins slowed and ceased during this interval and the distance between Australia and New Zealand has remained approximately constant since then.

A most significant mammal fossil, the first since the *Steropodon* jaw of 110 million years ago[177], was found at Murgon in southeastern Queensland in rocks 55 million years old. A minute tooth of a placental mammal, it created a stir among palaeontologists because it had been assumed, up to the time of its discovery, that placentals had not entered Australia in its Gondwanan cargo, and only did so when there was contact with Asia late in the Tertiary. This assumption had been used to explain why marsupials, deemed to be 'inferior' to placentals, had managed to thrive here. The Murgon site has revealed a few primitive marsupials; a primitive bat; a madtsoiid python which indicates that this snake group is Gondwanan; and a frog, turtle and crocodile. These fossils help to fill the blank in the fossil record between *Steropodon* and the vertebrate faunas of the latest Oligocene.

The Transantarctic Mountains in Antarctica were rising during the 50 to 55 million year interval, and very high mountains so near the South Pole would have encouraged accumulation of snow and ice, which were to become a major feature of Antarctica for the remainder of the Cainozoic. (Antarctica's gradual refrigeration resulted from its isolation when Australia had separated from it and the Drake Passage opened between South America and the Antarctic Peninsula probably between 40 and 30 million years ago, allowing a circum-Polar current to develop.)

A new global tectonic phase followed the 55-50 million year interval of re-adjustment around Australia. This phase would see Australia finally severing its connections with Antarctica by the movement of the trailing edge (Tasmania and the South Tasman Rise) clear of the Antarctic continental shelf. Australia's subsequent movement northwards as an island continent from 45 million years ago constitutes the Drifting section in this book. An account of the changing flora in Gondwanan lands during the Late Cretaceous and the Early Tertiary (Palaeocene and Early Eocene) follows, in order to establish the changes that occurred during that interval which preceded Australia's emergence as an island continent.

CHAPTER THREE

THE FORESTS OF GONDWANA

THE LATE CRETACEOUS TO PALAEOCENE FLORA OF SOUTHERN LANDS

(96 to 58 million years ago)

The origin of flowering plants in West Gondwana, and their radiation and spread north into Laurasia and south into Gondwanan lands, had taken place during Early Cretaceous times, from about 130 million years ago, within the Age of Conifers, Cycads and Ferns. The fossil pollen record shows that following the earliest 'generalist' pollen that cannot be assigned to any family[9], pollen of lilioid, magnolioid and lauralean type appeared, signalling the start of the Age of Flowering Plants (and the start of the Age of Mammals). Thereafter, early angiosperms progressively and rapidly adapted during the interval between 125 and 100 million years ago to fill all the niches in which competition was reduced. Many of the southern (Gondwanan) plant families had evolved by the end of the Cretaceous, as evidenced by pollen of PROTEACEAE, the banksia family; MYRTACEAE, the lillypilly, teatree and eucalypt family; RESTIONACEAE, rush-like plants; CASUARINACEAE, the river oak and desert oak family; and many others.

Until very late in the Cretaceous, forests were conifer-dominated. The emerging angiosperms filled the weed, fringe and disturbed niches available in the rapidly changing world. From about 100 million years ago, tree forms of flowering plants existed which were capable of competing directly with conifers as forest dominants, and by the end of the Cretaceous, flowering plants were increasingly important elements of the forest vegetation worldwide.

Each of today's southern lands has a fossil history that tells the story of the early steps towards the development of broadleaf Gondwanan forest.

AFRICA

After the initial dispersal of early angiosperms in the Middle Cretaceous, distinct northern and southern floral provinces became established in Africa. The northern province, with a warm tropical climate and seasonal aridity, became part of the *Palmae Province* that extended through all lands in low latitudes in the Late Cretaceous, following on an ephedroid flora (characterised by *Ephedra* of the family GNETACEAE) which had occupied the region originally.

The climate of the southern province was warm, humid, subtropical to warm temperate, and the southern conifer podocarp-araucarian forest that grew there in the Early and Middle Cretaceous was gradually invaded by increasingly dominant flowering plants. Because of Africa's geographic situation, its closeness to Eurasia and its isolation from South America and other Gondwanan lands after the South Atlantic opened, modern floral similarities to

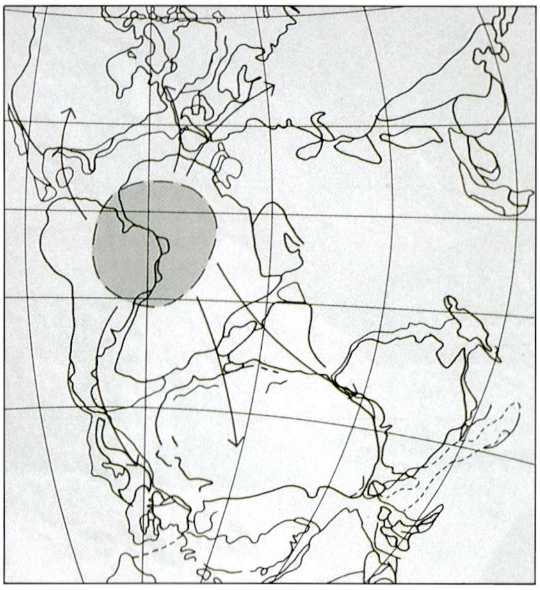

From a centre in West Gondwana the early families of flowering plants were able to spread into Northern and Southern Hemisphere lands while the supercontinents were more or less intact.

to which Australia remained attached for another 45 million years after Africa's total separation from South America.

The large number of southern families of flowering plants shared by Australia and southern Africa testify to the early appearance of these families. It is of interest to note that CASUARINACEAE, now absent from Africa, has a fossil record in that continent.[2]

India is also mainly outside the scope of this chapter by reason of its early flight from Gondwana, at which time it took with it a Gondwanan component of early angiosperms, which, like Africa, it had acquired in the first distribution of flowering plants across a still-connected supercontinent.

Australian plants result from the very early dispersion of ancestral southern flowering plants to all the lands when Gondwana was still more or less intact.

The South Atlantic was opening between South America and Africa from about 130 million years ago, starting in the south and propagating northwards, so that until about 90 million years ago there was still connection in the north. The story of the evolution of Africa's flora through time is thus of less interest as background to Australia's story than is information from South America and Antarctica,

SOUTH AMERICA

The evolution of a 'mixed' flora, described as the *Paleoflora Mixta* by the South American palaeobotanist Edgardo Romero[28], in Patagonia during the Late Cretaceous is of great significance. This flora is 'mixed' in that it contains elements which today occur in subtropical regions, and others which are plants of cold temperate modern floras. This flora evolved *in situ* from the warm Cretaceous vegetation of Patagonia, and by migration of subtropical species from the north and of more temperate ones from the south. The flora was

The wet tropical rainforests of northeastern Queensland are the closest living forests to the ancient Gondwanan forest that was widespread in all parts of the super-continent before Australia separated from Antarctica.

JIM FRAZIER

NOTHOFAGUS AS AN INDICATOR OF CLIMATE

Nothofagus is a significant plant in the fossil record because it is an indicator of climatic conditions. *Nothofagus* pollen is of several different types, each borne by a number of species, living and fossil. The *brassii* variety occurs in species which require warm temperate to subtropical and perhumid conditions without any seasonal aridity. Species with *fusca* and *menziesii* pollen types are cool temperate and can cope with a little seasonal dryness. The *Nothofagadites senectus* type of pollen, which is the earliest to appear in the pollen sequence, used to be equated to the *brassii* type. It is now considered to be an ancestral type and not the same as any living pollen group. No species can survive lengthy dry intervals, so as Australia progressively dried out *Nothofagus* became confined to refuges where its preferred conditions still existed or migrated into suitable areas (like the New Guinea mountains) when they became available.

DICK BURNS.

to develop further in the Palaeocene and would continue to thrive during Eocene times (58-36 million years ago). *Paleoflora Mixta* assemblages, vegetation of types that have no modern analogue, were to occupy all southern lands then still connected — South America, Antarctica, Australia and New Zealand — before declining and disappearing as entities in Oligocene times.

Late Cretaceous floras of South America show climatic zonation[29] and rich diversity. The northern province was characterised by American forms and among the flowering plants palms dominated, this region being part of the *Palmae Province* that was a feature of tropical regions worldwide in the early stages of flowering plant distribution. Also in this province were the early examples of well-known modern plants like *Ficus* of the fig family MORACEAE; *Nectandra* (LAURACEAE, the laurels); and *Theobroma fossilum* (fossil cocoa — STERCULIACEAE).

Late Cretaceous floras of southern South American provinces show the forests of podocarp conifers (*Podocarpus* and *Dacrydium*) with a rich fern component including *Hymenophyllum* (filmy ferns), *Gleichenia* and *Blechnum*, being progressively invaded by angiosperms. About 30 different species of leaf imprints, most of which have been

Palms, like these slender ones that form groves on Cape York Peninsula, are members of an ancient family of flowering plants, the ARECACEAE.

JIM FRAZIER

attributed to Early Cretaceous forms from North America, constitute the flowering plant element of this flora.

During the Palaeocene, the northern province of South America was still tropical with abundant palms, including *Nypa* the mangrove palm, and with *Bombax* (BOMBACACEAE, kapok trees), *Celtis* (ULMACEAE, the elm family), *Anacolosa* (OLACACEAE, the olive family) and *Psidium* (the genus to which the guava belongs, family MYRTACEAE) appearing in the pollen record.

A Late Palaeocene flora in northwest Patagonia dated at 59 million years ago contains nearly 100 species of leaves, of which 69 per cent have entire margins, a feature diagnostic of paratropical to seasonally dry rainforest. Several genera including *Annona* (ANNONACEAE) and *Cochlospermum* (COCHLOSPERMACEAE) that today form part of the present Neotropical Region, and others that now characterise the Antarctic Region are elements of this 'mixed' flora. (*Lomatia*, of family PROTEACEAE and *Peumus* of MONIMIACEAE are examples of genera of the *Paleoflora Mixta* that now occur in the Antarctic Region flora.) Some genera of eastern Gondwanan origin in the 'mixed' flora are *Dacrycarpus* (PODOCARPACEAE), *Dicranopteris* (a fern), and *Beauprea* (PROTEACEAE). *Gunnera* (GUNNERACEAE) is also present and may have had a northern Gondwanan origin. The Palaeocene vegetation type of southern South America approximates to the present-day vegetation of southern Brazil, some 2000 kilometres further north, confirming the subtropical climate of the times. Crocodile fossils of Palaeocene age from southern South America also testify to the subtropical conditions that prevailed while the 'mixed' flora flourished.

ANTARCTICA

Antarctica, which is still largely the mystery continent, hiding its information under its immense ice cap, is very significant in the story of flowering plant evolution. From latest Jurassic to the end of the Cretaceous, Antarctica's land vegetation comprised a series of evergreen coniferous rainforests. Similar forests of podocarp and araucarian conifers, like the forests in Australia, New Zealand, India and parts of southern South America at the same time, grew on the elevated land on the volcanic arc that now forms the Antarctic Peninsula.

Petrified forests of large tree trunks of Cretaceous age occur on Seymour Island.[15] These trunks have wood with consistently wide and uniform growth rings, indicating that during the time when the land was forested conditions were very favourable for tree growth.[16] (Other petrified wood of Eocene age on Seymour Island shows similar uniform growth rings, indicating uniformly favourable conditions then also.) The forest composition and the tree-ring patterns are comparable to those of cool- or warm-temperate forests like those of New Zealand, Tasmania or montane New Caledonia. However, some petrified wood dated at 65 million years old, at the transition from the Cretaceous to the Palaeocene, shows a radical short-term change in growth pattern, in accordance with evidence from other sources of an abrupt, temporary climatic change at this time.

The Antarctic Peninsula pollen record shows that the podocarp[5] and araucarian forest contained very large numbers of fern species[6], some mosses, liverworts and lycopods (clubmosses). The first evidence of flowering plants in the Antarctic Peninsula is pollen (*Clavatipollenites*, CHLORANTHACEAE) from 120 million years ago. Comparable pollen was widely dispersed, but not abundant, in eastern Australia at much the same time. Angiosperms were becoming increasingly common during the Late Cretaceous.

Pollen of southern or Antarctic beech — *Nothofagus* — occurs in the forest pollen assemblage for the first time about 80 million years ago, and fossil leaves of the plant are found in strata about 80 million years old also on Seymour Island, representing the oldest record of *Nothofagus* macrofossils. *Nothofagus* belongs to the same family as the beech trees of the Northern Hemisphere, FAGACEAE. (It is unfortunate that in Tasmania the living *Nothofagus cunninghamii* is known as 'myrtle' or 'myrtle beech' making it appear to belong to the MYRTACEAE. Common names, apart from being by no means consistent, are often misleading.)

The Antarctic Peninsula vegetation was richer in *Nothofagus* species than the contemporary vegetation of Australia and had less diversity of PROTEACEAE.[74] Tall open-forests were dominated by podocarps and the beeches, and dry-zone areas probably supported *Beauprea* and *Xylomelum* (woody pear) — members of PROTEACEAE. In wet areas *Sphagnum* moss (peat moss) and the water fern *Azolla* were abundant.

Pollen of the Huon pine appeared in the Antarctic Peninsula about 80 million years ago, having appeared in New Zealand some 10 million years earlier. MYRTACEAE pollen is present in the Late Cretaceous on the Peninsula, long before it became widespread in the Australasian region at the start of the Tertiary. (Some myrtaceous pollen has been recognised in latest Cretaceous sediments in the Bight Basin.)

Spore and pollen records from James Ross Island, and other smaller islands next to the Antarctic Peninsula, also show the podocarp-araucarian forests of mid Cretaceous times being modified by the introduction of *Nothofagus* about 80 million years ago, and thereafter the flowering plant component increases rapidly.

Palaeotemperatures for the Late Cretaceous in the Antarctic Peninsula[74] are estimated to have been 14°C (mean annual temperature), declining to 12°C by the end of the Period. While the pollen floras of the Antarctic Peninsula are similar to pollen floras of Australia and New Zealand, and to a lesser extent show similarity to forest floras of Patagonia, the Falkland Plateau region appears to have had a dissimilar flora through most of the Cretaceous. There climax vegetation of dry-zone and mangrove conifers of cheirolepidaceaen type (an extinct swamp-conifer group) was in sharp contrast to the tall forest vegetation on the Antarctic Peninsula until late in the Period. Then the podocarp-rich coniferous forests spread over the Plateau as well. It appears that there was pronounced regionalism within southern Gondwana, with a zonation of forest communities across the latitudes, and of understorey associations in an east-west direction.

NEW ZEALAND

New Zealand has a Middle Cretaceous macroflora in rocks[3] that were laid down in a coastal region of South Gondwana about 110 million years ago. The locality, in the Clarence Valley, South Island, was then within the Antarctic Circle. The flora contains thalloid and leafy liverworts, a horsetail, several ferns, several bennettitalean cycadophytes, a pentoxylalean cycadophyte, *Ginkgo* sp. (a maidenhair tree), araucarian conifers (both *Agathis*, kauri, and *Araucaria*), many podocarps, some pteridosperms (seed ferns), and angiosperm leaves. The angiosperm leaves are of at least 20 different dicotyledonous genera — a broadleaf component of temperate forest in high latitudes.

New Zealand's Late Cretaceous pollen flora[19] is similar to that of Antarctica and Australia.[25] Palaeotemperatures during the Late Cretaceous are estimated to have been between 15°C and 18°C[74] (mean annual temperature), declining to 14°C by the end of the Period.

The first appearance of *Nothofagus* pollen is in the Paparoa beds (dated at about 80 million years), and the three pollen types *brassii*, *fusca* and *menziesii* were soon established and *Nothofagus* was abundant. The flowering plant families PROTEACEAE, LORANTHACEAE (mistletoes), GUNNERACEAE, WINTERACEAE, LILIACEAE (lilies), CARYOPHYLLACEAE, EPACRIDACEAE, GNETACEAE (*Ephedra*), CHLORANTHACEAE, CASUARINACEAE and many others were present in the conifer-beech forests. A rich fern, moss, liverwort and lycopod flora was also present.

The Palaeocene flora is similar to that of the Late Cretaceous, though there was proportionately more podocarp pollen and less southern beech, and a number of tropical elements, including *Anacolosa* (OLACACEAE) and palms of several genera including *Mauritia*. The presence of palms of genera that today have tropical distribution, and other subtropical or tropical elements that coexisted with cool temperate forms identify this flora as a *Paleoflora Mixta* which has no modern counterpart. The family MYRTACEAE appears in the pollen record with both *Metrosideros* (rata) and *Leptospermum* (teatree) types, and TILIACEAE is also recorded.

THE LATE CRETACEOUS FLORA OF THE ANTARCTIC/AUSTRALIAN RIFT

Sediments deposited in the rift zone between Antarctica and Australia in the Late Cretaceous contain a pollen record of the flora of the times.[19, 6, 74] Rocks laid down on the Australian side of the rift in the Otway Basin

Above: Lomatia silaifolia *is a member of the* Banksia *family,* PROTEACEAE. *This species regrows from a lignotuber after fire or drought has destroyed its aerial parts. It has become an endangered plant because it has been over-exploited by cut-flower sellers, growing as it does on sandstone in the Blue Mountains region adjacent to Sydney.*

Right: The blueboy, Stirlingia, *one of Western Australia's endemics, is much sought after as a cut flower. It is an example of a sclerophyllous plant, adapted to poor sandy soils and periods of aridity. The origin of the genus in the rift zone between Australia and Antarctica late in the Cretaceous implies that conditions suiting plants with such adaptions existed then.*

KINGS PARK HERBARIUM

and in the Bight/Duntroon Basins, contain a record of the flora that fringed the northern margin of the embryonic Southern Ocean. On the Antarctica side of the rift the Late Cretaceous sedimentary rocks contain pollen from plants which grew on the southern margin of the Ocean. Evidence from the Otway Basin side of the rift showed that the rift valley was the cradle for *Ilex* (holly — AQUIFOLIACEAE) and for three members of PROTEACEAE — *Beauprea*, *Knightia* and *Gevuina/Hicksbeachia*. In other words the rift valley was a centre where diversification of proteaceous plants took place in the Late Cretaceous. *Beauprea* is today an endemic in New Caledonia; *Knightia* is now confined to forests in New Zealand and New Caledonia; *Gevuina/Hicksbeachia* pollen type occurs in the Tertiary in Australia and *Gevuina* ranges to Papua New Guinea and occurs also in Chile today while *Hicksbeachia* occurs only in rainforests of northeastern Australia.

Other pollen identified in the Otway sediments is of *Gunnera* (GUNNERACEAE), WINTERACEAE, EPACRIDACEAE, and TRIMENIACEAE. *Callitriche* (CALLITRICHACEAE, water weeds known as water starworts in Britain)[74] and aquatic ferns grew in the waterlogged areas. *Nothofagus* of the *brassii* pollen group type is also present. The podocarp conifers are represented by three genera: *Podocarpus* (plum pine), *Dacrydium* and *Dacrycarpus*. It seems likely that *Dacrydium*, like many members of PROTEACEAE and *Ilex*, may have had its origin in the rift, on the fringes of the infant Southern Ocean. Araucarian conifer pollen is also abundant. Mean annual temperature during the Late Cretaceous in the Otway Basin is estimated to have been between 16.6°C and 19°C.

On the southern side of the rift, recent investigations[7] have been carried out on marine muds from a range of localities[8] in the Ross and Weddell Seas fringing Antarctica. The muds result from the weathering of Late Cretaceous rocks. The pollen contained in the rocks has survived breakdown of the rock to mud, and supplies exciting new information. The pollen record for the southern margin of the rift, like that of other southern regions, shows podocarp-araucarian forests being invaded by angiosperms.

Both rainforest and sclerophyllous members of PROTEACEAE, in addition to the three forms identified in the Otway Basin, had their origins in this region: within the subfamily Proteoideae, the genera *Adenanthos* and *Stirlingia* (Western Australia's woollybush and blueboy), both of which are today restricted to the southern mediterranean-climate regions of the continent; in subfamily Persoonioideae, *Persoonia*, which today occurs in higher rainfall areas of eastern and northern Australia; in subfamily Grevilleoideae, *Grevillea* and *Telopea* (waratah), now found in eastern Australasia; in subfamily Carnarvonoideae, *Carnarvonia*, which is found in rainforest in eastern Australasia. *Macadamia*, a member of the PROTEACEAE cultivated today for its nuts, makes an appearance and confirms what its primitive floral characteristics had suggested to systematic botanists — that it was among the oldest of the proteaceous genera. Today *Macadamia* occurs in

The woollybush, Adenanthos, *is a Western Australian endemic. Pollen of the genus appeared in the Late Cretaceous in sediments that accumulated in the rift valley between Australia and Antarctica.*

ELIZABETH GEORGE

Lillypillies are fleshy-fruited members of the family MYRTACEAE to which the eucalypts, tea trees and paperbarks (all of which have dry fruit types) belong.

DENSEY CLYNE

The family WINTERACEAE is an ancient family, among the earliest to appear in the pollen record. It is closely related to the Magnolia group of families.

Above: Telopea truncata, *a waratah. Pollen of the genus occurs in the rift valley sediments between Australia and Antarctica.*

JEANETTE CLOSS

Left: Banksia grandis *growing in the Stirling Ranges of Western Australia.*

M.E.W.

Below left: Grevillea eriostachys *is a widespread species with varieties in semi-arid and sandplain regions.*

M.E.W.

rainforests of northeastern Australia, New Caledonia and eastern Malesia.

Pollen of the conifers *Microcachrys* and *Lagarostrobus* (Huon pine) is found in the muds on the Antarctic side of the rift with other podocarp and araucarian conifer pollen. Dredge samples from the Great Australian Bight adjacent to South Australia suggest that the vegetation in that part of the rift zone and probably in the south-central part of the continent was dominated by temperate (perhaps cool temperate) rainforest. Podocarps were abundant, particularly Huon pine. *Dacrydium*, *Microcachrys* and several species of *Podocarpus* together with araucarians grew in the forests where the understorey was of tree-ferns and mosses, including *Sphagnum*. The flowering plants present were *Ilex* and the proteaceous genera that had appeared in other parts of the rift zone in the Late Cretaceous, associated with *Gunnera* and rare *Nothofagus*.

The Gippsland Basin, outside the rift zone, has a Late Cretaceous pollen flora with less species diversity, and some of the PROTEACEAE appear later in the sequence than they do in the Otway Basin. The sclerophyll element appears to be absent, and the communities are perhumid forest with the canopy dominated by podocarps, and an understorey of ferns, clubmosses and mosses.

In the context of the modern Australian vegetation, the composition of the rift zone flora is particularly interesting as many of the taxa found in the rift sediments are today found only in the montane vegetation of northeastern Australasia. In addition, study of the rift flora and of fossil floras of the same age in other southern lands has made it possible to deduce where some taxa originated within Gondwana.

Evidence of the rift valley system acting as a locus for the evolution of new plants suited to all the niches that occur within rifts confirms the concept that rifting provided some of the environmental changes which were the driving force behind the evolution of the angiosperms. Charles Darwin had been among the first to see that environmental change is the driving force that promotes evolutionary change, with natural selection acting as the agent. He would have been delighted to see how well his concept worked in rift valleys during supercontinent break-up.

An interpretation of the vegetation types represented by the rift valley pollen flora is as follows[74]: A mosaic of communities is suggested. Tall open-forests had podocarps, araucarians and PROTEACEAE (like *Macadamia* and *Knightia*) in the canopy and other PROTEACEAE (like *Telopea*) with WINTERACEAE, TRIMENIACEAE, *Ilex*, *Gunnera*, and ferns, clubmosses and mosses in the understorey. The open nature of the forest was a requirement in the high latitudes of the times. A low-angle sun would not have supplied sufficient light for layered forest with understorey, and with the darkness/low light intensity during winter months it was necessary for the forest to be adapted to the conditions which prevailed so close to the Pole. (Closed forest/rainforest with high species diversity as we know it today would not have been viable in such high latitudes.) The conical-shaped crowns of the Cretaceous forest dominants in the tall open-forests allowed entry of enough light for a shrubby understorey and a ground cover of ferns and mossy plants.

Sclerophyllous PROTEACEAE (*Stirlingia*, *Adenanthos*) and EPACRIDACEAE probably grew on the fringes of the forest on soils which were nutrient-poor or waterlogged, and aquatic ferns and *Callitriche* waterweeds grew in the swamps.

Flora of the Palaeocene and Early Eocene in Australia

Podocarp conifers, including Huon pine, dominated the Early Palaeocene floras in the Otway and Gippsland Basins. The araucarian conifers *Agathis* and *Araucaria* were also abundant. *Nothofagus* was rare, and *Anacolosa* (OLACACEAE) and *Beauprea* occasional. High rainfall and cool to warm temperate conditions are suggested by the forest flora.

At other localities of Early Palaeocene age in mainland Australia and Tasmania, similar conifer forests occurred, indicating that high rainfall was a feature even of some inland areas round Lake Eyre where some of the localities are. *Cupania* (SAPINDACEAE) is present and pollen of MYRTACEAE becomes abundant at some localities.

The Bureau of Mineral Resources Muloorina 2 bore hole[101], at a locality just south of Lake Eyre and close to Lake Eyre South, contains a pollen record which spans the Late Palaeocene to Early Eocene interval. The Late Palaeocene vegetation here was dominated by the family CUNONIACEAE, of the section which has a tricolpate pollen type, and 67 per cent of the pollen sum in samples was of this kind. (Living representatives of this section of the CUNONIACEAE are *Pseudoweinmannia*, *Pullea*, *Caldcluvia* and *Vesselowskya*, which are all important canopy, sub-canopy and stream-side components of the eastern Australian closed forests.) Pollen of the other section of the family (dicolpate type) is present only in small amounts. Living representatives of this section of CUNONIACEAE are *Ceratopetalum* (New South Wales Christmas bush) and *Schizomeria*. (High proportions of tricolpate CUNONIACEAE pollen have been recorded from the Late Quaternary of northeastern Queensland. Their abundance there is taken to indicate an expansion of simple notophyll vine forest at the expense of 'drier' vine forest with araucarians and to be related to increase in rainfall. Mean annual temperature of 18–19°C and at least 1400 mm of rain are suggested conditions for this recent forest type. By inference such a climate must have prevailed in the Palaeocene/Early Eocene.) MYRTACEAE, araucarian and podocarp conifers, and native cypresses (*Callitris*) were fairly common and PROTEACEAE was slightly less abundant. CASUARINACEAE was present in very low frequency and *Nothofagus* was a rare component of the pollen flora. Tree-ferns were abundant in the Late Palaeocene but much less so in the Early Eocene.

The Early Eocene assemblage in the Muloorina Bore was similar to the Late Palaeocene assemblage though conifer pollen was less abundant and MYRTACEAE and PROTEACEAE were increasingly dominant. From the middle of the Early Eocene MYRTACEAE becomes co-dominant with CUNONIACEAE, and it has taken over dominance by the Middle Eocene.

In Western Australia the mangrove palm *Nypa* occurs in the Kings Park Formation[264], which is largely of Palaeocene age. A diverse proteaceous flora is associated, and podocarps and *Nothofagus* were not abundant. Ferns, however, were very abundant. This flora suggests that conditions locally were very humid and warm. An early Leeuwin Current bringing warm waters down along the Western Australian coast from the north is the probable reason.

In the Lake Eyre Sub-basin the Early Eocene forest was interrupted by floodplains where *Sparganium* (burr reed) was abundant and *Gunnera* grew in the marshes.[101] *Gymnostoma* (CASUARINACEAE) was common in some forest communities, while araucarians and podocarps or CUNONIACEAE and MYRTACEAE dominated in others. The presence of palms, EUPHORBIACEAE, *Anacolosa* and the xeromorphic proteaceous genus *Xylomelum* (the woody pear) in some assemblages is indicative of warm, possibly seasonal conditions. MALVACEAE, the *Hibiscus* family that loves warmth, was sometimes abundant.

An Early Eocene flora from Regatta Point in Tasmania[67] contains araucarian material of a new genus *Araucarioides* which has characteristics that make it intermediate between the living genera *Araucaria* and *Agathis*.

During the Late Palaeocene and Early Eocene widespread rainforest may have extended over most of the continent. Globally this interval was warm and humid Polewards in both hemispheres. The Early Tertiary was a 'golden age' of warmth and high rainfall. In Australia a wet and warm climate prevailed over the southern half, possibly a drier and warmer one over the northern half. It appears that different communities formed a mosaic then as now. Some areas were dominated by angiosperms. While CUNONIACEAE was very abundant in some, MYRTACEAE dominated others. Some forests in wetter areas had an understorey of tree-ferns, ferns and mosses, others in drier parts had only a small fern component. In some communities conifers and angiosperms were equally abundant, in others one or the other dominated. *Nothofagus* was sometimes present as a minor element, in others it was absent.

Thus we see that by the Middle Eocene when Australia became an island continent and its drifting phase began it was a well vegetated land with diverse forest communities forming a mosaic over much of the land. The forests were essentially *Paleoflora Mixta* types and *Nothofagus* was sometimes present though by no means ubiquitous or a major component.

Mesas of the breakaway country near Coober Pedy. The weathered profile grading from red-brown to white and the 35 million year old duricrust capping are striking features of this desert landscape.

M.E.W.

Occasional Acacia *bushes grow on the silcrete-capped ridges that rise above the saltbush plains of the breakaway country near Coober Pedy. The pallid zone of the weathered profiles in the mesas makes white splashes in the red gibber desert.*

M.E.W.

The silcrete capping of the mesa, product of an arid phase about 35 million years ago, has protected the regolith below it from erosion

M.E.W.

CHAPTER FOUR

WORN-OUT SOILS, HARD CAPS AND INVERTED LANDSCAPES

DEEP WEATHERING AND THE FORMATION OF DURICRUSTS

Many of the features which characterise our desert and arid landscapes today were in the process of formation during the very different Early Tertiary times when Australia had just become an island continent. It is surprising to realise that it was the conditions which applied in the warm, well-watered golden age, when forest made Australia a green and fertile land, that have resulted in major characteristics of our landscapes of stony deserts, gibber plains and coloured mesas. The duricrusts which have protected some of the weathered mantle of the land from erosion were the product of that green Australia, not the arid brown one which it has become. Deep weathering of the surface of the land under warm and wet climatic conditions set the stage and started to make today's scenery.

Deep weathering involves the alteration of rocks and minerals[86] to depths of hundreds, or even thousands, of metres. The chemical nature of the weathered material changes and soil and rock profiles can be divided into zones on their mineralogy and colour. The amazing and beautiful colouration of mesas near Coober Pedy and elsewhere in arid central Australia graphically illustrates deep weathering profiles. A red, oxidised upper zone grades through a mottled zone into a bleached zone (known as the pallid zone) which may be startlingly white so that mesa sides from a distance appear to have been mined.

The famous explorer E. J. Eyre[95] passed through the southwestern part of the Lake Eyre Sub-basin north of the Flinders Ranges on his journey northwards through South Australia in 1840. There he noted 'many very singular looking table-topped elevations from 50 to 300 feet in height and with steep, precipitous sides which were red, with the ironstone above and white, with a substance like chalk, below' — classic weathered profiles.

For a long time it was assumed by geologists that hot, wet climatic conditions were required for deep chemical weathering.[86] It is now generally accepted that all that is required is deep groundwater and long periods of time. Chemical weathering to great depths has occurred under different climatic conditions, some cold-climate, in different parts of the world. However, there was an interval in the Late Cretaceous to Early Eocene when global conditions were warm

AUSTRALIA'S CAINOZOIC SEDIMENTARY BASINS

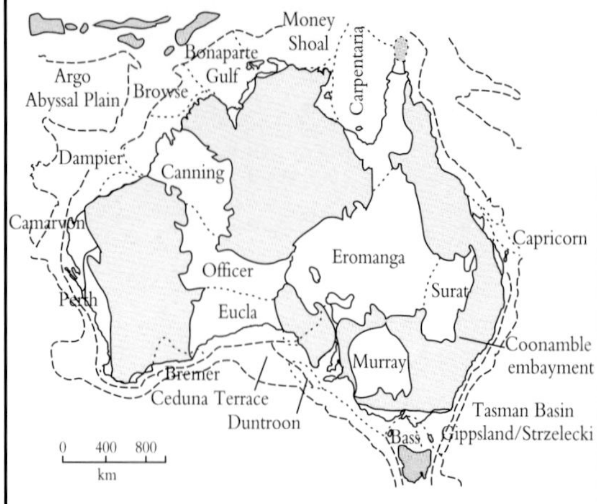

and wet and this was a period of deep weathering in many parts of the world.

It has been suggested that at this time when supercontinents were fragmenting and a great deal of associated volcanic activity was occurring, acid rain would have been a global phenomenon. The weathering potential of acid rain would obviously be far greater than that of uncontaminated rain, and this might account for the unusually deep weathering during this interval.

In Australia a deep blanket of weathered material was produced over much of the continent. Such a blanket of soil and weathered rock is known as regolith.

In order to understand the deep weathering processes, it has to be remembered that groundwater is held in weathered rock and that the zone which is saturated is topped by a watertable which separates an upper oxidised zone from the saturated zone below which lacks oxygen.

The formation of the Cretaceous and Early Tertiary weathered profiles resulted from a scenario like this:

Above the watertable, groundwater percolating through the material leached it of iron and other elements, carrying them downwards to the watertable. (Many economic mineral deposits were produced by the concentration of elements such as copper, gold, uranium and opal at this horizon.) Chemical reactions occurred within the silicate minerals, which form the bulk of continental crust, breaking them down into clay minerals. Groundwater redistributed the clay-forming elements downwards, resulting in the decrease in colour from iron-stained, through mottled, to the white clay pallid zone in profiles.The relationship of opal to the pallid zone in deeply weathered regolith in Australia is of great significance. Deep weathering was required to free silica, and a clay layer was required to trap it and allow it to precipitate as opal. The mottled zone, through which the groundwater percolated, has, as its name implies, an unequal distribution of iron. Mottles resulted from the segregation of iron oxides. Small hematite crystals coalesced to form nodules of different sizes in this zone.

The upper part of the saturated zone in the basement rock on which all the weathered material lies is the zone of groundwater discharge. Groundwater moves towards springs or seepage sites in this zone, carrying weathering products including silicic acid (dissolved silica) in solution. It also carries iron when the regolith through which it percolates is iron-rich. The removal of weathering products maintains a chemical gradient and results in further weathering. In the stagnant lower part of the saturated zone the products cannot simply be washed away by groundwater flow like those from above the discharge zone. They diffuse upwards by a process of osmosis resulting from a gradient in concentration towards the zone and are removed, promoting further weathering. Where ferrous iron is the element that diffuses upwards in this way it is converted to ferric hydroxide when it reaches the oxidising zone of the water-discharge zone and in some cases it may be precipitated in that zone as massive or nodular ferricrete. In the Yowah opal field of western Queensland, 'Yowah Nuts' — ironstone concretions which sometimes contain precious opal as their core — occur. They result from a combination of available silica in concentrated solution as a gel and the precipitation of iron to form nodules in the watertable region in a deeply weathered profile. Boulder opal, occurring in cracks and fissures in massive concretionary ironstones also in western Queensland, has a similar origin.

It has been calculated that a zone of ferricrete a metre thick can be produced in 10,000 years by diffusion of iron from the saturated zone when the underlying rock is iron-rich.[86]

Iron in solution carried away in the groundwater becomes concentrated in low-lying areas where flow is restricted or sluggish, and there it impregnates soils, sediments or permeable rock layers to create different forms of ferricrete. In these sites the ferricrete layer is younger than the weathered profile[107] on which it forms and it is not part of the local deep weathering sequence. (It is described by geologists as being separated from the underlying sequence by an unconformity.) Later, when erosion lowers the landscape, the ferricrete duricrust protects the strata below it and a typical profile capped by a duricrust is produced. The ferricrete that was formed in the lowest part of the landscape ends up as the highest stratum in the 'inversion of relief' that results from erosion of the softer material which surrounds it.

As with the dissolved iron, the free silica produced in the chemical weathering process becomes concentrated in groundwater where flow is impeded or very slow due to little topographical gradient. There it becomes a cement for sediments and soils in low-lying areas, forming silcrete. It may also infiltrate porous rock strata making them very resistant to erosion. When later the silcrete protects the strata below it from erosion, it, like ferricrete in similar circumstances, ends up capping a profile on which it rests unconformably.

For deep weathering to occur, the rate of alteration and breakdown must exceed the rate of erosion. Flat topography therefore favours deep weathering, and Australia was a very flat land in Late Cretaceous and Early Tertiary times when high rainfall provided the groundwater that was necessary for the process. The deep regolith that resulted was subsequently stripped when the rate of erosion overtook the rate of formation.

Old river gravels show the progressive stripping of the weathered material that occurred when topography became more varied with elevation of some areas and rejuvenation of drainages, demonstrating a steady change in composition. The early gravels are of quartz-dominant material, the only material available for erosion when the entire landscape was deeply weathered; later gravels contain a wider range of rock types as more fresh rock was exposed in the landscape. The change from quartz-rich to mixed gravels occurs in Miocene times. (Old river gravels — known as deep leads — are mined because they are often rich in the gold and other elements that were concentrated at the watertable in the weathered profile before erosion stripped the land and left only the more resistant gravels and heavier mineral component in the river beds.)

Because deep groundwater is the criterion necessary for deep weathering, arid areas may be sites where it occurs in spite of the lack of surface water. Weathering products are less likely to be removed if there is no external drainage, so deposits of gypsum, other

FERRICRETE FORMATION FROM IRON DISSOLVED IN GROUNDWATER

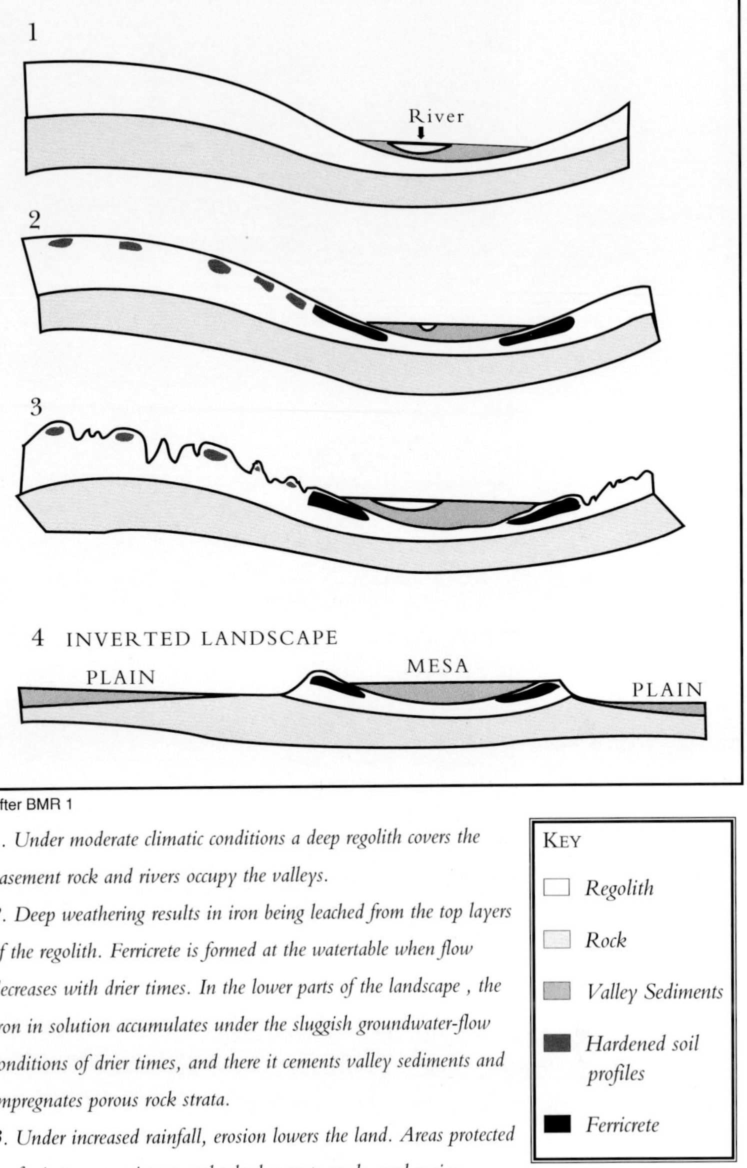

After BMR 1

1. Under moderate climatic conditions a deep regolith covers the basement rock and rivers occupy the valleys.
2. Deep weathering results in iron being leached from the top layers of the regolith. Ferricrete is formed at the watertable when flow decreases with drier times. In the lower parts of the landscape , the iron in solution accumulates under the sluggish groundwater-flow conditions of drier times, and there it cements valley sediments and impregnates porous rock strata.
3. Under increased rainfall, erosion lowers the land. Areas protected by ferricrete are resistant and take longer to erode, and major ferricrete zones in low parts of the landscape are protected by build-up of valley sediments from the eroding higher ground.
4. Erosion strips the land, leaving the larger areas that are protected by ferricrete as mesas. In this inversion of relief the ferricrete which was formed in lowest parts of the landscape becomes the capping on the highest features.

The breakaways in the weathered Bulldog Shale, duricrust capped, at Painted Hills, Arckaringa, South Australia.

N.F. ALLEY

A vast silcrete gibber plain, about 60 kilometres from Tibooburra.

M.E.W.

precipitates and sometimes opal may result.

The Great Artesian Basin[86], which occupies a great deal of the eastern half of the Australian continent, is an example of a mainly arid region in which deep weathering is occurring. Groundwater extends to a depth of 3000 metres, so chemical weathering can also extend to this depth. The zone where most recharge of stored waters occurs is situated in an area of humid climate along the Great Divide on the eastern rim of the Basin. Groundwater is lost at discharge zones and mound-springs mainly on the western rim. The different chemical content of the artesian waters in different areas and of the water in mound-springs depends on the nature of the chemical processes that are occurring underground and the types of rock through which the water has travelled in its extraordinarily slow journey across the Basin.

Some ferricrete has developed at the surface rather than at a sub-surface watertable and this process has been on-going right up to the present.[97] One type of ferricrete is composed of small round nodules cemented together. Because of their pea-like form the small nodules are known as 'pisolites'. Their formation is believed to have occurred in the following way: Where the ground surface was progressively lowered by erosion of an ancient weathered profile, the mottles and concretions of the mottled zone accumulated at the surface.

Left: The silcrete cap on the weathered Bulldog Shale profile near Arckaringa.

N.F. Alley

A quarry where pisolites are mined for road construction. Agonis *shrubs grow out of this desiccated material.*

M.E.W.

Here chemical alteration took place within small nodules. (It has been suggested that fire has sometimes played a part in the chemical alteration of some nodules.) In addition, some of the iron-rich material exposed on the ground surface broke down and soluble iron was carried into lower parts of the landscape by runoff. Over time the nodules also were transported into lower parts of the landscape, becoming rounded into pisolites on the way. Their cementing together by iron from groundwaters resulted in the formation of pisolitic ferricrete, made up of multi-millions of small round balls. Pisolitic ferricrete weathers again into free pisolites when the softer iron cement breaks down.

In parts of Western Australia pisolites are quarried from weathered duricrust and used as road gravel, making walking on some roads rather like walking on marbles and creating problems for cars when brakes are suddenly applied.

In the Pilbara region of Western Australia, in the Hamersley Ranges[96], high-grade pisolitic iron ore deposits are mined.[99] The Yandicoogina and Marillana deposits occur as Eocene infilling of ancient drainages and as small mesas adjacent to the palaeochannels. Similar palaeochannel deposits in the Robe River, and lesser ones in the Duck Creek and Beasley River systems, are also pisolitic.

The distribution of silcrete and ferricrete duricrust[85] over the whole of Australia has been mapped, and where detailed mapping of smaller areas has been carried out, the duricrust follows dendritic patterns of ancient drainages. In Queensland ancient 'surfaces' of duricrust are recognised, in which duricrust of a particular age and character occurs over very large areas. The Morney and Canaway profiles,[93] of the Late Cretaceous/Early Tertiary and Late Oligocene/Early Miocene respectively, are products of two major phases of duricrust formation on the Winton and Mackunda Formation sedimentary sequence of the Great Artesian Basin. (The Morney duricrust is about 50 million years old, the Canaway about 25 million.) The duricrust did not form a continuous surface which capped the entire landscape, though wide flat flood plains in the featureless land often provided extensive areas for deposition of silica- and iron-rich material. Breakdown of the duricrust in the course of erosion has resulted in the spread of gibbers over intervening areas between the production sites, increasing the impression of continuity over vast tracts of land.

It seems that major episodes of silcrete/

PRECIOUS OPAL—A PRODUCT OF DEEP WEATHERING

Australia is the world's largest producer of opal.[122] The only other country that produces a significant amount is Mexico. Deposits of precious opal in Australia occur on the edge of the Great Artesian Basin or within it, and they are associated with the pallid zone in deeply weathered regolith profiles.

Opal was formed by the gradual concentration of entrapped groundwater containing colloidal silica (silica gel). The silica was produced during the weathering process when smectite clay was altered to kaolinite, and possibly by corrosion of quartz grains. It is believed that opal formation occurred in one unique climatically controlled event about 30 million years ago. Concentration of silica in groundwater to produce colloidal silica, its accumulation in cracks and voids, and its crystallisation into precious opal over possibly hundreds or thousands of years required special conditions — hence the rarity of opal.

Precious opal consists of equal-sized, regularly packed silica spheres (when seen under a scanning electron microscope) and its fire results from diffraction of light by the spheres. Their size and how closely they are packed in any one stone will determine the flash and fire of the gemstone.

Australian precious opal is of three main types which are typical of three distinct regions, all in the Great Artesian Basin:

1: On the southwestern margin of the Great Artesian Basin the Coober Pedy, Andamooka and Mintabie opal fields produce 70 per cent of total world production — mainly '*white opal*' with milky background and fine colour.

2: Lightning Ridge and White Cliffs in the southeastern and south-central part of the Basin are famous for '*black opal*' which has brilliant fire originating from a dark grey to black siliceous background.

3: The opal fields in the central portion of the Basin in western Queensland produce '*boulder opal*', the least common type which accounts for only 3-4 per cent of Australia's total production. This kind of opal is basically a dark or black variety with colours emanating from the dark brown or black ironstone background. The boulder opal deposits are scattered throughout a 300-kilometre wide belt which extends from the Queensland-New South Wales border northwards for about 900 kilometres to the vicinity of Winton. Within this zone are several hundred sporadically worked deposits, including the Yowah opal field and others in the Eromanga district. Boulder opal is ironstone-hosted.

At Andamooka the opal occurs in a deeply weathered Cretaceous sequence of the Bulldog Shale, a shallow-marine deposit that was laid down in the Eromanga epicontinental sea. At the surface Quaternary soil rich in gypsum and mantled with gibbers creates the empty desert landscapes. Some of the massive quartzose silcrete and porcellanite (silicified 'kopi', which is patchily developed at the top of the weathered profile) occurs as outcrops and some is below ground, under the gibbers and soil.

Several layers are recognised in the weathered Bulldog Shale. An upper white, sandy clay layer, up to 30 metres thick in places, is known as 'kopi'. At its base is an extensive sandy boulder bed which the miners call 'concrete'. The pebbles, cobbles and boulders in this layer are of rocks older than Cretaceous. Below this conglomerate band is a layer of 'mud' which may reach a thickness of 4 metres. Contact between the conglomerate and the mud (which is of the same composition as kopi but with less quartz sand in the clay) is the 'opal level'. Here most of the good opal occurs. Below the mud layer is a zone which miners call 'bull dust'. It consists of up to 4 metres of pale yellow to red and brown pebbly silt which may contain large boulders.

Within the opal level the precious opal is randomly distributed and rare, and a great deal of the level has to be worked to find

DISTRIBUTION OF OPAL DEPOSITS IN AUSTRALIA

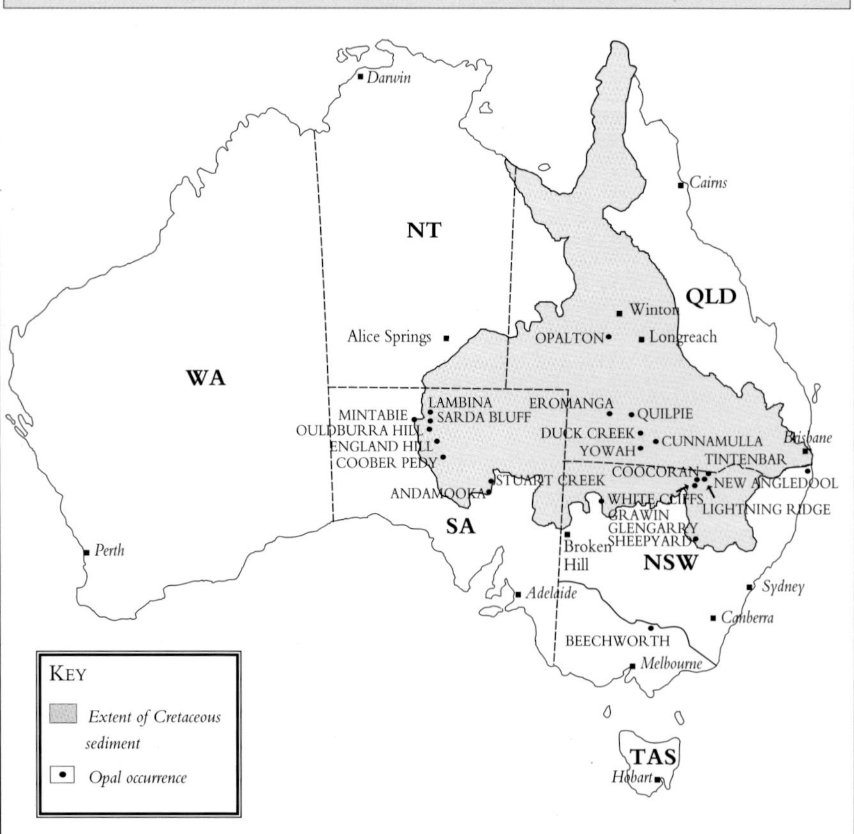

After Barnes and Townsend 122

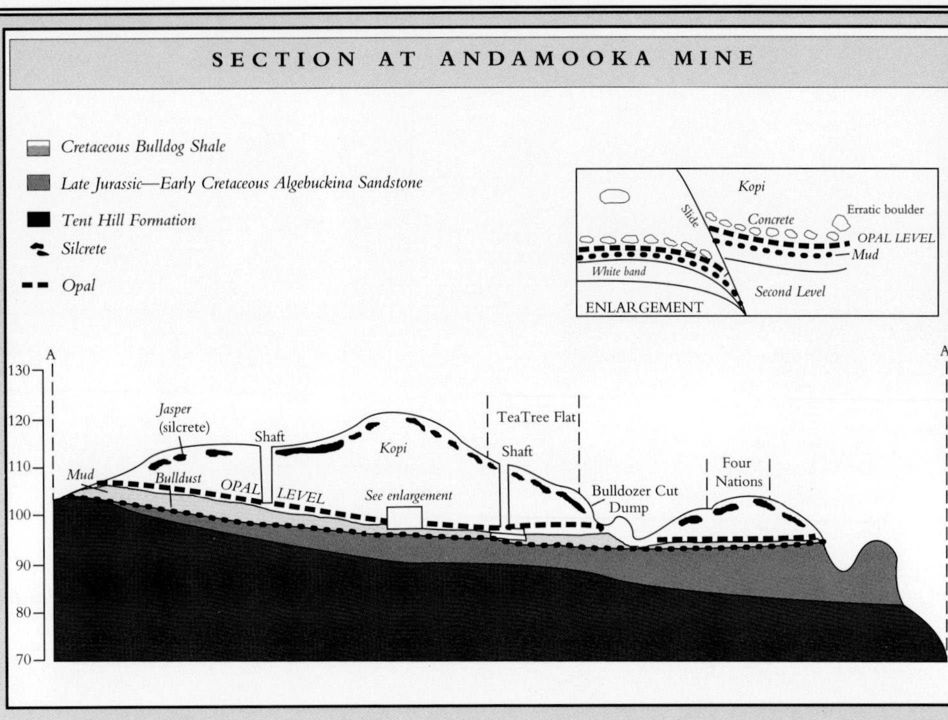

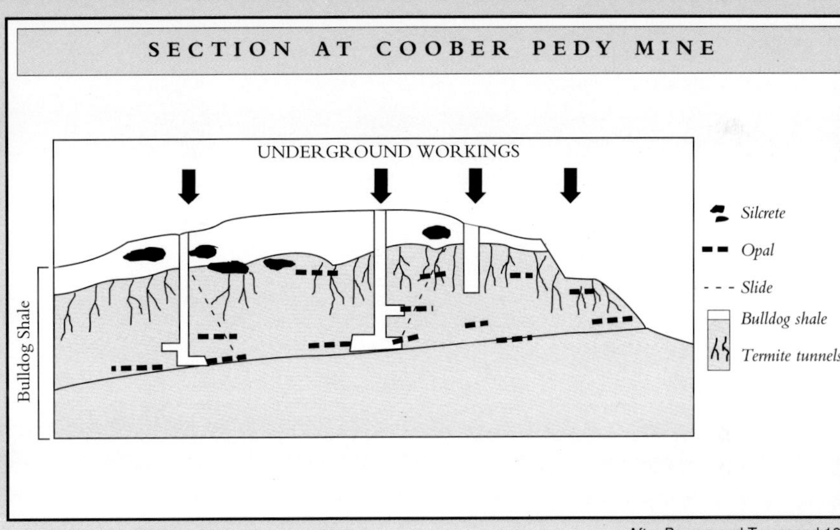

After Barnes and Townsend 122

stones. Where limestone boulders have been impregnated with opal 'matrix opal' is produced, and 'painted lady' stones result from the filling of cracks in quartzite boulders.

At Coober Pedy the precious opal also occurs in weathered Bulldog Shale. The weathering extends to depths in excess of 50 metres. The uppermost 20 metres of white to pale mauve porous claystone is called 'sandstone' by miners, and opal is found throughout this zone but is concentrated in a band about 5 metres above its contact with a darker-coloured claystone that lies below, and in the upper 1 to 2 metres of that clay layer. The precious and potch opal is found as irregular veins infilling cracks and joints and replacing fossil shells. Opal veins are usually in 'levels' where weathering products such as gypsum and iron oxides are concentrated and where there are numerous red-brown tubules. These are termite burrows, and the 'levels' represent old groundwater tables.

The weathered zone is cut by steeply dipping 'slides' which are small faults which resulted from change in volume during weathering. These fracture lines created cavities in which weathering products, including the silica required for opal formation, could accumulate.

At Lightning Ridge opal occurs in deeply weathered sediments of the Rolling Downs Group, equivalent to the Bulldog Shale, of Early Cretaceous age. These sediments were also laid down in the epicontinental sea and today they outcrop in low ridges and mesas that rise 15 to 20 metres above the surrounding black soil plains. Tertiary gravels and sands, in some places silicified, overlie the Cretaceous sediments on the ridges. The weathered sequence consists of: 2 to 4 metres of cream to white claystone at the top, silicified in places to form porcellanite which is known by miners as 'shincracker'; underlain by 4 to 6 metres of fine to medium-grained clayey sandstone. This lower layer contains a discontinuous silicified layer up to 30 centimetres thick, known as 'steelband', and 'Finch clay' lenses of soft grey claystone — the 'opal dirt'. Opal is mainly found in irregular nodules ('nobbies') or in thin seams within Finch clay lenses. Several levels of opal dirt can be found within any one area, most between 6 and 18 metres from the surface. Opalised shells, wood and reptilian bones are occasionally found.

White Cliffs was Australia's first-discovered opal field and was essentially worked out between 1890 and 1915. The host rock is a weathered Cretaceous sequence, capped by Tertiary silcrete, which outcrops as low mesas. Up to six clay horizons separated by silicified zones and layers of boulders produce opal. As well as infillings in joints and cavities, 'painted ladies' occur where quartzite cobbles have opal in fractures, and 'opal pineapples' are produced by replacement of glauberite crystals. Shells and plant material are also replaced by opal. A complete opalised skeleton of a plesiosaur was discovered by miners at White Cliffs and reassembled by the Australian Museum. Within the stomach region of the skeleton opalised bones of a fish were found — remnants of the creature's last meal.

The boulder opal occurrences of Yowah, Duck Creek, Eromanga, Opalton and other sites in western Queensland are all within a weathered profile of Winton Formation of Middle Cretaceous age. Ironstone formed from iron leached out of upper layers and precipitated near the watertable is the host rock in boulders.

At Yowah a layer of small to large ironstone concretions lies at the base of a pink sandstone. Below the ironstone layer is white or yellow mudstone. 'Yowah Nuts' are golf ball-sized spheres of ironstone in which the formerly hollow centre has been partly or completely filled with opal (if you are lucky — most are filled with clay). The concretionary ironstone may have veins and cracks between the concentric layers of iron filled with opal.

Fossil termite burrows, infilled with iron-rich sediment, create a pattern in the upper section of the weathered profile at Coober Pedy. M.E.W.

The weathered Bulldog Shale profile with opal diggings into the pallid zone at Coober Pedy. N.F. ALLEY

A moon-landscape with opal diggings at Coober Pedy. N.F. ALLEY

*Colours of the leached profile can be seen in the side of the ridge at this boulder opal site. The gidgee trees (*Acacia cambagei*) grow out of the desert duricrust and each is surrounded by a grey circle of its fallen leaves.* M.E.W.

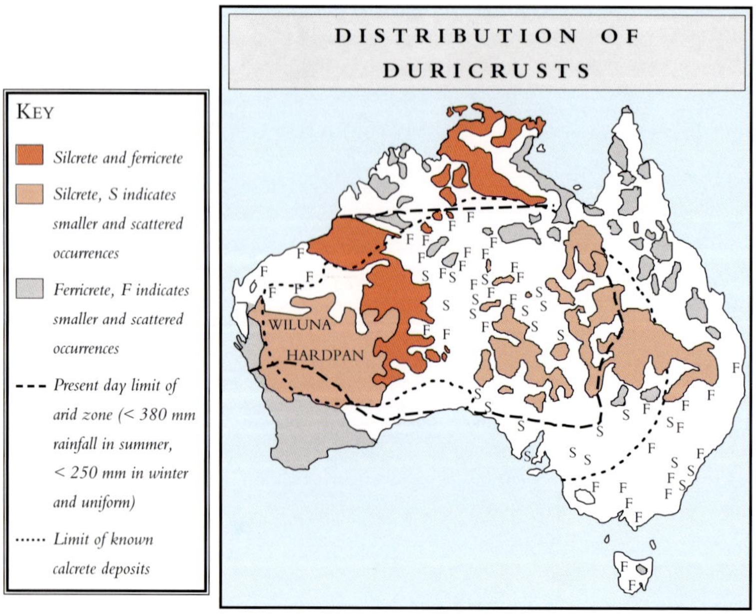

After Ollier 100

ferricrete formation (when 'surfaces' were formed) relate to episodes of climatic change. There is an approximate correlation between the times of major duricrust formation and global sealevel and temperature changes which are attributed to short episodes of Antarctic cooling. If sudden relative aridity at these times slowed or stopped the groundwater percolation, the concentrated mineral matter at the old watertable would have been likely to solidify in one episode over perhaps 10,000 years. (The Queensland duricrusts have been dated by palaeomagnetic means and evidence of a polar reversal within the time of formation of the Morney surface over wide areas establishes its genesis within a limited interval.) However, there appears to be no necessary association between duricrusts and widespread

Far left: A conglomerate cemented with ferricrete. When iron-rich groundwater accumulates in low lying areas an arid phase causes it to precipitate and cement gravels and sediments together in this manner.

JIM FRAZIER

Left: A lump of pisolitic ferricrete showing the round pea-like ironstone nodules that are cemented together with softer ironstone.

JIM FRAZIER

Massive pisolitic ferricrete in Western Australia.

M.E.W.

Below far left: Wreath Lechenaultia *grows happily on the hot and completely dry pisolitic ferricrete surface in Western Australia. Its delicate flowers contrast with its harsh surroundings.*

M.E.W.

Below left: This boomerang or cow-horn trigger plant grows on the pisolitic ferricrete surfaces, thriving in the seemingly inhospitable environment.

M.E.W.

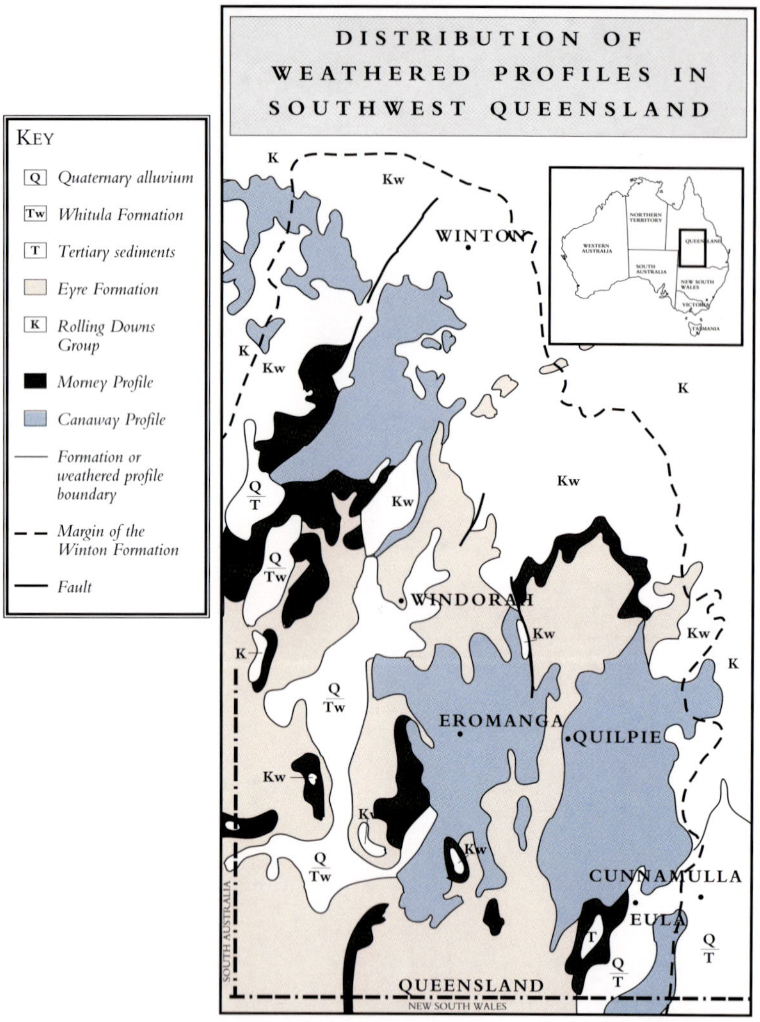

After Senior 92

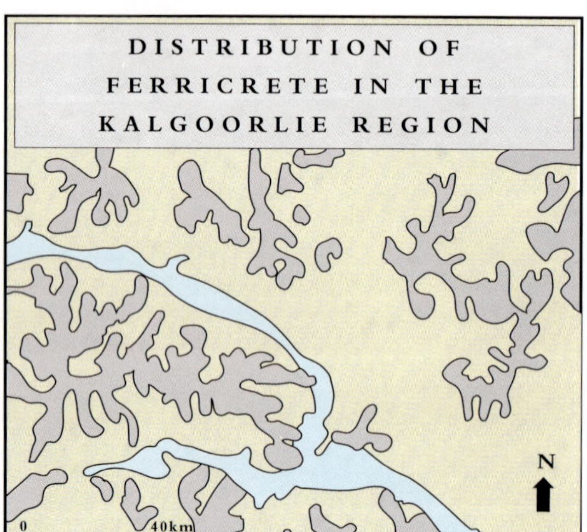

The dendritic pattern of ferricrete remnants supports the idea that ferricrete was formed in palaeovalleys. Landscape inversion has caused the ferricrete to be situated on plateau tops today.

After Ollier 85

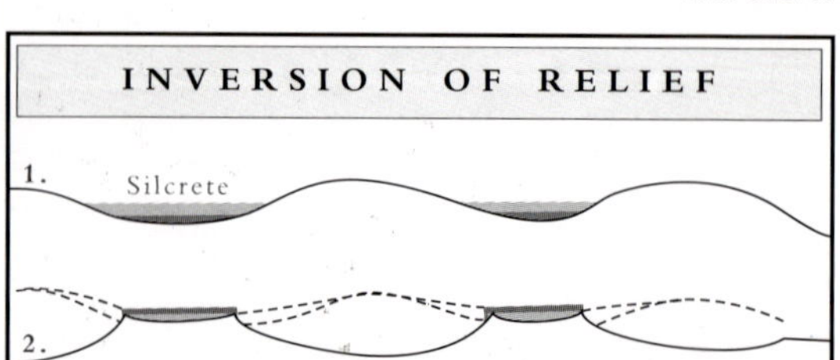

The protective effect of silica-indurated palaeovalley-floor sediments results in silcrete-capped mesas in modern landscapes.

After Twidale & Milnes 94

aridity. Both silcrete and ferricrete can develop in many different environments, and deposits of different ages are scattered on the continent.

Cementing of soils and rock detritus by iron is on-going, and near Darwin old vehicle engines are 'ferricreted' into the rock platform on the coast and have become 'fossils'.

Calcrete and gypcrete duricrusts are produced where soil water is heavily laden with lime and gypsum. As with silcrete and ferricrete, the calcium and gypsum are deposited in low-lying or stagnant areas and there they cement the soil and sediments and form resistant layers in profiles. Where erosion has subsequently lowered the landscape and inversion of topography has occurred, they may eventually become the capping on mesas or be exposed in plateaux breakaways.

The Wiluna Hardpan on the ancient Yilgarn Block of Western Australia[91] is an area where a siliceous layer is developed in the red-brown soil. It looks like soil but is a barrier to digging. Its presence in young alluvium suggests that it is still being formed.

Bauxite, the ore of aluminium, used to be considered a simple product of deep weathering.[368] It was thought that when practically all the other elements had been removed by intense leaching under wet and warm conditions, the aluminium-rich residue was concentrated — usually on plains. It is now known that bacteria play an important role in the concentration of the aluminium and it is their activities on the deeply weathered material that results in bauxite deposits.

At Weipa in north Queensland the largest single occurrence of bauxite in the world is the result of bacterial activity on the Late Cretaceous and Early Tertiary deeply weathered regolith. At Gove in the Northern Territory the deposits have a similar history. The Jarrahdale bauxite in Western Australia results from concentration of aluminium by bacteria in deeply weathered material derived from ancient Precambrian rocks during the same Late Cretaceous to Early Tertiary interval. The bacterial activity and the production of bauxite is a continuing process, right up to the present. (Mineral deposits that result from bacterial activity are known as biogenic deposits. Concentration of gold is often biogenic, and the immense banded ironstone deposits of Precambrian times also owed their origins to bacteria.)

CHAPTER FIVE

Rivers Beheaded, Broken and Bent

A legacy of rifting

What happens to rivers when the supercontinent lands across which they have flowed for millions of years are separated by rifts? When continental margins are upwarped after rifting, how does the changed topography affect the drainage? When raising of continental margins results in the sinking of the centre of the continent and enormous inward-draining regions result, what does that do to the hydrology of the land? When volcanic activity and tectonism create new topographical highs and rivers are diverted around them and new watersheds are created can we see evidence of the superseded drainage in landscapes?

By 'reading' the ancient drainage patterns and observing the paths taken by modern rivers some of their history can be interpreted. Modern technology such as aerial photography and remote sensing from satellites enables us to see the ancient drainage over parts of the continent. Lines of salt lakes can be seen to lie along palaeodrainage channels and sediments within the channels can tell us when last the rivers ran.

Australia is unique in that only its western, eastern and northern marginal provinces drain to the ocean while its centre and the middle sector of its southern margin are inward-draining.

The manner in which tributaries join rivers can tell us whether they have always flowed in the same direction or whether alteration to the topography and creation of new drainage divides has changed their behaviour. When the course of a river is broken by a rift zone the part on the headwater side of the rift will cut down

INWARD AND EXTERNAL DRAINAGE OF AUSTRALIA

After van de Graaff *et al.* 128

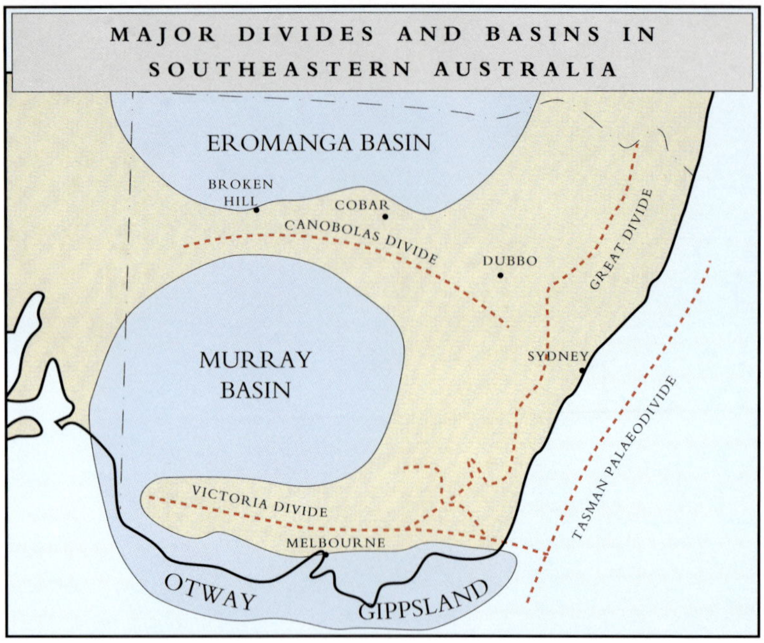

After Ollier *et al.* 123

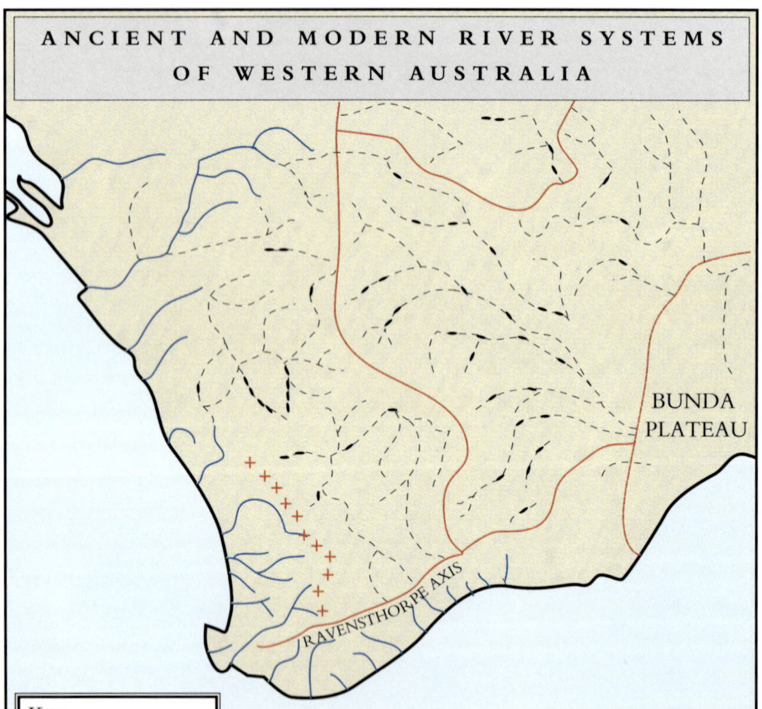

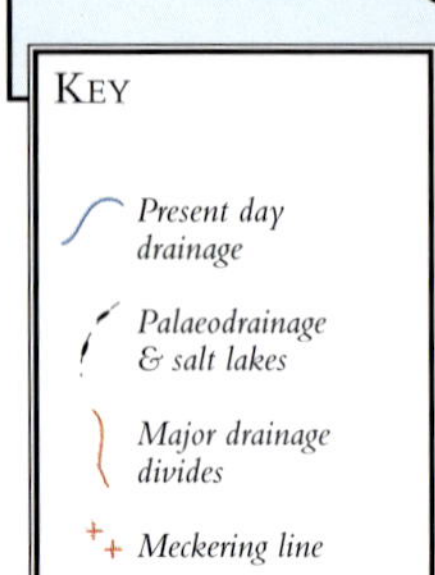

After van de Graaff *et al.* 128

through the rising rift margin to reach the new ocean that forms in the rift. The other part of the river that is stranded on the far side of the rift will have to reverse its flow if it is to carry its waters from the new rift-margin watershed to the sea. It will be possible to see where this has happened because the tributary streams that meet the now-reversed river form a barbed pattern (running 'up' in relation to the catchment to meet the river) in contrast to the dendritic pattern of normal rivers where the feeder streams run down away from the catchment to meet the river.

The palaeochannels of many of Australia's rivers tell of changes to drainage that have resulted from rifting along the continent's margins. The course of some of the major present day rivers supplies evidence of the upwarping and subsiding of parts of the land that changed drainage patterns. In addition, the sedimentary rock sequences in basins provide evidence of the direction and productivity of sediment supply.

Rifting on Australia's margins proceeded from the northwest, down the Western Australian coast when Argoland and India moved away; round the southern margin as the Southern Ocean opened between Australia and Antarctica; and northwards along the east coast when the Tasman Sea opened separating the New Zealand Subcontinent from eastern Australia. Thus changes to drainage patterns related to rifting occurred first in Western Australia, later in South Australia and southern Victoria, and later still along the New South Wales and Queensland coastline.

Rivers of Western Australia and the Southern Margin of Australia

In Western Australia the Darling Fault bounds the Tertiary[128] and Quaternary deposits of the Perth Basin, separating them from the ancient Yilgarn Block which is one of the most ancient land surfaces in the world. Uplift of the Block on this fault and repeated marine incursions onto the down-thrown coastal strip have obliterated any evidence of palaeorivers that once ran between India and Australia before the Indian Ocean opened.

However, in the southwestern province of Western Australia a system of interconnected valleys arises near the southern coastline and heads inland in a northerly direction before linking with the present course of the Swan River. This valley system is dry now, and chains of salt lakes occupy the drainage channels. The width of the valleys indicates that major rivers used to run in them. These rivers had their headwaters in Antarctica, and these pre-rift valleys must be among the oldest drainage features preserved anywhere on Earth.

The rift between Australia and Antarctica had started from the west and these rivers were some of the first to be beheaded — possibly 150 million years ago! There has been upwarping across the palaeodrainage lines along the Ravensthorpe Axis, a new watershed that formed when the continental margin rose following rifting. Subsequent rejuvenation of the drainage has occurred south of this axis of upwarp.

Further to the east along Australia's southern margin palaeodrainage of the Bunda Plateau (the Nullarbor Plain) has not been entirely obliterated by the marine incursions of Eocene and Miocene times. Palaeochannels can be seen from satellites using remote sensing. The large rivers that flowed from the Musgrave Block and the Stuart and Gawler Ranges during the Cretaceous cut deep channels. The Eocene epicontinental sea penetrated far up these channels infilling them with Pidinga Formation sediments. There is evidence that the southern continental margin was not yet uplifted in the Eocene when the eastern and western margins were already raised. Had such elevation already taken place, a drainage divide would have resulted, stopping flow up rivers from the sea. Elevation of the area only occurred in Late Miocene times when the limestones on the bottom of the Miocene sea were raised to form the Bunda Plateau.

The palaeorivers in the desert north of the Nullarbor are now chains of salt lakes. When water flows in them after rain at irregular and unpredictable intervals it disappears into the limestones of this vast karst desert where all the water is subterranean.

South Australia east of the Nullarbor owes much of its relief to block-faulting in Tertiary times. Such faulting has created the internal drainage of the Lake Eyre Sub-basin, the uplift of the Mt Lofty-Flinders Ranges and the formation of the St Vincent and Spencer Gulfs. Experts do not agree on whether the mountains were elevated in the Late Miocene, or in the Palaeocene and Early Eocene, or even if there may have been earlier Flinders Ranges during Cretaceous times. The Murray River makes a sudden bend at Morgan, turning sharply to the south. Some say this was due to its westward path being impeded by uplift of the Flinders Ranges, whenever that occurred. There is no geological evidence that the Murray ever crossed the area where the mountains are to reach the Great Australian Bight, though a 'western mouth' still has appeal to some experts.

Before the Lake Eyre Sub-basin subsided, rivers to its south ran to the sea in the Bight. Some streams now run backwards in ancient beds to drain into Lake Eyre when they have any water to run.

RIVERS AND PALAEODRAINAGE OF SOUTHEASTERN AUSTRALIA

Throughout the Jurassic and the first half of the Cretaceous,[123] the New Zealand Subcontinent formed the easternmost province of Gondwana and rivers that drained its highlands on the Tasman Divide carried sediment to the Eromanga-Surat Basin (which corresponds roughly to the Great Artesian Basin). The Otway, Torquay and Gippsland Basins were separated from the Eromanga Basin by the Victoria Divide, an east-west palaeodivide which was roughly coincident with the present-day divide. They accumulated sediment (with a high proportion of volcanic detritus) from the New Zealand Subcontinent during the Cretaceous.

Drainage from the Victoria Divide carried sediment northwards into the Eromanga Basin and southwards into the Otway-Gippsland Basins. The Victoria Divide was apparently fairly flat and had developed along a plain which was probably a feature of the landscape during the Triassic. Valleys were cut into the plain and a great escarpment probably existed on the edges of this divide, similar to that on the Great Divide today. (Erosion has since removed most of the evidence of these ancient scarps.) Remnants of the palaeoplain exist today as the major plateau in the Kosciusko region and the isolated smaller plateaux of the High Plains.

The ancient drainage of southeastern Australia was as follows:

During the Mesozoic Era, the Kiewa and Mitta Mitta Rivers flowed northwards to join the Ovens and Goulburn Rivers in a northwesterly flowing system. The Upper Murray, Tumut and Murrumbidgee united into a major system flowing northwards to the southern part of the Surat Basin. Volcanic activity in the Eocene, when a major shield volcano was centred on Brown Mountain and many subsidiary vents erupted over a wide area, dammed some of the north-flowing tributaries and created local radial drainage patterns.

Corridors of life, the lines of river red gums follow the usually dry watercourses into the arid lands. They depend on water stored at depth in the river channels for survival and provide refuges for birds and other inhabitants of the desert. Northern Flinders Ranges district, South Australia.

M.E.W.

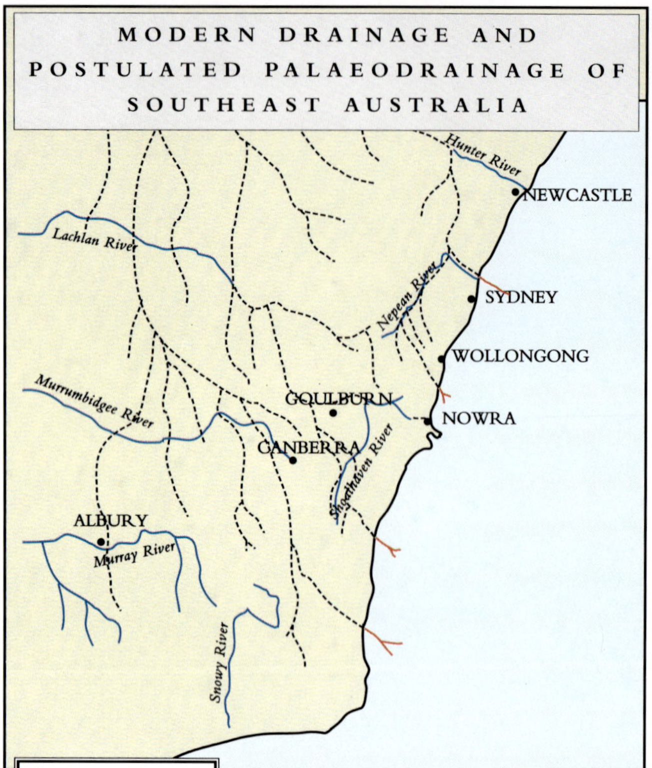

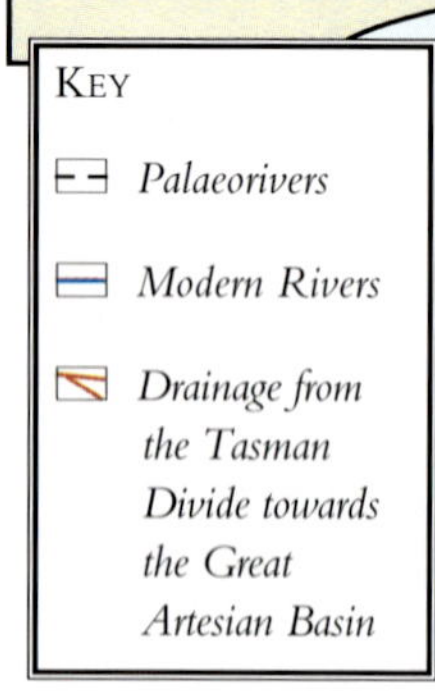

After Ollier 123

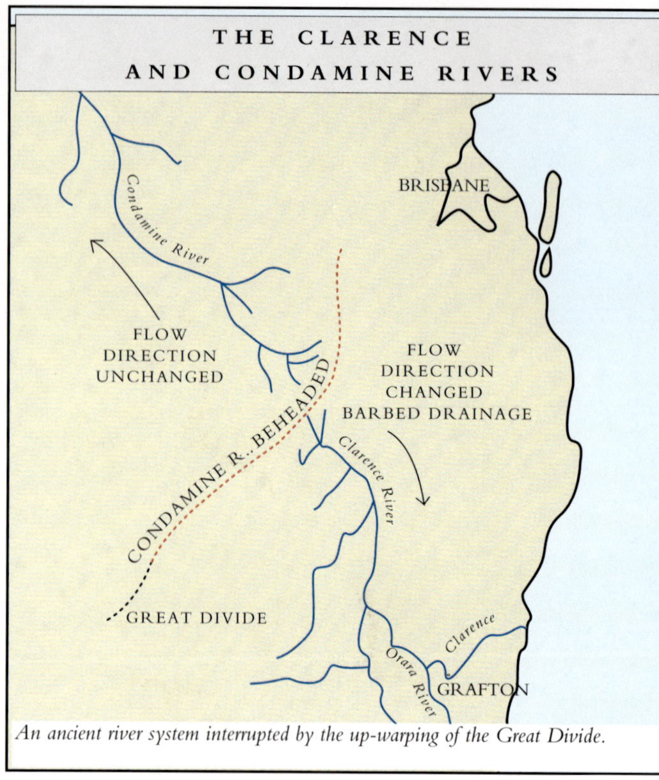

An ancient river system interrupted by the up-warping of the Great Divide.

After Ollier 125

Drainage from the south had to bend round this Monaro Volcano to enter the Snowy River.

Later, in the Miocene, faulting elevated the Kosciusko area and emplacement of more basalt tilted the palaeoplain to the southeast, again reversing some rivers. These then had to bend round the Monaro Volcano to enter the Snowy or the Murrumbidgee.

The Lower Shoalhaven ran via the Lachlan to the Macquarie, and eventually to the Surat Basin. The Upper Shoalhaven, Deua, Moruya and Tuross were tributaries to this system, as was the system comprising the Wollondilly, Abercrombie, Coxs River and the Tuross.

Two major rivers followed the Hawkesbury and Hunter Valleys to the west, joined by the Nepean and Wollombi. This system drained the Sydney Basin, carrying sediment that was to be deposited in the Surat Basin as the Pilliga Sandstone (of Jurassic age).

About 80 million years ago the Tasman Sea entered the rift zone between the New Zealand Subcontinent and Australia. Uplift of the continental margin and downwarping of the rift zone created the Great Divide. The rivers that had flowed from the Tasman Divide on the New Zealand Subcontinent were beheaded. This cut off the supply of coarse river-borne sediment to Australian basins. Rivers to the east of the Great Divide were reversed and the Great Escarpment was formed and started to retreat inland as erosion proceeded. Even today the Great Divide separates an area of simple dendritic drainage to its west from a region of complex drainage to its east. The picture has been complicated by the fact that the Divide itself has been moved in places by volcanism (which emplaced basaltic masses along it) and affected by the westward retreat of the Escarpment. This scarp retreat resulted from river headwaters eroding back, uniting and creating the massive gorges of the 'ranges'.

The Murray Basin started to subside at the start of the Tertiary[125] (about 65 million years ago) and this created a divide — the Canobolas Divide — between it and the Eromanga Basin. This effectively cut off supply of sediment to the Great Artesian Basin from the south. Rivers from the Victoria Divide which had previously flowed northwards started to flow to the west, like the modern drainage of the Murray Basin. Traces of ancient north-flowing rivers crossing the Canobolas Divide can still be detected.

In the southeastern corner of Australia volcanism during the Tertiary with emplacement of basalt, and faulting during Miocene times complicated the topography. The Eastern Highlands were produced by elevation of fault-bounded blocks hundreds of metres above the surrounding terrain. The Mt Bogong-Omeo area is a fault block with 800 metres of uplift at its northern margin and 200 metres at its southern margin.

Though faulting can raise blocks, areas can also be depressed by faults. Lake George near Canberra is an example of a fault-angle depression which was formed in the Tertiary. There, lake sediments have accumulated since Mesozoic times, providing a valuable record of changing vegetation and climate.

EAST COAST RIVERS MODIFIED BY RIFTING

The Clarence-Condamine drainage of modern Australia[125] is interpreted as being an example of drainage that has been altered by rifting and the creation of the Great Divide. The segment that is east of the Divide has been reversed as a consequence of back-tilting of the area between the Divide and the new coastline. During the Mesozoic, drainage was from east to west, flowing towards the the Great Artesian Basin. Barbed drainage patterns of tributaries of the Clarence show reversal of direction of flow, and the Condamine River on the west of the Divide has dendritic arrangement of feeder streams.

The Hunter River is another that has had its flow reversed. It has flowed on its present course since Miocene times.

The Shoalhaven River has complex drainage patterns. The lower Shoalhaven has barbed entry of the Kangaroo River and other tributaries indicating a reversal of flow. Before it was reversed it probably ran to the Upper Wollondilly and thence to the Lachlan.

RIVERS OF NORTHERN AUSTRALIA

In northern Australia, subsidence resulting from tilting of the edge of the continent has resulted in drowning of the northern ends of some rivers. Apsley Strait between Bathurst and Melville Islands, and Dundas Strait which separates Melville Island from the Coburg Peninsula represent a northern extension of the Darwin, Adelaide and Alligator River systems. The large swampy areas of estuarine plain on Melville and Bathurst Islands and on the southwestern Coburg Peninsula result from this regional tilting and the lowering of the land.

RIFTING COMPLETED—THE DRIFTING BEGINS

THE STAGES OF RIFTING SUMMARISED

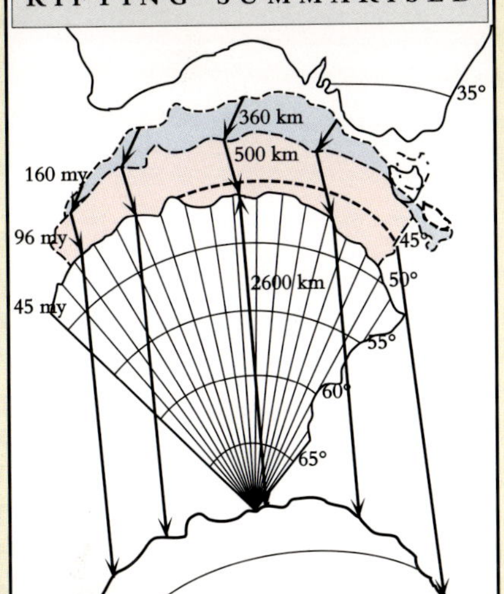

After Veveers 37

When the trailing edge of Australia — Tasmania and the South Tasman Rise[37] — finally cleared Antarctica and island continent status was achieved, it was as though the brakes were released and the continent was free to move northwards at a steady, much increased rate (6-7 centimetres per year).[77]

Its initial preparations for rifting from Antarctica by continental extension had been painfully slow, and its subsequent slow drift while India was making its rush towards Asia was still only millimetres a year. While India's movement slowed drastically as it bored into Asia between 55 and 50 million years ago, spreading of the Tasman and Coral Seas on Australia's eastern margin had slowed and ceased. Then for a further five million years, during a period of global stress readjustments, Australia had a slightly accelerated rate of movement which saw the continent, as it were, getting into gear for its independent journey northwards, when it would speed up to six or seven centimetres a year.

The rifting processes which had occurred along Australia's western, southern and eastern margins had created some of the topography, geological features and drainage patterns that characterise the modern continent. Australia was essentially a flat land then as now. Deep weathering had produced a leached regolith and soils of low nutrient status. The Gondwanan forests had both sclerophyll and broadleaf flowering plants, of mixed cool and warm temperate kinds, and southern conifers were important components of forests. Global climates were warm and humid, with little polar ice.

RELATIVE MOVEMENTS OF AUSTRALIA AND INDIA DURING RIFTING

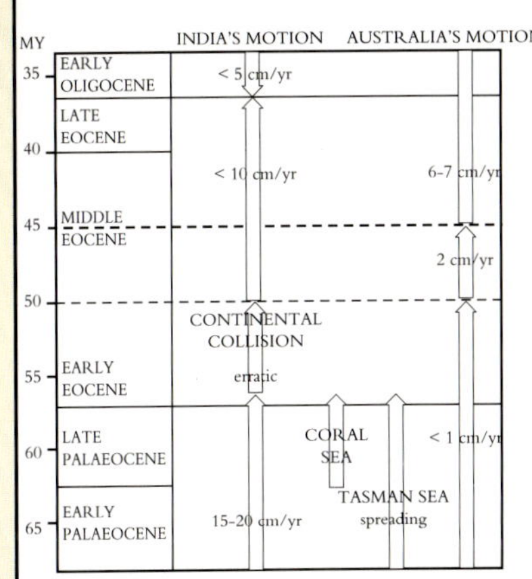

After MacGowran 77

PART TWO

DRIFTING

(FROM 45 TO 2.4 MILLION YEARS AGO)

INTRODUCTION

Contrasts in South Australia's Desert

Cool green forests locked in hot desert rocks

During the 45 million years since Australia severed its last connections with Antarctica and started moving northwards, the continent has changed from a green, forested land to the driest vegetated continent on Earth. Only Antarctica, whose water is locked up as ice, rivals it for dryness.

The epitome of arid landscapes in Australia is the sandridge deserts, the gibber plains and the salt lakes of Central Australia. A large proportion of South Australia's land is blown-sand and gibber desert and Lake Eyre and the multitude of smaller salt lakes and pans that surround it are within that State. South Australia has fossil floras and faunas that tell the story of evolution of the biota through time, making it an ideal region from all points of view to focus upon. The evolution of its landscapes and biota through the last 45 million years of geological time epitomise the 'browning' of Australia.

The contrast between South Australia today and what the State's area of land must have been like at the time of the start of drifting of the island continent is immense. A journey through the intensely arid below-

Lake Eyre South across a gibber plain with sparse saltbush in the dead heart of Australia.

M.E.W.

LAKE EYRE IN AUSTRALIA'S DEAD HEART

Above left: Gypsum crystals growing in the water of Warriner Creek, near Lake Eyre.

N.F. ALLEY

Above right: A huge dune of Quaternary lake gypsum at Lake Eyre North. Gypsum blown from the dry lake floor during an acutely arid phase comprises these dunes.

Left: Small shoreline benches eroded during filling of Lake Eyre South.

N.F. ALLEY

sealevel region round Lake Eyre, across blown-sand dunes of the Great Victoria Desert, the Strzelecki Desert and the Simpson Desert which form the northern half of South Australia gives no intimation of a green and forested past. Travelling across the Nullarbor Plain, a considerable part of which forms the southwestern part of the State, it is hard to imagine times of high rainfall when the Eucla Basin was a sea. The desolation of the limestone karst-type treeless desert tells only of aridity.

Nowhere is one more forcibly struck by the contrast between the arid present and the green and forested past than at two of the three plant fossil localities in South Australia chosen to describe the Gondwanan forest that existed when Australia had just become an island continent about 45 million years ago. These two localities are Nelly Creek and Stuart Creek near Lake Eyre South, and the third is Golden Grove near Adelaide. At Stuart Creek, in particular, where silcrete rocks form ridges in the gibber landscape, the leaves of the ancient forest are marvellously displayed among the small, hardy desert herbs and grasses that exist precariously in an almost waterless land.

CHAPTER SIX

Flame Trees Illuminate the Warm, Wet Forests

The Island Continent's original Gondwanan flora

The vegetation in Australia at 45 million years ago when the independent continent began its northwards movement at about 7 centimetres a year can be reconstructed from information supplied by fossil floras in South Australia. They, and the more or less contemporaneous floras of Anglesea in Victoria and Nerriga in New South Wales, tell us that the Middle Eocene forests were very similar to those that today occur in wet-tropical refugia in northeastern Australia.

The Nelly Creek flora Lake Eyre Sub-basin, South Australia

Nelly Creek is the easternmost arm of Welcome Creek[56] and it runs into Lake Eyre South on Muloorina Station in one of the driest, most desolate landscapes seen anywhere in Australia. To get there the starting point is Marree, which can be reached by the Strzelecki, Birdsville or Oodnadatta tracks. From Marree, the nearest town with a general store, postal agency and a hotel, one takes the tourist road towards Lake Eyre as far as Muloorina's gate. After that it is by four-wheel-drive vehicle across the open gibber plains, following a track along the Dingo Fence that bisects the property. (The fence allows sheep to be run on the half of the property that is within the fence, and cattle on the other — in years when there is rain and the stony desert can support such life at all.)

The only relief in the flat landscape is the breakaway edge of a low plateau where gypsum capping (the hallmark of a desiccated landscape) has protected the Bulldog Shale from eroding away to the uniform flatness of the plains. The Crowsnest Bore, with a clump of foreign palms planted long ago and a narrow *Acacia*-fringed drain running like a thin living line into the arid distance, interrupts the featureless land. Here nitre bushes grow on the gibbers surrounding the stream, indicative of the salt- and mineral-laden watertable. Thereafter there is nothing but gibbers until the saltbush of the sandplain round Lake Eyre South and Nelly Creek is reached. A few *Acacia* bushes struggle to survive on blown-sand dune ridges beside the river.

Lake Eyre South lies 15 metres below sealevel. Only in years when Lake Eyre North is full does it receive any water. It is usually a

Left: The plateau breakaway near Crowsnest Bore where the gypsum capping on the Bulldog Shale plateau results in it resisting erosion and forming low mesas. Heavy rain a week earlier led to a green tinge on the gibber plain.
M.E.W.

Middle left: The digging at Nelly Creek. Wet, much-consolidated mud was encountered at a depth of about a metre below the sand behind a bar of silcrete rock, salt-whitened, that lies in the river bed. The mud is full of mummified leaves.
M.E.W.

Middle right: A lump of mud with mummified leaves, dug out of Nelly Creek.
M.E.W.

Bottom left: Gypsum roses, growing in this radial pattern on the salt-encrusted soil surface at Nelly Creek.
M.E.W.

Bottom right: Patterns in the salt crust in Nelly Creek. The iron-stained mud segmented by the pattern of cracks is partly fluffed up by the expansion of gypsum within it, acting rather like yeast in dough. Walking on the almost dry surface is hazardous because the salt and cracked mud overlies a slippery wet layer.
M.E.W.

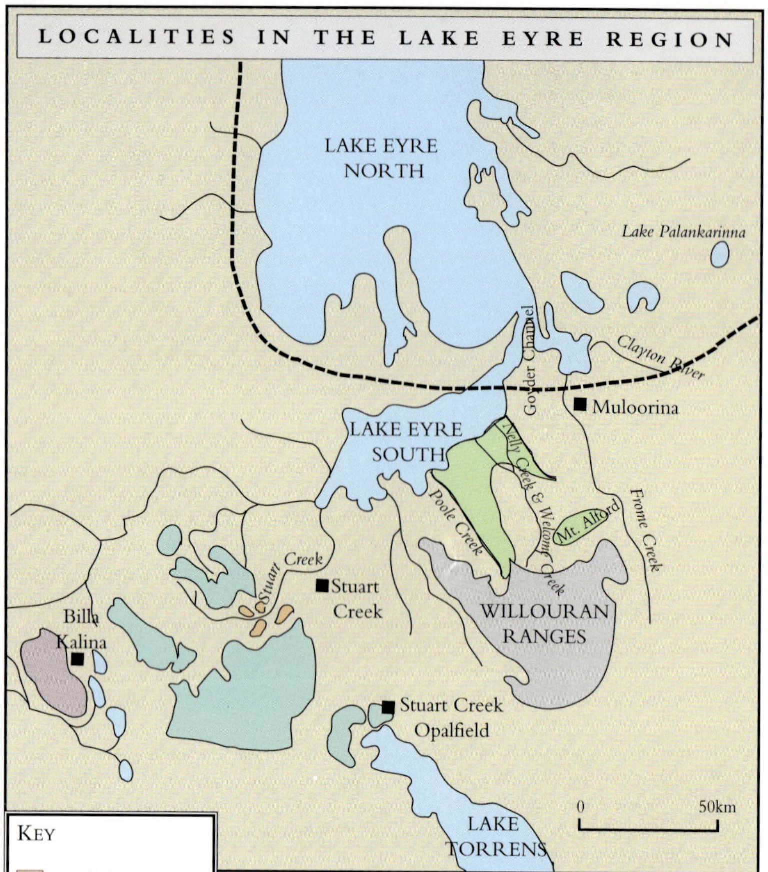

After Callen *et al* 113

vast saltpan, glaringly white, ringed by samphires and surrounded by saltbush sandplain that merges into gibber desert. Nelly Creek runs briefly when the rare rains fall, holds pools of salty water for a time, and dries again to be a salt-edged channel where salt-crack patterns beautify its bed. About a kilometre from its entry into Lake Eyre South a salt-whitened strip of silicified Eyre Formation rock forms a bar across the channel. It dams a salty pool after rain and maintains sub-surface dampness when the last surface water has gone.

Coloured samphires in Nelly Creek.

M.E.W.

Here in this most unlikely place is evidence of a forest that spread its cool green shade round meandering rivers about 45 million years ago. Then, in the middle of the Eocene epoch, Australia had just become an island continent and started its independent journey northwards, away from Antarctica. Its Gondwanan forest was already diverse, with different communities adapted to the different soils and conditions found on such a large landmass. No modern analogues exist for the *Paleoflora Mixta* forests which had elements that today are sifted into temperate and tropical forests, or sorted into broadleaf and sclerophyll communities.

The Nelly Creek fossils occur in a carbonaceous mud layer[55] that lies in the river bed below the silicified Eyre Formation rock that forms a bar across the channel. One might legitimately ask how such a fossiliferous layer, hidden under the salt crust and sand in a river bed in the middle of nowhere came to be found. Geologists, in the course of regional mapping and trying to sort out the stratigraphy in parts where there is just about no outcropping rock, visit places never examined by anyone before. Sometimes they dig where a soft surface allows such activity to try to find clues to the rock sequences below the surface. Thus the black, damp, soft mud behind and below the rock bar in Nelly Creek was found, and the mummified leaves that have survived within it for 45 million years were brought to light. The plant material is not altered by mineralisation — the leaves have outer cuticles, with veins and some tissue preserved between, and the bits of woody stem are like carbonised wood, not permineralised.

Examination of some of the wood by officers of the Wood Technology Branch of the Forestry Department at Beecroft has resulted in an identification of *Ackama* (*Caldcluvia*) of the family CUNONIACEAE. (Modern species are corkwoods in wet tropical rainforests of eastern Australia.) In addition, microscopic examination of the wood has revealed the presence of brown rot fungi, whose delicate threads have survived 45 million years of mummification.

It is possible to cut lumps of the wet mud with a spade. The leaf and other plant material can be separated out in a laboratory and mounted on glass for examination.

Not only are the Nelly Creek macrofossils capable of identification to plant family level

PLANTS OF THE ANGLESEA (VICTORIA) AND NERRIGA (NEW SOUTH WALES) MIDDLE EOCENE FOSSIL LOCALITIES

The fossil flora from **Anglesea** has been compared in composition[51] to the extant rainforests that are found at Noah Creek, north of the Daintree River in north Queensland. There forests have been classified as complex and simple mesophyll vine forest types, with streamside vegetation modified towards complex notophyll vine forest. The Anglesea forest is assumed to have existed under similar conditions of rainfall and temperature to those that apply today in the wet tropical rainforests of Queensland.

Identified at Anglesea are:

Macrofossils:

LAURACEAE — *Ediandra, Litsea, Neolitsea, Cryptocarya, Cinnamomum* (LAURACEAE, the family comprised more than 30 per cent of the flora.)
PROTEACEAE — *Banksieaephyllum, Musgraveinanthus* (flowers), *Orites*[50]
EBENACEAE — Flowers and leaves, *Austrodiospyros*
SAXIFRAGACEAE — *Quintinia* (possum wood)
STERCULIACEAE — *Brachychiton* (flame trees, kurrajongs)
CASUARINACEAE — *Gymnostoma* (twigs and fruits)
Leaves probably of CUNONIACEAE, MONIMIACEAE, ELAEOCARPACEAE
MYRTACEAE — *Myrtaciphyllum*
Cycads *Bowenia* and *Pterostoma*[61]

Pollen Record:

The pollen record shows high percentage of *Nothofagus* pollen[51], but no leaves have been found. *Nothofagus* pollen is notorious for its long distance dispersal and may not have been growing locally in any significant quantity. Pollen of MYRTACEAE and PROTEACEAE is abundant, yet leaf fossils of these families are sparse. Araucarian and podocarp conifers are common. The podocarps are represented by at least six species from four modern genera: *Dacrycarpus, Decussocarpus, Falcatifolium,* and *Podocarpus*, whereas only *Podocarpus* and *Prumnopitys* are present in the wet tropical forests of Queensland. (Pollen of *Prumnopitys* cannot be differentiated from other podocarp pollen.)

Families represented by pollen and not identified by leaf fossils are SAPINDACEAE, CHLORANTHACEAE (*Ascarina*), SANTALACEAE (quandongs), AQUIFOLIACEAE (holly), PANDANACEAE/LEMNACEAE (swamp plants), LORANTHACEAE (mistletoes), MALVACEAE, and JUNCAGINACEAE (aquatic water-ribbon plants).

An analysis of the Noah Creek flora and the fossil flora at Anglesea shows that virtually all the modern equivalents of the Eocene flora are present at Noah Creek. [48]

At **Nerriga** in New South Wales 44 per cent of the leaf fossils[66] are assigned to LAURACEAE.[70] Altogether 25 different leaf taxa occur there and apart from the lauraceous examples, *Casuarina* and three species of cycad, they have been classified numerically and not assigned to families or genera.

BRUSH BOX FOREST

This brush box forest in the Lamington National Park of the Border Ranges between northeastern New South Wales and southeastern Queensland seems to be similar to the forest that is fossilised at Nelly Creek (and Stuart Creek). A climate similar to the subtropical conditions under which flame trees, brush box, eucalypts, and rainforest including southern beech in suitable protected gully sites occur there today is implied for the Lake Eyre region 45 million years ago.

The silcrete ridge fossil site on Stuart Creek Station. The line of trees in the distance marks the Stuart Creek channel in the grassy floodplain which merges into gibber desert away from the river.

M.E.W.

Above: A cast of a large gumnut with its cap preserved is seen in this specimen (which had to be photographed in the field as the huge block of silcrete is immovable.) Though Eucalyptus *fruits are known from this fossil site, examples in which the cap is preserved were previously unknown.*

N.F. Alley

Above right: Large blocks of conglomerate, containing silcrete pieces, occur at the base of the Willalinchina Sandstone which is silcreted and forms the fossiliferous ridges at Stuart Creek.

M.E.W.

Right: Blocks of silcrete with pronounced ripple marks show that the sandstone (which was later silcreted) was laid down under quiet water conditions.

M.E.W.

Fossiliferous silcrete on the ridge at Stuart Creek. The 'dead finish' Acacia *trees and sparse desert grasses and forbs are the only vegetation.*

M.E.W.

and sometimes to genus, but the mud layer has also preserved within it a pollen and spore record. Thus the age of the flora is known and it is possible to establish that the forest 45 million years ago here at Lake Eyre South was approximately contemporaneous with the Middle Eocene forest fossils preserved in the Golden Grove[51] and Maslin Bay localities near Adelaide.[70] (The fossil floras at Anglesea in Victoria and at Nerriga in New South Wales are also Middle Eocene in age, and probably slightly younger than those at Nelly Creek. They contain similar mummified material capable of close identification.)

Blocks of puddingstone silcrete, patterned like Aztec carvings, occur at the base of the silcrete ridge. How such ropy silcrete forms is a mystery.

M.E.W.

The forest communities in all the localities mentioned above are believed to represent very warm temperate vegetation types at least, and they may even be subtropical or tropical.

Among the leaves identified in the original sample collected[56] at Nelly Creek are members of PROTEACEAE, MYRTACEAE, LAURACEAE, araucarian conifers (including *Agathis*), podocarp conifers, CASUARINACEAE (*Gymnostoma*), and *Brachychiton* (flame trees or kurrajongs, STERCULIACEAE). In all, 269 leaves with sufficient area preserved were examined and 64 per cent belonged to one taxon which cannot be identified as having the characteristics of any extant family, and which may therefore be of a family that is now extinct. The recent visit to Nelly Creek by Dr Neville Alley and the author resulted in a very large sample being collected, and work on the fossils by Dr Christophel at Adelaide University will add greatly to knowledge of this flora.

The pollen and spores identified in the Nelly Creek mummified[131] leaf horizon include: *Anacolosa* (OLACACEAE, the olive family); many different PROTEACEAE including *Banksia*, *Xylomelum* (woody pear), *Gevuina/ Hicksbeachia*, *Grevillea* and *Beauprea*; *Cupania* (tuckeroo, SAPINDACEAE); *Casuarina* and *Gymnostoma*; grasses; RESTIONACEAE; MYRTACEAE including *Eucalyptus*, *Baeckea-Backhousia*, *Tristania* (brush-box, water gum); *Ilex* (holly family, AQUIFOLIACEAE); SAPOTACEAE; CUNONIACEAE; ERICACEAE; lilies; MALVACEAE (*Hibiscus* family); palms (ARECACEAE); *Santalum* (quandong or

Abve left and top right: A hole through a large block of silcrete is patterned on its inside in the same way as the tree-fern trunks seen as impressions. The hole is the result of the decay of a trunk, leaving only a cast of its more enduring outer layers.

Above middle: The great number of leaves on this block of silcrete gives an indication of the richness of this fossil site.

Bottom left: A fossil of a Brachychiton *leaf, from a flame tree or kurrajong that grew at Stuart Creek when the region was forested and had lakes and meandering streams where now there is only desert, epitomises the drying of Australia.*

Bottom right: The colouration of the leaves varies on the silcrete blocks.

M.E.W.

native pear, SANTALACEAE); mistletoes; *Austrobuxus* and *Dissiliaria* (EUPHORBIACEAE); POLYGALACEAE; *Ascarina* (CHLORANTHACEAE); araucarian and cypress conifers; many ferns; and mosses, including *Sphagnum*. The *Nothofagus* beeches (with *fusca* and *brassii* pollen types) and *Podocarpus* form only a minor part of the pollen spectrum. The beech pollen may have come from far away, perhaps from the ancestral Flinders and Willouran Ranges. *Nothofagus* is a notorious long-distance contaminator of pollen assemblages.

The nature of the sediments comprising the Nelly Creek rock sequence indicates that rivers meandered through a flat landscape much like that of today in Middle Eocene times, but it was a forested land, or at least there were gallery forests on valley floors. Drier areas may have supported *Casuarina* woodland and more sclerophyllous vegetation.

About a kilometre upstream from the Nelly Creek mummified[55] leaf site is a river bend where the contact between the Cretaceous Winton Formation and the base of the Tertiary Eyre Formation is exposed in the eroded river bank. A zone of shiny, rounded quartz pebbles, grey and white and small (maximum diameter 0.5 cm) is exposed in basal Eyre Formation. Below it the Winton shale has carbonaceous lenses in which masses of carbonised wood are preserved. The wood is like charcoal, not permineralised. (A sample of wood was examined by a wood technologist but unfortunately showed no structure as a result of compression.) A spore and pollen flora is preserved, dating the horizon and identifying the vegetation as conifer forest with fern understorey and only two angiosperm taxa. *Microcachrys* conifer pollen forms 60 per cent of the pollen sample, other podocarps, araucarians and ginkgoes were present but not abundant and the ferns were mainly ground ferns, not tree-ferns. A cool temperate climate is suggested by this Cretaceous flora, and the nature of sedimentation implies that the region was already flat with low relief.

On the banks of Nelly Creek today the sunshine glints on pieces of platy gypsum that look like broken glass, and in some areas 'gypsum roses' occur — beautiful structures where crystals grow in triangular segments from a central point. Rain had fallen two weeks before our visit to this starkly beautiful place and the bladder saltbush, pale yellow-green with inflated fruits like polystyrene packing bubbles; samphires in an extraordinary array of colours from orange-red to pink and mauve; *Frankenia* saltbush with pink flowers and looking superficially like sea-thrift; and a few small ephemerals including a bright blue *Erodium* (GERANIACEAE) formed a bright garden on the river banks... A very different picture from the forests that grew here 45 and 65 million years ago.

THE STUART CREEK SILCRETE FLORA

The Stuart Creek fossil flora occurs on silcrete ridges on Stuart Creek Station.[47] This pastoral holding is managed together with Anna Creek Station[55], forming the largest rangeland property in Australia. Gibber desert, low plateau breakaways, saltbush, patches of bleached grass; only the ephemeral watercourses tree-lined — some of the most empty landscapes imaginable. Like the rest of Australia's amazing central deserts, the vast emptiness comes alive after rain. Grass and ephemeral herbs grow in haste to complete their cycle and set seed before they dry up and die. Even the gibber plains take on a green tinge.

A stark contrast exists between the present-day gibber plains and ridges covered in silcrete and the evidence presented by the silcrete fossil flora of the very different conditions that prevailed here when the silcrete was formed. Although the age of the silcrete and its flora is not agreed upon by all the scientists who are trying to establish the stratigraphy in this out-of-the-way region, it would seem that a Middle Eocene age is the least contentious. This would mean that the plants in the silcrete were growing here at about the same time as the Nelly Creek and Golden Grove forests existed, and therefore they would be roughly the same age as the Anglesea and Nerriga floras (of Victoria and New South Wales respectively).

The silcrete capping of the ridges is broken up into boulders and slabs of all sizes, and into scree that covers the flanks and merges into the gibbers on the surrounding plains. Many of the boulders and slabs have surfaces covered in leaf impressions, as do many of the bits of scree. Often the impressions are coloured orange-red or shades

A large seed (diameter 1.75 cm) probably of PROTEACEAE.

The large leaf, upper right, is probably of MYRTACEAE.

A transverse section of a Banksia*-like inflorescence.*

This leaf has not yet been identified.

This leaf with narrow lamina is identified as close to Orites*, a member of the* PROTEACEAE.

PHOTOS: JIM FRAZIER

The broad leaf at the top in this specimen, length about 7 centimetres, is identified as of the family MYRTACEAE.

These leaves have not yet been identified.

This magnificent fossil specimen with a large Brachychiton *leaf is richly coloured and has acquired a desert varnish from exposure to sun and sand-laden wind, in the same way as desert gibbers acquire their patina.*

This example of a large leaf of a flame tree, a species of Brachychiton*, shows an unusual colour range in the fossil.*

Frond of a palm, from Eyre Formation near Woomera.

A conifer twig, length 6 cm, from Woomera Plateau.

A frond which may be of a fern close to Gleichenia.

Part of a big leaf of Brachychiton *is seen in this Haggard Hill specimen.*

A leaf resembling Banksia *from Mt Eba.*

A latex mould of the inside of a hole in a specimen from Haggard Hill reveals a cluster of gumnuts probably of Eucalyptus *sp. The technique of using a latex solution and filling cavities that were left when a plant fragment rotted away to obtain a cast that shows the impression of the outer surface of the fragment is very useful. In a matrix as fine and hard as silcrete a great deal of detail is visible on the latex mould. Because latex is elastic the mould can be pulled out revealing the form of the object that left the hole.*

A latex pull from a hole in another Haggard Hill specimen revealed a perfect Casuarina *fruit.*

Leaves like this with pointed, narrow segments are one type of Brachychiton *(family* STERCULIACEAE*) common at Stuart Creek.*

PHOTOS: JIM FRAZIER

Above: Ruins of a shepherds' hut on Stuart Creek station, the walls made of blocks of silcrete with fossil leaves on their surfaces.

Left: A latex mould of two elongated, segmented cavities in a specimen from Island Lagoon near Woomera has enabled the identification of the organs which left the cavities as male catkins of Gymnostoma, *a genus close to* Casuarina. Gymnostoma *was abundant in Australia while the climate was warm and wet in the Early Tertiary. Today it is found in New Guinea, having migrated off mainland Australia as the continent dried.*

Jim Frazier

of pink and yellow, and some have a high shine (or patina) from the oxidation of the surface by weathering under a desert sun.

An amazing diversity of 45 million year old leaves is preserved and on display on the stony ridges near Stuart Creek. Today's occasional small, thorny trees of *Acacia tetragonophylla* grow among the blocks of silcrete, and small desert plants survive as best they can among the scree...the desert present rooted among a warm, wet forest past. (*Acacia tetragonophylla* is known as 'dead finish', being the only survivor in prolonged droughts and appearing to be dead before reviving when the rains come.)

Among the fossil leaves in some areas gumnut-type fruits and conifer twigs and branches are found. One large block of silcrete has a cylindrical hole (diameter 9 cm, length 1 metre) through it, made by a trunk that had rotted away leaving only a cast of its outer form. Parts of the cast show a pattern which can only have been made by a mass of adventitious rootlets clothing the stem. The only plants that have such a zone of rootlets clothing their trunks are tree-ferns. Impressions of narrower strips of the same pattern are seen on some of the silcrete blocks. They can only be parts of trunks, not thin stems covered in the same rootlets, as tree-ferns do not have branches and young plants do not have rootlet-covered stems. The hole made by a tree-fern trunk, narrow stem-like impressions with roots attached and an assortment of leaf fossils were photographed *in situ* in the field.

All silcrete fossils in South Australia are protected and the outcrops have been declared geological monuments, so collecting can only be by permit to obtain specimens for museums and scientific institutions where work on fossil floras is taking place. The South Australian Museum and the Mines Department kindly lent specimens from their collections for photography under studio conditions. A sample of Stuart Creek fossils could thus be illustrated for the first time in a way that enables their beauty and scientific importance to be widely appreciated.

Among the plants identified so far in the silcrete flora, which still requires a great deal of scientific work, are the following taxa: Conifer shoots of PODOCARPACEAE (*Dacrycarpus* and *Dacrydium*), ARAUCARIACEAE and CUPRESSACEAE (*Papuacedrus*). Fruits of *Eucalyptus* and other capsular-fruited MYRTACEAE.[60] Abundant leaves of MYRTACEAE (*Eucalyptus* and/or *Melaleuca*). Leaves of *Brachychiton* (flame trees, kurrajongs, lace barks, STERCULIACEAE) and PROTEACEAE (*Banksieaeformis* and *Orites*). Cone-like fruits of CASUARINACEAE (*Gymnostoma* and *Casuarina*).

This described plant association and the impression of the vegetation that is gained from a study of the silcrete flora in the field appear to represent a *Paleoflora Mixta* type of forest, possibly with paperbark communities in adjacent swamps. The presence of large silcrete blocks with ripple marks indicates that the conditions for deposition were quiet. The terrain is believed to have been one of low

A leaf of Brachychiton from Golden Grove.

JIM FRAZIER

relief where meandering rivers with gallery forest and oxbow lakes with swamps were probably features of landscapes. The climate was probably warm and humid, but so far inland may have been seasonal with dry intervals.

The silcrete at Stuart Creek is silicified Willalinchina Sandstone, equivalent in age to the Eyre Formation at Nelly Creek. At its base is a layer of conglomerate in which rough pebbles and pieces of silcrete (from a previous phase of silcretisation) are cemented together. The top of the unit is composed of 'reed mould' silcrete whose method of formation is not understood. This silcrete has vertical channels or hollow tubes through it. Some amazing examples of blocks of 'pudding stone' silcrete with surface patterning that looks like Aztec carvings are found among the boulders on the ridge near the Willalinchina shepherds' hut which was built from blocks of silcrete, many with plant fossils decorating them. *Eucalyptus* fruits were recorded from the ridges near the hut, but none remain there now. Before the silcrete floras were protected (and maybe since!) collectors have stripped the choice specimens from the area.

OTHER SOUTH AUSTRALIAN SILCRETE FOSSIL LOCALITIES

The **Poole Creek** palaeochannels on the plateau near Lake Eyre South[113] contain a flora which includes fruits of *Eucalyptus* in a sandstone layer which is equivalent to the Willalinchina Sandstone of Stuart Creek. Two sandstone levels are present at the Poole Creek site and the gumnuts are confined to the lower, while *Casuarina* and *Grevillea*-type fruits and a few leaves characterise the upper sandstone. This upper level may be equivalent to Miocene Etadunna Formation.

In the **Woomera district** and within the **Woomera Prohibited Area** silcrete floras are found at a number of localities. These include Island Lagoon and Freds Swamp just west of Woomera; Shell Lagoon and Koolymilka; Mt Eba, Haggard Hill, Jack's Hill and Hallam Rise on 'The Twins' station near Millers Creek northwest of Woomera near the Stuart Highway. Although dating of these localities is uncertain, they appear to be part of the same Middle Eocene silcrete province as Stuart and Poole Creek.

THE GOLDEN GROVE FOSSIL LOCALITY[130, 126]

An enormous sand and brick-clay quarry in Golden Grove, an outer suburb of Adelaide, is the locality for a Middle Eocene fossil flora that is contemporaneous with the Nelly Creek flora and with a similar one at Maslin Bay.

The North Maslin Sand that has been mined out of the pit at Golden Grove contains clay lenses in which the fossils occur. Erosion of the walls of the quarry has created a theatrical landscape like a mini-Grand Canyon, intricately dissected and richly coloured. Swallows have made their bottle-shaped clay nests in the canyon caves.

In the Middle Eocene the region was a plain in a drainage basin[59] with braided streams meandering across it. Small oxbow lakes developed from meanders which became cut off from the streams, and in these occasionally flooded lakes the clay sediments with their plant inclusions accumulated. A forest surrounded them or was close by on the edges of the basin.

The fossils are either preserved as brown impressions on a beige matrix, in which case no plant tissue is preserved, or as mummified material in a grey, less-oxidised matrix. The mummified leaves and plant fragments can be removed from the clay just as the Nelly Creek fossils can be, and their diagnostic features of cuticle structure and venation make their identification possible.

The commonest leaves have serrate margins and are identified as belonging to the

Golden Grove Fossils

Left: A very large leaf similar to leaves of MORACEAE, the fig family, and more than 20 centimetres long was dug out of wet grey claystone in the quarry.

M.E.W.

Right: Large unidentified leaves.

Below left: A small fragment of podocarp conifer foliage, bottom left, and unidentified leaves.

Below right: Banksiaephyllum, *a leaf closely resembling those of some modern species of* Banksia.

Jim Frazier

The Golden Grove fossil locality is in an enormous sand quarry in the outer suburbs of Adelaide.

M.E.W.

family ELAEOCARPACEAE[54], genus *Sloanea* or *Elaeocarpus* (blueberry ash). *Banksieaephyllum* leaves, similar to those found in the tribe Banksieae of the PROTEACEAE, are also abundant. They may represent the leaves of the proteaceous flower *Musgraveianthus* which has been found at Golden Grove. (This flower was first described from the slightly younger Anglesea deposit in Victoria.) Pollen of the sort identified in *Musgraveianthus* is found loose in Golden Grove samples.

Laurels, an important component of Early Tertiary floras, are represented by at least two taxa at Golden Grove. STERCULIACEAE (*Brachychiton*, kurrajongs and flame trees) and MYRTACEAE (*Myrtaciphyllum*) are common. Recently some leaves were discovered which may prove to be *Eucalyptus*. They are still to be described. The conifers *Decussocarpus* and *Podocarpus* (PODOCARPACEAE) and *Lygodium*, a climbing fern, have also been identified.[49] Very large leaves up to 20 cm long and 10 cm wide, with rounded bases, were found on a recent collecting visit. They appear to be similar to *Artocarpidium* (MORACEAE, the fig, breadfruit and jak fruit family) described by Ettinghausen in his classic 1883[73] treatise on Eocene plants from Vegetable Creek in northern New South Wales.

The Golden Grove spore and pollen record gives a more complete picture of the vegetation. Ferns were particularly abundant; conifers represented a minor component with podocarps the commonest; among the angiosperms *Nothofagus* species and CASUARINACEAE were present but not abundant; PROTEACEAE was well represented with high species diversity in some samples.

The Maslin Bay fossil locality, which is also in North Maslin Sand[79], has palm pollen (ARECACEAE) in addition to the plants identified at Golden Grove. *Agathis* (kauri pine) is also present.

Outliers of the North Maslin Sand occur in the One Tree Hill[127] area, St Vincent Basin. Palynology indicates a Middle Eocene age for the flora, the same as for the Sand at Golden Grove. CYPERACEAE (sedges) are a component of the Middle Eocene floras of the St Vincent Basin and occur in the latest Eocene[132] in the southeastern Eucla Basin. (The intracontinental St Vincent Basin[133] was formed in the Middle Eocene when Australia's final separation from Antarctica occurred and the drift northwards accelerated.)

Similar Middle Eocene pollen floras to those in Maslin Sand occur in the western Murray Basin though the proportions of the component species vary considerably. *Nothofagus*, for instance, is usually abundant, particularly the *brassii* group.

Some of the leaves of LAURACEAE and ELAEOCARPACEAE in the Golden Grove flora (as well as some from Anglesea)[53] have domatia ('mite houses') on their leaves. Oribatid mites occupy these structures and maintain leaf hygiene. Some adult mites have been found preserved in the 45 million year old samples. They ate the algae and lichen that grew on the leaves, as do modern leaf-cleaning mites that occupy similar domatia on ELAEOCARPACEAE and LAURACEAE in modern Australian rainforests.

CHAPTER SEVEN

AUSTRALIA ADRIFT IN THE 'GOLDEN AGE' EOCENE

(45 to 38 million years ago)

Forty-five million years ago Australia lay between 60° and 30° South, the mainland already separated from Antarctica by 10° of latitude. The development of fully marine conditions to its south and the sagging of passive rift margins resulted in the sea invading the Bremer and Eucla Basins, and the Carnarvon Basin region of Western Australia was also inundated. With the onshore part of the Bremer Basin flooded, arms of the Eocene sea penetrated far into southern Western Australia along valleys, as far north as Norseman. The Eucla Basin became a wide embayment of the Southern Ocean and marine limestones were deposited beneath its warm waters.

The southwestern part of the Murray Basin[146] was invaded by the sea and the Buccleuch Beds were laid down during this Late Eocene to Early Oligocene event. Recent evidence suggests that the marine influence of this interval in the Murray Basin was much more widespread than had previously been thought.[147] The Murray Basin was sinking throughout the Tertiary. As a result, the land surface that was exposed in Late Cretaceous times is now 500 metres subsurface; the Buccleuch Beds are 200 metres down and the Miocene Dubbo Limestone is 80 metres below today's ground level.

The sea had not yet created Bass Strait in the Eocene — most of the area was a swampy plain connecting Tasmania to the mainland. In the Gippsland Basin the sea reached its present limits. Channel, point-bar and barrier sand deposits formed in its offshore section creating reservoirs for some Gippsland oil and gas deposits.

It has been suggested, on the basis of nanoplankton evidence[184], that during the Eocene surface-water temperature decreased progressively in an easterly direction along the southern margin of Australia. A similar west to east gradient in temperature existed during Middle to Late Oligocene times. An explanation involves the intermittent presence of a strong, warm proto-Leeuwin Current

DRIFTING—45 MILLION YEARS AGO

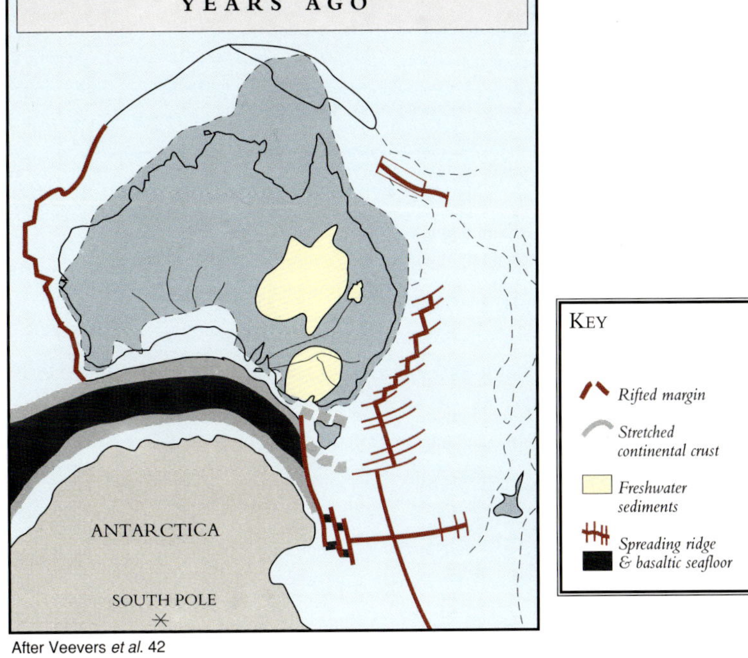

After Veevers *et al.* 42

East Mt Barren showing the basal platform that represents sealevel height during the Eocene.

J.S. BEARD

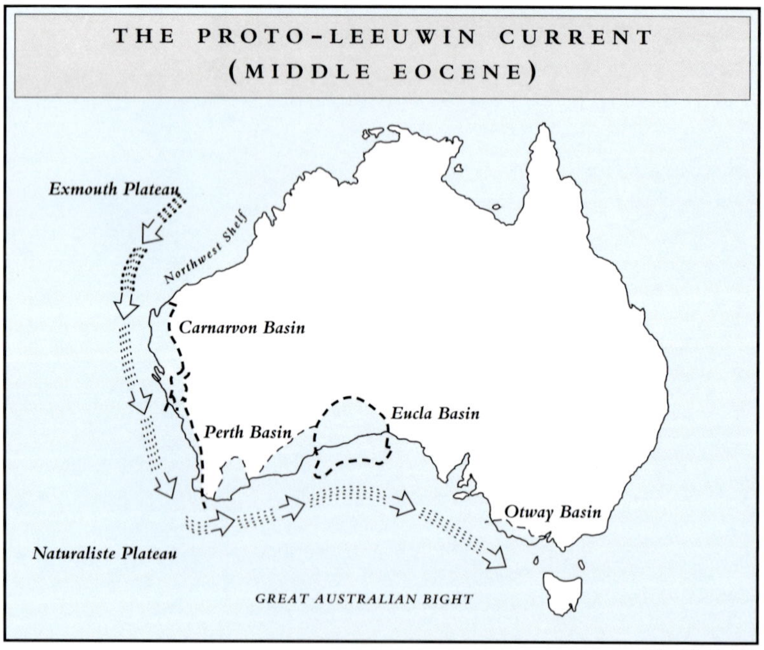

After Shafik 184

The warm proto-Leeuwin Current was initiated during the Middle Eocene. It flowed southwards along the western continental slope and turned eastwards round Cape Leeuwin to flow into the Great Australian Bight. The temperature of its waters decreased progressively southwards and eastwards.

travelling down the Western Australian coast from the north and sweeping round the southwestern corner of the continent into the Great Australian Bight Basin. Surface waters in the Otway Basin further east would be expected to be cooler than those in the Bight Basin under this influence and nanoplankton differences confirm this hypothesis. The Leeuwin Current today contributes to the occurrence of a fauna of tropical benthic (bottom-dwelling) invertebrates and several tropical species of pelagic tuna in the Bight.

Had it not been for a warm current flowing from north to south along the western margin of Australia during the time when it was still close to Antarctica, it might have been expected that a progressively colder current would have flowed along that coast from 35 million years ago when Antarctica started to cool. In South America and in Africa the cold currents that have flowed along west coasts have resulted in the formation of west coast deserts, intensely arid and inhospitable, dating from the time when Antarctica's real refrigeration began. Early aridification of Australia's western margin with regions suitable for the development of an arid-adapted flora would have been the expected scenario. Floral, faunal and sedimentological evidence shows that this was not the situation, strengthening the case for the presence of the proto-Leeuwin Current.

In Western Australia today one can see landscape features created by the marine invasion of the Bremer Basin. The Barren Ranges, east of the Stirling Ranges between Bremer Bay and Hopetoun, have a basal platform about 90 metres above sealevel from which the peaks rise. This 90-metre terrace represents a wave-cut platform eroded by the Eocene sea. The peaks of the Barren Ranges and other promontories in the Albany-Frazer province were islands in the Eocene sea, and comparable with those of the Recherche Archipelago today.

Spongolite breakaway with the Bremer River below.

M.E.W.

Far left: The spongolite quarry at Twertup where Eocene marine rock full of sponge spicules is cut into building blocks.

M.E.W.

In the adjacent Fitzgerald National Park low mesas of spongolite (marine sediment with siliceous sponge spicules) have been left standing out of the flat landscape, their tops all the same height, where the plateau of Eocene marine sediments has been eroded away. At Twertup the river has cut down through the plateau (which was once a flat sea bottom) forming a breakaway. Spongolite and the Pallinup siltstone in which it forms layers is quarried for building stone.

Abundant, varied and beautifully preserved lithistid sponges[249] have been described from the Late Eocene spongolite[250] as well as the first sphinctozoan sponge to be recorded in the Tertiary. Plant fossils are also found in some of the deltaic sediments in the sequence. These include strap-like leaves with parallel veins of rushes that grew on the water margins, some medium-sized leaves with crenulate margins which resemble ELAEOCARPACEAE and many that are too poorly preserved to be identified. A small araucarian cone was present in a collection kindly supplied by Bill Moir of Borden, Western Australia.

The Eocene Sea reached its greatest extent in early Late Eocene and then retreated. As it withdrew from the valleys, terrestrial sediments were deposited in them, and it is in these post-Middle Eocene sediments that plant

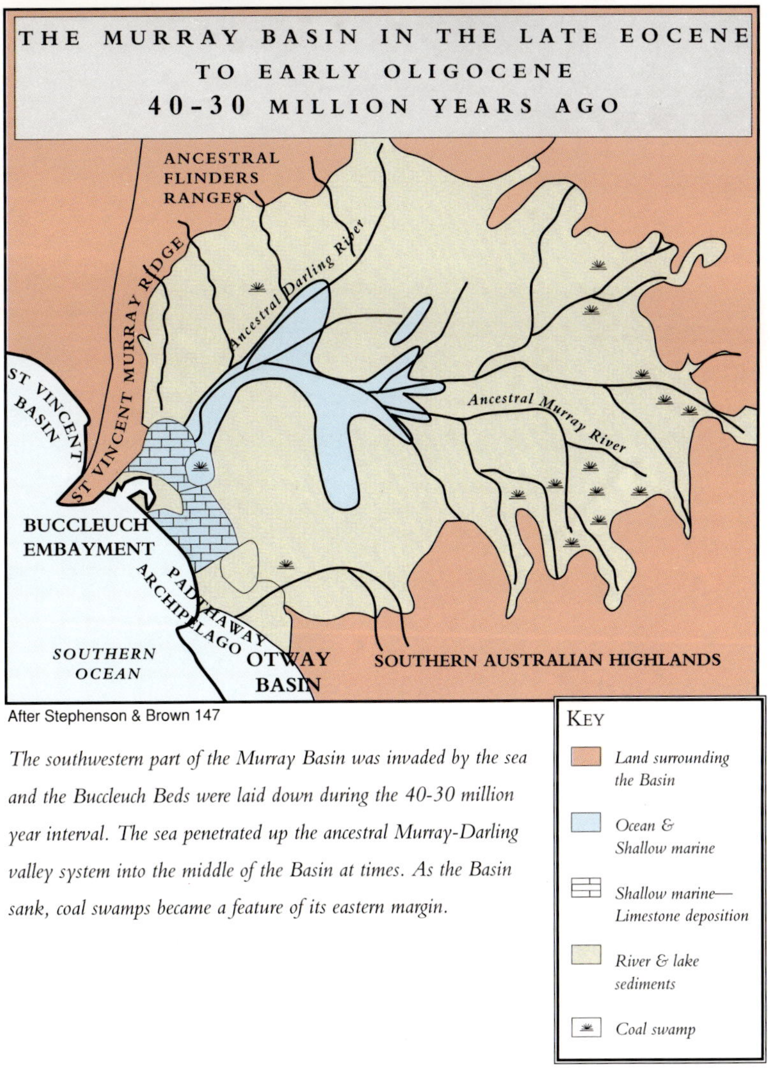

After Stephenson & Brown 147

The southwestern part of the Murray Basin was invaded by the sea and the Buccleuch Beds were laid down during the 40-30 million year interval. The sea penetrated up the ancestral Murray-Darling valley system into the middle of the Basin at times. As the Basin sank, coal swamps became a feature of its eastern margin.

A modern cypress pine — Callitris.

A Nothofagus *leaf from the Kojonup Sandstone.*

STEVE MCLOUGHLIN

A leaf of a member of the PROTEACEAE, similar to Grevillea.

STEVE MCLOUGHLIN

fossils are found, adding to our knowledge of the vegetation. Post-Eocene uplift of the ancient land and its drainage channels has resulted in their being 300 metres above sealevel today.

In the Eucla Basin, the sea flooded into a far larger area than in the Bremer Basin — covering what is now the Nullarbor Plain. The marine sediments laid down in it are today covered by a Miocene limestone which formed during another major marine incursion 15 million years ago. The Bremer Basin was not invaded again by the sea after the Eocene episode as the edge of the continent in southern Western Australia was upwarped along what is known as the Ravensthorpe Axis. This watershed follows a a line that runs parallel to the coast from Augusta near Cape Leeuwin, through Kojonup and Ravensthorpe to the edge of the Eucla Basin. Rivers to the south of this upwarp were rejuvenated and some reversed their direction in beds of ancient drainages in which rivers had flowed from Antarctica into Australia before rifting.

Plant macrofossils in the silicified Kojonup Sandstone[62] (part of the Plantagenet Group) at Muradup where they occur on the drainage divide include: *Nothofagus*; podocarp and

THE VAN DIEMEN SANDSTONE OF MELVILLE ISLAND

The Van Diemen Sandstone type locality, Melville Island.

BRIAN SENIOR

Left and middle: Leaf of a species of Grevillea. *Right: A long, slender leaf of unknown affinity, and a leaf of* Grevillea.

JIM FRAZIER

araucarian conifers (including *Agathis*); *Livistona* palm; *Ficus* (fig trees, MORACEAE); APOCYNACEAE; STERCULIACEAE (flame trees, kurrajongs); several PROTEACEAE, including *Lomatia*, *Banksia* and *Grevillea*; *Eucalyptus*; *Gymnostoma* (CASUARINACEAE); and rushes (*Typha*).

Mangrove pollen has been identified in Plantagenet Group sediments and in the Kings Park Shale of the Perth Basin.[264] From the present day ecology of tropical mangroves and their Tertiary occurrence in Australia it is reasoned that hot, wet tropical conditions characterised the Middle to Late Eocene in southern Western Australia.

At Calingiri, on a major drainage divide, blocks of silcrete have a similar flora to that at Muradup, including *Eucalyptus*, *Lomatia* and *Apocynophyllum*. The silcrete here is underlain by a conglomerate and some silcrete blocks have ripple marks, duplicating the situation at the Stuart Creek locality in South Australia, though the age of the fossils at Calingiri and Muradup is probably somewhat younger.

Other Eocene plant fossil localities in Western Australia occur at Lake Cowan near Norseman, near Coolgardie and at Darkin Swamp. Abundant tree-ferns and ferns; *Nothofagus*; *Banksia*; *Araucaria*; *Podocarpus*; *Nypa*, a tropical mangrove palm; mangroves and other plants are found in these deposits in which the plant material is solidly layered. A new fossil locality on the eastern Darling Range at Westdale has well-preserved plants in a lake deposit. This flora is currently being studied.

In the Fortescue River valley, north of the Hamersley Ranges, a rich assemblage of *Nothofagus* and other typical Early Tertiary taxa occurs in shales which are believed to be Middle to Late Eocene in age.

The Merlinleigh Sandstone of Late Eocene age in the Carnarvon Basin[140] contains silicified wood and a *Banksia* 'cone' just like those of living species. This type of fructification, which today gives *Banksia* an advantage and enables it to survive bushfires, was already developed long before regular fires became a feature of environments.

The pollen and spore record from a Late Eocene lignitic sequence[64] in Zanthus 6 Coal Project bore on the western margin of the Eucla Basin near Balladonia shows a rich flora of conifers, flowering plants and ferns. Ten different pollen species of *Nothofagus* are present, dominating the assemblage and comprising 35-58 per cent of samples. The *Nothofagus* species are predominantly of the *brassii* pollen type; *fusca* type are rare and *menziesii* type are absent. Gymnosperms, including celery-top pine (*Phyllocladus*), and CASUARINACEAE comprise up to 30 per cent; PROTEACEAE are very diverse and comprise 4-16 per cent. *Acacia* and grass pollen are both present, and SPARGANIACEAE (burr rushes), CYPERACEAE (sedges) and horsetails (*Equisetum*) indicate swamp environments. The assemblage is similar to one from the Wilkinson 1 bore in the Pidinga Formation[137] in western South Australia. (*Acacia* pollen does not appear until the Late Oligocene in the Murray Basin, or until the Miocene in the Gippsland Basin. Sedges do not appear in the record in eastern Australia until Miocene times, yet they were widespread in the southeast Eucla Basin, in the St Vincent Basin[132] and the Lake Eyre Sub-basin of South Australia[133], and at Napperby in Central Australia in the Eocene.)

Other bores in the Eucla and Bremer Basins which intersect the Eocene lignites show the same abundance of *Nothofagus* and the great diversity of PROTEACEAE. Of particular interest is the presence of sclerophyll members of PROTEACEAE including *Sterlingia* and *Adenanthos* (Western Australia's blueboy and woollybush), genera that evolved in the rift zone between Australia and Antarctica.[63] Today 33 of the 39 extant species of *Sterlingia* are restricted to southwestern Australia associated with dry sclerophyll or heath vegetation.

The mixture of cool temperate, subtropical, sclerophyll and broadleaf angiosperms and the high proportions of conifers and ferns make the forests 'mixed' with no modern analogue. The abundance of southern beech and very different proportions of other components accord with evidence of climatic change in the latter part of the Eocene. While the Middle Eocene was subtropical, ocean temperatures around Australia, both of surface and bottom water, decreased by about 5°C during the Late Eocene, though there were probably fluctuations. A warmer reversal of the cooling trend is known to have occurred late in the epoch.

In southeastern Australia the same increase in *Nothofagus*[136] abundance indicates the cooler conditions. Rainforest with year-round humidity and a rainfall of 1500-2000 millimetres a year is suggested by the vegetation type. In the southern part of South Australia this sort of forest with abundance of tree-ferns, ferns and mosses (including *Sphagnum,* the peat moss) occurred during the Late Eocene.

Further inland, near Lake Eyre, *Nothofagus* beech[47] and podocarps were far less prominent, CASUARINACEAE were abundant and the climate was probably seasonal and drier. Significant values for pollen of CUPRESSACEAE (*Callitris,* the native cypress pine)[139] occur in the Middle to Late Eocene in Central Australia, although such pollen does not appear in southeastern Australia until the Miocene/Pliocene. It is possible that *Callitris-Acacia* woodland may have been a much more widespread vegetation type in drier areas of the continent in the past. In the Torrens Basin up to 10 per cent of *Eucalyptus* pollen is found in samples; CUNONIACEAE has patchy distribution, abundant in some areas; PROTEACEAE was abundant but species diversity was not high. The abundance of burr rushes and RESTIONACEAE indicates claypan and swamp habitats, and grass pollen in low frequencies tells of opening up of forest into woodland as rainforest became restricted to valley floors.

Throughout the Eocene, the rivers south of Lake Eyre probably drained to the sea through Stuart Creek and across the northern Gawler Ranges into the Eucla Basin, or via the Torrens Basin to Spencer Gulf.

In the Hale River Basin, northeast of Alice Springs[82], the pollen record in the Ulgnamba Lignite of Middle to Late Eocene age contains abundant rainforest taxa which suggests high, evenly distributed rainfall. Among the pollen is that of a sundew (DROSERACEAE) closest to the Venus fly trap, *Dionaea,* which today occurs only in the southeastern United States of America. Grass pollen may reach 2 per cent of samples, suggesting that patches of grassland existed in Central Australia by this time.

In a palaeovalley between Ayers Rock and the Olgas, the Late Eocene section of a pollen sequence shows a different forest type. Conifers were very abundant; PROTEACEAE were common but diversity was low; MALVACEAE, *Quintinia* (possum wood) and *Gyrostemon* (the desert poplar of today's arid lands) were present.

In Tasmania *Nothofagus*-dominated forests with high diversity[65] of angiosperms, gymnosperms and ferns occurred in the latter part of the Eocene.

The Vegetable Creek flora from Elsmore[73], near the New South Wales-Queensland border, is similar in composition to the silcrete floras of South Australia. Very little is known about vegetation of the northern half of Australia in the Eocene (or, in fact, throughout the Tertiary). The Rundle Oil Shale deposits[49] of Queensland indicate that similar vegetation, rich in LAURACEAE and therefore approximating the floras of Nerriga and Anglesea, occurred that far north. It was microscopic algae which produced droplets of oil in their cells that were partly responsible for formation of the oil shales. The oil permeated the mud in the bottom of lakes which was later converted into shale. (These lakes, between Gladstone and Proserpine, had formed in troughs created by faulting on the coast during the 55-50 million year period of readjustment of rifting regimes while India's northwards movement was being slowed down by its collision with Asia.)

Near North West Cape in Western Australia a Late Eocene spore[173] sequence in a recent mineral exploration bore indicates that rainforest was in the area in the Eocene. Diverse PROTEACEAE, *Casuarina, Nothofagus* of *brassii* pollen type, podocarps and possible CHENOPODIACEAE (chenopods, saltbush) are identified.

The Van Diemen Sandstone of Melville Island has a macrofossil flora[124] which includes *Grevillea* and *Elaeocarpus.*[190] The age of the flora is not yet certain but it is probably within the range of Eocene to Middle Miocene.

South Australia has a rare vertebrate fossil site of Eocene age. At Port Noarlunga, in the Late Eocene Blanche Point Formation, fossils of giant penguins occur. Fossils of the same birds occur in rocks of the same age in New Zealand and in Seymour Island.

At the end of the Eocene epoch a sudden cooling (the Terminal Eocene Event)[47] saw the formation of glaciers on West Antarctica. Westerly winds over the southern part of the Australian continent are inferred from the direction of ancient coastal dunes of the Eucla Basin. Sealevel fell by 125 metres as a result of the capture of water in ice on Antarctica. Before this, at a time still to be agreed upon by scientists, the Drake Passage had opened between the tip of South America and the Antarctic Peninsula, allowing the development of a circum-Polar current. The immediate effect of the current was to uncouple high and low latitude temperatures for a while, so that Antarctic cooling had little effect on areas close to the Equator. The strengthening of this current and the widening of the distance

Ayers Rock and The Olgas

The Ayers Rock arkose and The Olgas conglomerate are inselbergs[72] — isolated outcrops of rock[119] — that rise like islands abruptly out of a sea of plains.[98] The ancient sedimentary rocks were formed about 600 million years ago. Their sediments came from the erosion of high ground that had been produced by the Petermann Ranges mountain-building event (orogeny). The sediments were subjected to great compacting forces and later in the Alice Springs Orogeny they were folded. This complex history resulted in alteration and toughening of the rocks, making them extremely resistant. The folding of the layered sequence of sediments is clearly seen on Ayers Rock, where strata which were flat-lying when they were deposited are now almost vertical, giving The Rock its characteristic ribbed appearance.

Latest investigations by the Australian Geological Survey Organisation[121] determined that there was a difference in the formation history of the arkose that forms The Rock and that of the conglomerate that forms The Olgas. As a result a new unit name — Uluru Arkose — has been proposed for the former while the latter is still designated Mt Currie Conglomerate. The direction of flow in the palaeochannels that provided the sediment for the Mt Currie beds was from the west and southwest, while the Uluru sediment was brought in from the south.

The outer shell of the inselbergs is hardened by weathering and derives its rich colouring from the iron concentrated in it. The fresh rock seen inside some of the caves on Ayers Rock which penetrate beyond the hardened skin is pale greyish-white or pinkish and is coarse-grained and poorly sorted — the sediment which formed it was of various grain sizes, mixed together. Fine striations due to cross bedding add to the textured appearance of the fresh rock.

The reason for the survival of the inselbergs when they rise up out of a large area where the same rock has eroded away and is now sub-surface around them is not fully understood. The most likely explanation is that the compartments of rock which have survived were on elevated ridges of substrate (palaeodivides) so that they were subject to less chemical weathering by groundwater.

A deep valley (a palaeochannel) has been discovered by drilling between Ayers Rock and The Olgas. It has an infilling of about 100 metres of Late Cretaceous to Eocene sediments, which implies that probably as much as 75 million years ago[120] the rock massifs, of which today's inselbergs are a part, created elevated features in the landscape. The uplands probably took the form of a soil-covered dome in the case of Ayers Rock and a complex of low rock domes in the case of The Olgas.

Forests of podocarps with PROTEACEAE (including *Beauprea*) grew upon and around the rocky domes. As the rock was progressively exposed by lowering of the ground surface the forest became angiosperm-dominated, as happened generally over central Australia during the Palaeocene and Early Eocene.

A cave at the base of The Rock reveals the pale pinky grey colour of the fresh rock that exists under the red outer shell.

M.E.W.

On the shady side of The Rock early in the morning colours are soft.

M.E.W.

Top and right: Bleached tussock grasses and mulga form a thin cover on the red sand round the glowing orange-red beehive-shaped rock massifs of The Olgas.

M.E.W.

Above: The Olgas in the distance across a plain dotted with desert Casuarina *trees and mulga take on a purple hue unlike their close-up orange-red colour.*

M.E.W.

Throughout the Eocene, forests grew on the deep regolith that was a feature of the times and The Rock and The Olgas were not dramatic landscape features though they affected the local topography. With the stripping of regolith from Oligocene times onwards they came into their own, and as forest was replaced by dry plains vegetation as the centre of Australia dried out they became islands in a sea of sand.

Deep sub-surface weathering was breaking down much of the the Uluru Arkose and Mt Currie Conglomerate in the region and the compartments of more resistant rock on the drainage divides were undergoing surface weathering, rounding the edges and etching the ridges. Erosion of the caves that form a horizon[72] at about 60 and 30 metres up the sides of Ayers Rock today started while these levels were sub-surface and represent areas of scarp-foot weathering that relate to the height of surrounding slope at the time. 'Wave caves' at the foot of the inselberg today are the result of a lowering of the surrounding plains by three to five metres during recent times, probably during the extremely arid intervals of the Pleistocene ice age.

1. Early Mesozoic: The Uluru arkose forms a low soil-covered rise (a drainage divide) in the landscape.
2. Late Mesozoic-Early Cainozoic: Erosion lowers the terrain on either side of the divide and the rock is exposed on top. The sediment produced by erosion forms valley deposits. In the valley between Ayers Rock and the Olgas a lake forms and lake sediments contribute to the valley sediments. As the slope steepens, erosion starts to undercut the iron-hardened capping of the exposed rock and chemical weathering in the water run-off region round its base weakens the underlying rock.
3. Erosion at soil level and sub-surface undercuts the exposed rock and block collapse increases the steepness of the inselberg.
4. Late Cainozoic: As the Rock takes shape, valley floor sediments cover the lake and climatic change sees the creation of a surrounding sandplain.
5. Today the undercutting of The Rock at groundlevel continues and fallen blocks sit on the platform which was produced by deflation of its cover by the dry and windy glacial maxima of the Pleistocene Ice Age. Sub-surface erosion continues.

THE FORMATION OF AYERS ROCK

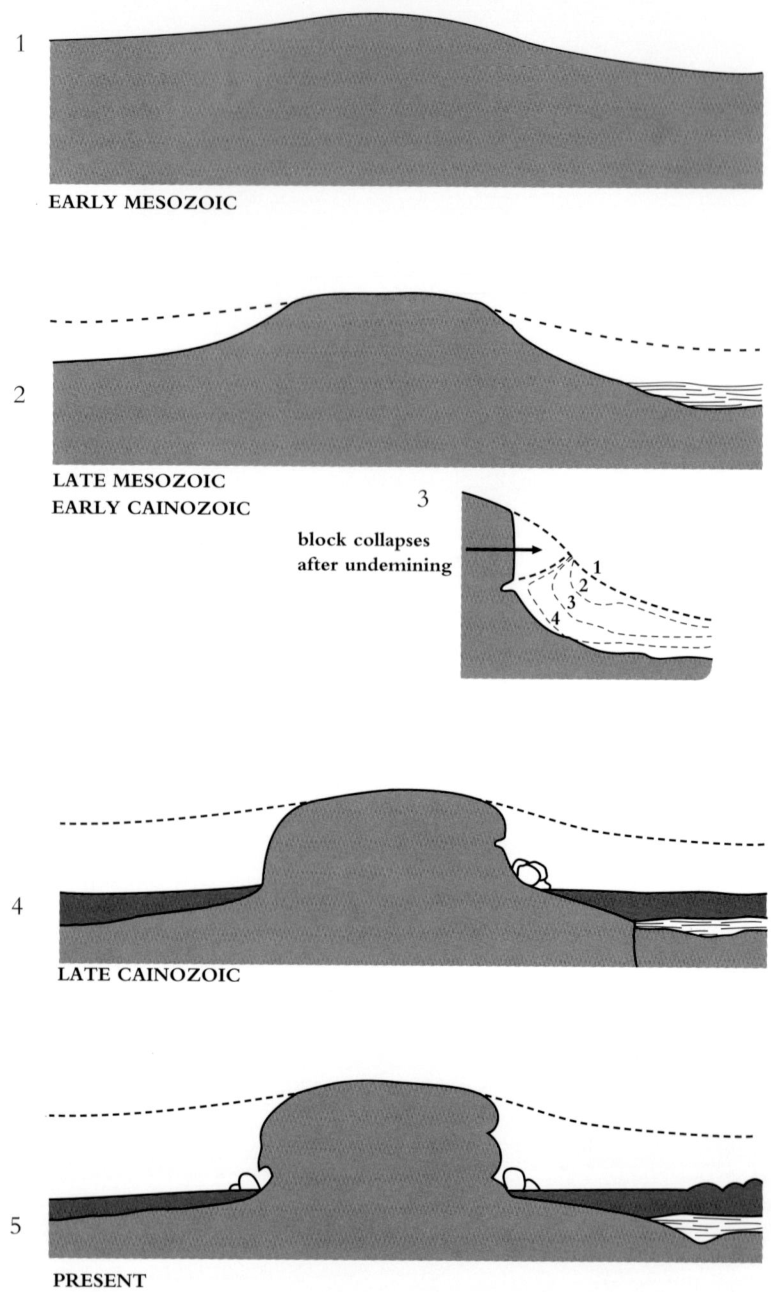

After Twidale & Bourne 119

between Antarctica and other lands as continents drifted northwards led to increasing cooling of Antarctica. From the time when the circum-Polar circulation first started, the Earth was set on a path towards an ice age, and this would mean progressive drying of the Australian continent. The Southern Ocean surface water temperature during the Terminal Eocene Event is estimated to have been 7°C and the duration of the Event 75,000 to 100,000 years.

A global tectonic event affecting Australia's and New Guinea's future development occurred at about 40 million years ago. Movement of the Pacific Plate changed direction from north to west-northwest. This would eventually result in collision with the Australian Plate in New Guinea and mountain building there from the Late Oligocene onwards. Subduction of the Australian Plate under Sundaland would also result from the changed direction of movement of the Pacific Plate.

HISTORY OF THE NULLARBOR

From the Middle Jurassic (160 million years ago) and through Cretaceous times up to 96 million years ago rifting was occurring between Antarctica and Australia by continental extension. The Nullarbor region was subsiding during early rifting and braided river systems flowed across it to enter the sagging rift valley that was evolving along Australia's southern margin. Early Cretaceous conglomerate, siltstone and shale were deposited in the adjacent rift during this phase. (During the Mesozoic the basin formed by the sagging rift and its onshore section which underlies the Nullarbor was known as the Bight Basin. As it continued to subside, the part below the Nullarbor and its offshore section became the Eucla Basin and site for deposition of Tertiary sediments.)

When global sealevel rose dramatically to peak at about 116 million years ago, much of Australia was flooded by the Eromanga Sea, including the Nullarbor region (the onshore Bight Basin). It became a seaway connecting all the other flooded continental basins to the proto-Southern Ocean (which had been forming since the Indian Ocean opened at 132 million years ago). The Nullarbor region then was connected to the flooded Officer Basin and the west coast via the Canning Basin. (The Cretaceous eustatic sealevel rise was the highest in the last 400 million years and much land worldwide was flooded.)

The first Cretaceous marine sediments to be deposited in the Nullarbor part of the Bight Basin were those that accumulated in the Eromanga Sea in the area during the flooding of the continent.

When the epicontinental sea retreated after peaking at 116 million years ago, seas retreated from the marginal basins including the Nullarbor region. Most of the continent was again well above sealevel. The narrow seaway in the rift valley system along the southern continental margin was progressing eastwards; a major depocentre developed in the Bight Basin — the Ceduna depocentre — and all drainage from the eastern part of the continent and from the southern margin was towards this centre. The Nullarbor region during this phase was again a floodplain with braided rivers.

At 96 million years ago a second phase of rifting started. (During the first continental extension phase, 360 kilometres of extended crust had been produced, separating the continents.) Sea-floor started to spread in the centre of the rift, and there was a global

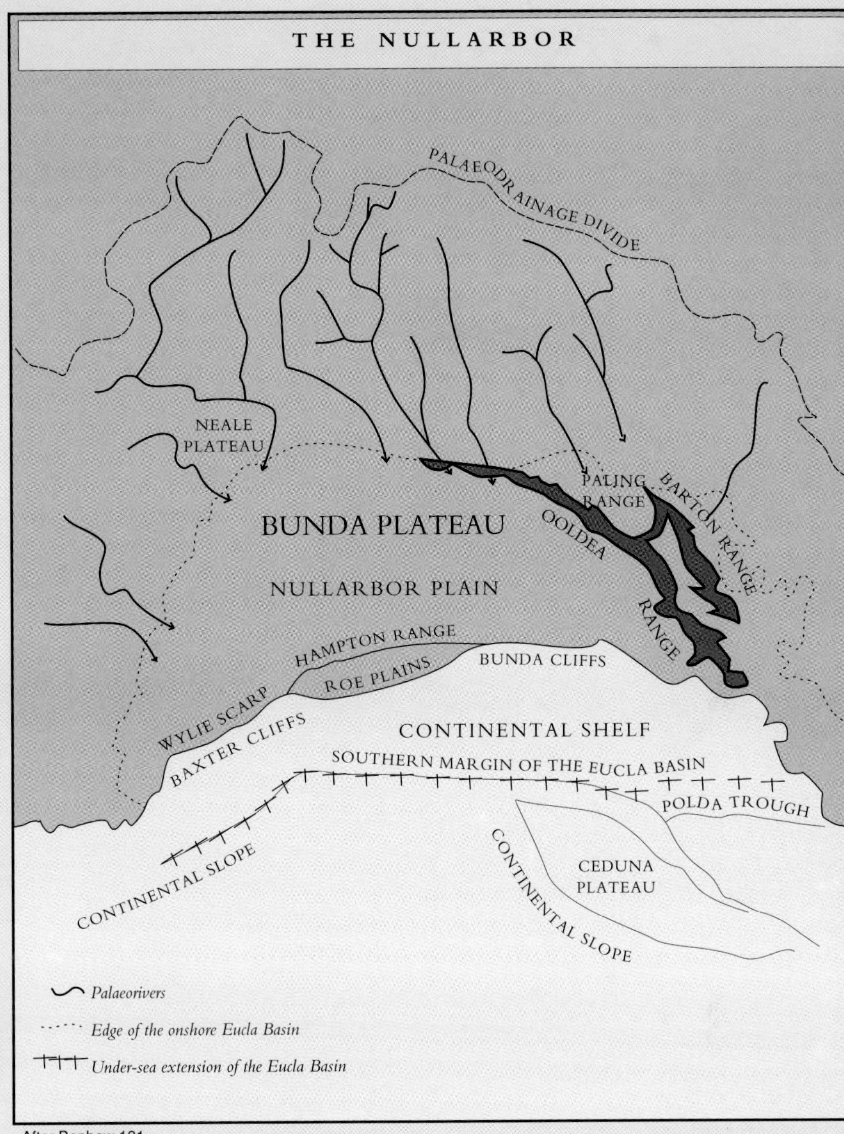

After Benbow 181

rearrangement of plate activities, setting up new tensions. The easternmost province of Gondwana (the New Zealand Subcontinent) started to rift and the Great Divide rose. The southern continental margin might have been expected to rise in this rifting phase but there is no geological evidence that this was the case in the Nullarbor section. Drainage of the continent was strongly north to south as though the land was tilted, and the Nullarbor remained a flat land traversed by rivers, its rocks eroding and the sediment deposited off the land in the rift valley system. The extensive and large river systems that flowed from the Musgrave Block and the Stuart and Gawler Ranges cut deep drainage channels and these palaeochannels can be seen on satellite images of the Nullarbor today in spite of the deposition of limestones in the Tertiary which today form the almost-flat Bunda Plateau.

About 50 million years ago there was another change in plate tectonics. India collided with Asia, the Tasman Sea widening ended and sea-floor spreading between Australia and Antarctica speeded up. The Eucla Basin (Tertiary successor to the Bight Basin in the Nullarbor region) underwent another phase of downwarping and was again invaded by the sea. Marine sandstone and marl were deposited forming the lower part of the Wilsons Bluff Limestone, the upper part being deposited late in the Eocene under quiet water conditions as the sea shallowed. The sea penetrated far inland up the large river valleys towards the Musgrave, Stuart and Gawler Ranges. Pidinga Formation Eocene sediments, of alternating marine and

Abrupt cliffs edge the flat Nullarbor plateau where it meets the ocean in the Great Australian Bight. The Nullarbor used to be the bottom of a sea during the Eocene and again in Miocene times. J.S. BEARD

brackish to freshwater types, in these palaeochannels testify to the fluctuating sealevels of the times.

It has recently been recognised that the Ooldea and Barton 'Ranges'[181] on the northeastern margin of the Nullarbor are ancient coastal dune ridges[182] which mark the extent of the Eocene sea. The Ooldea Range is a sand-range which stands out clearly from the flat Nullarbor Plain and runs for 650 kilometres. It lies from 25 to 300 kilometres from the present coast. The Barton Range occurs within the dunefield of the Great Victoria Desert and was only recognised as a separate feature in aerial photographs. To the southeast of the Ooldea Range further beach ridges — the Paling Range — have been recognised from LANDSAT imagery, and possible beach ridges have also been detected near Lake Maurice. The Eocene sea was far more extensive than had previously been thought.

The ancient coastal dunes provide evidence about the climate of the time of their formation, and since. Westerly winds, which must have been of some intensity, resulted in their formation in the easterly quarter of the Eocene seashore. During the Late Eocene when they were forming a temperate climate is implied, which agrees with other estimates of climate in that interval. The fact that sand dunes have survived for 34 to 37 million years is surprising and suggests a degree of aridity for much of the rest of the Cainozoic. This evidence of aridity in the Nullarbor region is earlier than any in other parts of southern Australia. Plant cover on the dunes and the permeability of their quartz sands have been factors contributing to their survival. The mineral-sands potential of these ancient dunes is considerable and no doubt in time they will be mined.

By the end of the Eocene (36 million years ago) the sea had again retreated from the continental part of the Basin and was not to re-enter until the Early Miocene — about 20 million years ago. Then the Abrakurrie Limestone was deposited in the centre of the Basin and as the area invaded by the sea widened the Nullarbor Limestone was laid down over the whole area, grading into calcareous sandstone on its northern margin, where river-borne sediment from the land mixed with the marine material.

The Eucla Basin was uplifted to become land by the end of the Miocene, its limestone floor forming the top of a plateau. This Bunda Plateau attains an elevation of 240 metres today, sloping gently southwards and eastwards, with high cliffs along its southern edge.

During the Pleistocene, the position of sealevel in relation to the cliffs of the Bunda Plateau appears to have been higher by a maximum of 36 metres, partly representing tectonic change locally and partly the fluctuations in volume of ice caps and hence in the height of global sealevel. Coastal lowland plains were carved into the Tertiary limestone cliffs by marine erosion. The Israelite Plain, Roe Plain, and Yalata Swamp result from this raised sealevel activity. The thin layer of Roe Calcarenite was deposited on the eroded ledges during high sea stands.

The Nullarbor is one of the largest arid to semi-arid karst regions in the world — more than 250,000 square kilometres. It has acted as a barrier for plant migration between southern Western Australia and the eastern States, increasing the endemism there.

An aerial view of the northeastern margin of the Nullarbor shows dunes of the Great Victoria Desert overriding the featureless limestone plateau. MARK BENBOW

The empty Nullarbor landscape. The saltbush vegetation has been seriously degraded by grazing of stock and rabbits. MARK BENBOW

CHAPTER EIGHT

A Cooler and Drier Australia in the Oligocene

(36.6 to 23.7 million years ago)

By the start of the Oligocene, about 10 million years after Australia had become an island continent, a broad band of oceanic crust formed the sea-floor between Australia and Antarctica within the original rift. Spreading ridges to the west and east of the rift zone between the two lands had developed so that the processes of sea-floor spreading now extended into the Indian Ocean and round Tasmania to cross the Tasman Sea below the Campbell Plateau (initiating the Antarctic-Pacific Ridge). All of the Australian continent was emergent and continental shelves were exposed because sealevel was lowered due to the glaciation of Antarctica that had marked the transition from the Eocene to the Oligocene.

The start of Antarctic refrigeration in the Terminal Eocene Event[141] was the beginning of the making of modern oceans. Filling of ocean basins with cold water as Antarctica cooled would convert them from a greenhouse state with salinity-dominated deep-ocean circulation to the icehouse state where circulation was dominated by temperature. Today icy cold Antarctic water travels as an abyssal current northwards, taking 400 years to travel to Greenland via the Atlantic and 1500 years to the North Polar regions via the Pacific Ocean. Then, slightly warmed by its journey, it returns as a middle layer, and when it reaches the descending colder water round Antarctica it wells up and promotes snowfalls which keep the South Pole frozen. A cold-driven engine in perpetual motion.

In the oceans around Australia altering marine faunas show the changing ocean climate over several million years around an extinction peak between 37 and 36.5 million years ago. Antarctica still supported forests during the Oligocene[17] where they co-existed with the early stages of glaciation. A Deep Sea Drilling Project core in the Ross Sea has a pollen record for the Late Oligocene which shows dominant *Nothofagus*, mainly of *fusca* pollen type, with podocarp conifers, MYRTACEAE and PROTEACEAE. Whether the vegetation type was tall forest, or whether it was more like the stunted, open, deciduous forests that grow in Patagonia today is unknown. The probability has to be considered that remnant forests continued to exist in refuge areas in Antarctica until about 3 million years ago, as *Nothofagus* fossils (petrified twigs and leaves) are known from the Sirius Formation of about that age just prior to the formation of the northern Polar ice cap.

In the Middle Oligocene, about 30 million years ago, the Pacific and Australian Plates started to collide in the New Guinea region. From then onwards mountain building in New Guinea accelerated. Late in the Oligocene, about 25 million years ago[165], the Sepik Province of New Guinea became attached. This province is a terrane — a fragment of continental crust that moved from elsewhere and eventually collided with, and became part of, another landmass. Volcanic islands appeared in the Tasman Sea from this time as well, perhaps making island chain migration routes to

New Caledonia. Except for the last of these islands — Lord Howe, which is only 6 million years old — all the islands are now submerged and form seamounts in the Tasman Sea. Movement on the Alpine Fault in New Zealand also began at this time, starting the uprising of the land.

Perhaps related to the colliding of plates in the New Guinea region[113], and resulting from the stresses that developed within the continent as a result, changes occurred to drainage patterns in Central Australia in the Lake Eyre region. South of the present-day Lake an upwarp formed the Stuart Range and the Willouran, Peake and Denison Ranges, interrupting southward drainage. Further warping and faulting resulted in the creation of the Lake Eyre inward-draining Sub-basin, and started its sinking, so that now the centre of the basin is many metres below sealevel. The original depocentre was north of its situation today, and Lake Eyre has migrated southwards through time with the development of the present north-south continental gradient. All sedimentation in the Lake Eyre Sub-basin is post-Oligocene, dating the tectonism that cut off flow to the sea from what had become an inward-draining region. South of the Stuart Range divide another basin was segregated and became the Billa Kalina Basin with its southern margin determined by the rising ground of the Gawler Ranges.

During the latest Oligocene and throughout most of the Miocene, a large lake filled the Billa Kalina Basin. The lake centre was in the vicinity of today's Lake Younghusband. Ancient shorelines of this lake can be seen in the modern desert landscape in the form of arcuate ridges that lie on the eastern side of the area once occupied by the lake. Their presence only on the eastern side indicates that strong westerly winds[187] were a feature of the Miocene climate as it became drier and the lake contracted.

The sudden decrease in rainfall and introduction of climates with seasonal aridity in the Oligocene which resulted from ice build-up in Antarctica caused sluggish conditions in watertables in flat areas and palaeodrainages in many parts of the continent. This led to the formation of ferricrete and silcrete duricrusts in suitable areas (the Cordillo Surface). This process was probably repeated each time there was a major cooling event with its attendant aridity after an interval with better rainfall. Such fluctuations characterise the lead up to the ice age through the Miocene and Pliocene and are much exaggerated in the last 2.4 million years of geological time when both Poles were glaciated and the 'Pleistocene Ice Age' had arrived.

The Oligocene climate overall was considerably cooler and drier than the Eocene had been. Such vegetation as can be reconstructed from the few fossil sites and pollen records of Oligocene age reflects the climatic changes.

Pollen floras that span the Eocene to Oligocene transition[132] occur in the St Vincent Basin of South Australia[133] and in the Murray Basin[47]. *Nothofagus* beech and CASUARINACEAE are increasingly dominant. The *Nothofagus* is predominantly of *brassii* pollen type, with a little *fusca* and *menziesii*. Flowering plant diversity is decreased compared to Eocene floras and the MYRTACEAE, CUNONIACEAE and PROTEACEAE are less abundant. Ferns and mosses are rare, indicating a drier forest type; sedges, burr rushes (SPARGANIACEAE) and RESTIONACEAE are common, indicating open marsh and clay-pan areas.

In Western Australia, in the North Stirlings district, lake and swamp deposits of probable Early Oligocene age contain leaves of *Banksia* and of sassafras (*Cinnamomum*, LAURACEAE) as well as petrified wood of *Banksia* and of *Barringtonia*, a freshwater mangrove now confined to tropical Australia.

By the Early Oligocene[49] a change in the composition of *Nothofagus* pollen, which some palynologists designate the '*Nothofagus* shift', sees the *brassii* group diminished and the *fusca* and *menziesii* groups more prominent. This reflects the climate change. The *brassii* species require non-seasonal, perhumid conditions, the others prefer cooler climates and can tolerate some seasonal dryness. The diversity of flowering plants is generally reduced in the *Nothofagus*-dominated forests and sclerophyllous taxa become more prominent.

Interpreting a general picture for Australia from the limited and patchy evidence available suggests that increase in the areal cover of *Nothofagus* forests meant reduction in areas of more complex rainforest. This latter, original Gondwanan, mixed forest probably maintained its character and species diversity in suitable areas which, even then, were virtually refuges.

There was also undoubtedly variation in

Dodonea viscosa, the hop bush, is widely distributed in Australia in arid regions. The same species occurs in New Zealand today. The family SAPINDACEAE includes Atalaya, the whitewood, another resident of arid places.

JEANETTE CLOSS

the nature of the *Nothofagus* forests, according to the rainfall and soil conditions in different places. For instance, the Oligocene forest at Pioneer in Tasmania was much more complex than today's cool temperate beech forests in Tasmania, and was also much more complex than the drier type forests of the Oligocene mainland. A reconstruction of the Pioneer forest suggests that conifers were emergent from the *Nothofagus* canopy; abundant angiospermous trees, including MYRTACEAE, CUNONIACEAE, SAPINDACEAE, *Quintinia,* and *Eucryphia* (leatherwood) and *Ceratopetalum* (New South Wales Christmas bush) formed a substratum or grew as tall shrubs; abundant tree-ferns, ferns and mosses formed lower strata. This Pioneer forest is similar to modern cool-temperate rainforests in the mountains of the Barrington Tops and Dorrigo Plateau in New South Wales.

Eucryphia lucida from the Frankland River, Tasmania.

JEANETTE CLOSS

In the Latrobe Valley in Victoria the fossil floras are microphyllous (characterised by small leaf size, a response to cooler climatic conditions). MYRTACEAE is abundant and sclerophyll genera like *Leptospermum* are evident. EPACRIDACEAE appears for the first time in a mainland assemblage. (It was found in the Australia-Antarctica rift in the Late Cretaceous sediments. The epacrids are southern heaths, closely related to ERICACEAE, the family of heathers of the Northern Hemisphere.) *Banksia* becomes a prominent component of the flora; CASUARINACEAE is represented by *Casuarina* and *Allocasuarina* rather than *Gymnostoma*; and *Callitris*, the native cypress pine, is present with podocarps (*Dacrydium*, *Dacrycarpus*) and with *Agathis*, the kauri pine (ARAUCARIACEAE). Some broadleaf angiosperm taxa are still present, including abundant LAURACEAE at some localities, and several reed-like monocotyledons are present. This vegetation type has been likened to stunted 'heath forest' of Borneo, but might just as well represent a mosaic of vegetation units in which heathland coexists with scrub and mixed woodland in the catchment area. Extensive swamps in the Murray Basin appear to have supported low-diversity forests lacking many of the taxa which now characterise relict rainforests.

At Glencoe, near Capella in central Queensland[149], a fossil flora dated by basalt as about 30 million years old contains fruits of *Sarcopetalum* (MENISPERMACEAE). *Sarcopetalum* is a genus with only one extant species, which has a range from southern New Guinea, down the Australian east coast from Queensland to Victoria. The Glencoe flora also contains *Athertonia* (PROTEACEAE) which has one extant species in the wet tropical rainforests of Queensland; *Cissus* and *Ampelocissus* (VITACEAE, the grape family), genera with a high concentration of species in northern rainforests; and *Pleigynium* (ANACARDIACEAE), a genus now confined to New Guinea.

Fossil wood at Glencoe shows evidence of damage by termites[148] and termite frass is preserved. Similarity of the frass to that produced by *Neotermes*, the extant wood-eating Isopteran, suggests that such termites had a Gondwanan ancestry.

By late in the Oligocene, small-leaved *Nothofagus* species had developed in Tasmania and sub-alpine vegetation had appeared at high elevations. At Monpeelyata the winter-deciduous Tasmanian endemic *Nothofagus*

Mt Warning seen from a lookout in the Border Ranges with the dissected basalt plateau in the foreground.

M.E.W.

Left: Richia scoparia, *a showy member of the Epacridaceae in Tasmania.*

Jeanette Closs

Right: Nothofagus gunnii *a deciduous beech which is a glorious sight in its autumn colour. It grows in Tasmania's mountains where winters are very cold.*

Dick Burns

OLIGOCENE, MIOCENE AND PLIOCENE VEGETATION OF NEW ZEALAND [174, 175, 80]

Oligo-Miocene floras of New Zealand are similar to the Latrobe Valley coal floras of Victoria, and in composition are close to the extant lowland forest of North Island (which exists today as only a few islands following Maori burning in pre-European settlement days and subsequent clearing for farming).

The Gore Lignites in South Island, of latest Oligocene to Early Miocene age, are the product of swamps in a forest in which podocarp and araucarian conifers are co-dominant with *Nothofagus* beech. The forest contained five taxa of PROTEACEAE; WINTERACEAE (bush peppers); *Ceratopetalum* (New South Wales Christmas bush, CUNONIACEAE); *Elaeodendron* (CELASTRACEAE); *Gymnostoma* (CASUARINACEAE); LAURACEAE; *Apocynophyllum* (APOCYNACEAE); *Dracophyllum* (a genus with many New Zealand species and including the 'pandani' of Tasmania, EPACRIDACEAE). Marsh and swamp plants of RESTIONACEAE, JUNCACEAE and TYPHACEAE (bulrushes) were abundant. An extinct species of *Phormium* (New Zealand flax, AGAVACEAE) also grew in these swamps.

The Oligocene in New Zealand was a time when erosion had so lowered the remnants of the once-large Subcontinent that marine transgression further reduced the land area. All that remained was an archipelago of small low, swampy islands. It is somewhat surprising that so much of the Gondwanan *Paleoflora Mixta* had managed to survive the steady degrading of the Subcontinent. But the remaining islands were refugia, much like the Gondwanan rainforest 'archipelago of refugia' — Dr Len Webb's description of relict rainforests of northeastern Australia — is today. The diverse flora that survived there had all the elements which would be able to expand into newly emerging environments as climates changed and modern New Zealand was built up by activity on the Alpine Fault.

The Early and Middle Miocene flora had the same rich diversity and mixture of subtropical and temperate elements. Many of the taxa have become extinct in New Zealand and are now restricted to Australia, New Caledonia and New Guinea. These include *brassii* type *Nothofagus*; CASUARINACEAE; *Acacia*; *Melaleuca* and *Acmena* (MYRTACEAE); and *Eucalyptus*. Coconut palm fossils (*Cocos*) are recorded from near Auckland indicating that warm conditions prevailed.

The Late Miocene vegetation reflects wet conditions in the north (*brassii* beech forest), cooler wet westerly conditions in the south (*Podocarpus* and *Dacrydium* forests), and drier conditions in the southeast in the lee of mountains that arose during the Miocene (*fusca* beech forests). Both *Eucalyptus* leaves and *Eucalyptus* pollen are recorded from Otago, and in central Otago near Gimmerburn chenopod saltbush, restios, and sedges are locally abundant and daisies are a significant component.

The Pliocene vegetation differed from that of the Miocene with the 'tropical' taxa disappearing and the selection of plants adapted to montane and cold temperate environments over much of the country, particularly South Island, by the end of the epoch.

The north of North Island remained warmer and more favourable. There *Eucalyptus* pollen in some Early Pliocene assemblages reached 50 per cent of the pollen count in samples[80]. The genus survived in New Zealand until the early Pleistocene, being present among a coastal MYRTACEAE-dominated flora in northwestern North Island. Westerly winds and coastal dunes are believed to have created special conditions for a flora which contained *Eucalyptus*, *Metrosideros* (rata), *Leptospermum* (MYRTACEAE); *Acacia*; PROTEACEAE, CHLORANTHACEAE, RUBIACEAE; *Dodonaea viscosa* (hop bush, SAPINDACEAE); and *menziesii* beech and podocarps.

gunnii, which now characterises Tasmanian sub-alpine regions, had arrived.

At Bacchus Marsh in Victoria, in beds of Oligocene-Miocene transition age, fruits of *Nothofagus* and ELAEOCARPACEAE are found in a low-diversity flora which also includes a small-leaved MYRTACEAE genus, *Araucaria*, and a few small-leaved LAURACEAE. (Small leaves are a characteristic of cool temperate floras.)

Late in the Oligocene, temperatures rose slightly and sealevel[83] started to rise with ice-melt in Antarctica[146], heading towards the Early Miocene peak when it would be warmer than at any time since the Eocene. The Oligocene-Miocene Geera Clay in the Murray Basin was laid down when the Basin was invaded by the sea, and in the oldest part a pollen flora shows *Araucaria* prominent and *Nothofagus* with *brassii* type pollen reduced in importance, reflecting a drier type of rainforest growing under a mildly seasonal rainfall regime. *Acacia* pollen is present in the Late Oligocene part of the Geera Clay.

Thus we can see that in response to the cooler and drier Oligocene climate, vegetation had changed considerably from the 'golden age' Eocene forest that had characterised Australia at the time of its separation from Antarctica.

The Late Oligocene and Early Miocene was a time of active volcanism[176] in southeast Queensland and northeastern New South Wales. The Glasshouse Mountains north of Brisbane and the basalts to their north in the Maleny region (dated at 24.8 million years) are eroded centres of eruption of Late Oligocene volcanoes. The Focal Peak and Mt Warning volcanic complexes (23.6-22.7 and 23.1-22.4 million years respectively) which today are important features in the Border Ranges-Tweed districts further south, show the southward migration of volcanic activity at the rate of between 60 and 70 millimetres per year. This migration represents the rate of movement of the Australian continent across a 'hot spot' on its journey northwards.

Volcanism along the Great Divide had commenced about 70 million years ago and moved southward as the continent travelled northwards. In Australia where soils are mainly old, leached and nutrient-deficient, the production of basalt and its weathering to make rich soils in many places is of great significance. The areas where basalt had erupted became islands of fertility. The ability of basalt soils to hold water was an extra bonus. In the warm Atherton Tablelands in northeastern Queensland today red basalt soils support more mesic vegetation than adjacent soils on non-volcanic rocks. The chain of basalt outcrops in eastern Australia could have acted as a pathway for the spread of moisture-loving species as drier conditions started to affect the continent.

In the Late Oligocene, vertebrate fossil deposits containing mammals at last start to appear on the continent to tell us about faunal change. (The *Steropodon* jaw of 110 million years ago and the Murgon fossils with the mammal tooth of 55 million years are tantalising and fascinating glimpses in an 80 million year blank interval.)

Late Oligocene sites at Geilston Bay in Tasmania have a marsupial fauna of phalanger-like possums, early diprotodontids and a pygmy possum. Near Wynyard in northern Tasmania at Fossil Bluff on Bass Strait the skull of *Wynyardia bassiana* was found in Oligocene-Early Miocene marine sediments. The dead wynyardiid had been washed out to sea in river sediments.

In Late Oligocene-Early Miocene sedimentary rock in Victoria the skull of a small baleen whale is the first evidence of whales in Australia. This whale was about 3 metres in length. At Lake Palankarinna in the Lake Eyre region in South Australia there is a sequence of fossils from Late Oligocene through Miocene. In the oldest section, in the Ditjimanka local fauna, bones of an insectivorous bat have been found.

Riversleigh, in the Mt Isa district of western Queensland[177] is proving to be one of the world's greatest fossil sites. Its faunas, preserved in limestones, span a time interval from latest Oligocene through Miocene (with little Late Miocene) to Pliocene and Pleistocene-Recent. In its Oligocene-Miocene sites an amazingly diverse marsupial fauna and other animals are being revealed. The gallery forests of the times were obviously luxuriant and it is regrettable that so far no pollen flora has been found in the limestone to give a factual picture of the composition of the rainforest. The limestones in which the animal fossils occur are outcropping and weathered, and it is in the nature of limestone to be porous and subject to penetration by groundwater, which results in unsatisfactory conditions for preservation of pollen. A few fossil leaves of broadleaf dicotyledon type have been found but they give no idea of the forest. Therefore it has to be assumed that the vegetation would have been similar to the gallery forests of the Lake Eyre Sub-basin at this time.

Thousands of fish and frog fossils; turtles, ordinary and horned; a variety of crocodiles; many different lizards; pythons, poisonous snakes and blind burrowing snakes; birds, including abundant passerines (perching birds), dromornithids (large running birds) and emus; enormous numbers of bat bones (but no evidence of fruit-bats); a complete skull of *Obdurodon*, an early platypus; and a host of marsupials make up this fauna. The high diversity of species implies rainforest conditions with high rainfall.

GONDWANA DISPERSION AND CLIMATIC ZONES AT 30 MILLION YEARS

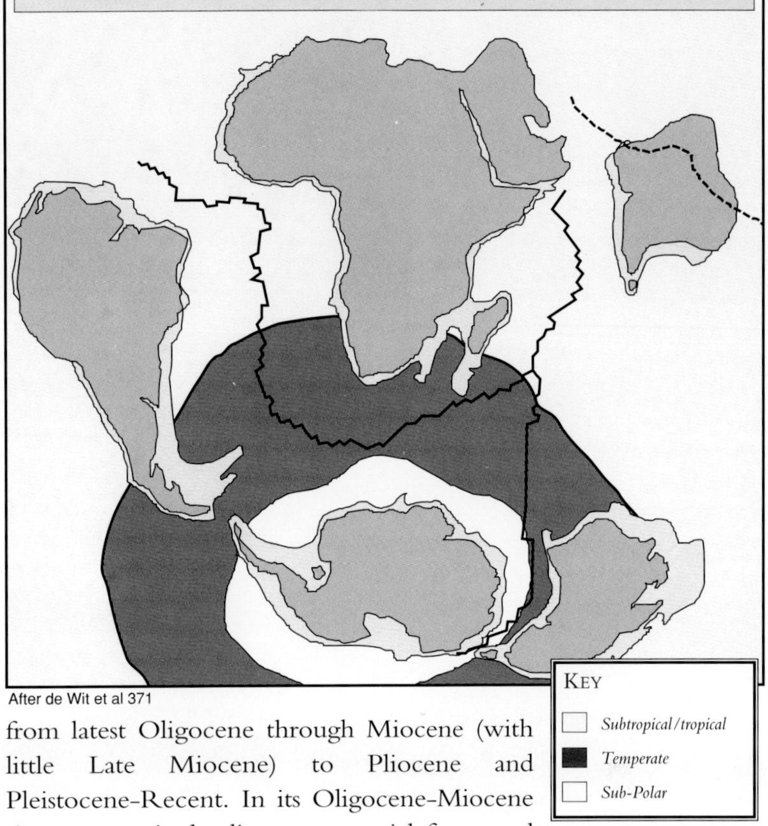

After de Wit et al 371

With the development of a circum-Polar current Antactica's refrigeration began. A Sub-Polar to Polar climatic zone developed progressively, moving the temperate zone, which had surrounded Antarctica previously, further north.

CHAPTER NINE

THE SWINGING MIOCENE

FROM WARM AND WET TO COLD, DRY AND FIERY

(23.7 to 5.3 million years ago)

The Miocene Epoch was a time of great significance in the evolution of modern vegetation and fauna in Australia. After Early Miocene warming, when temperatures were higher than at any time since the Eocene, cooling recommenced and the East Antarctica ice cap formed. A rapid decline in global temperature followed, culminating in the Terminal Miocene Event between 6.5 and 5 million years ago. The temperature decline was accompanied by increasing aridity as more of the Earth's water budget was tied up in ice. Antarctic ice was more voluminous six million years ago than it is now, resulting in a sealevel drop of 100 metres.

PALAEOGEOGRAPHY OF THE LATE TERTIARY (20 MILLION YEARS AGO)

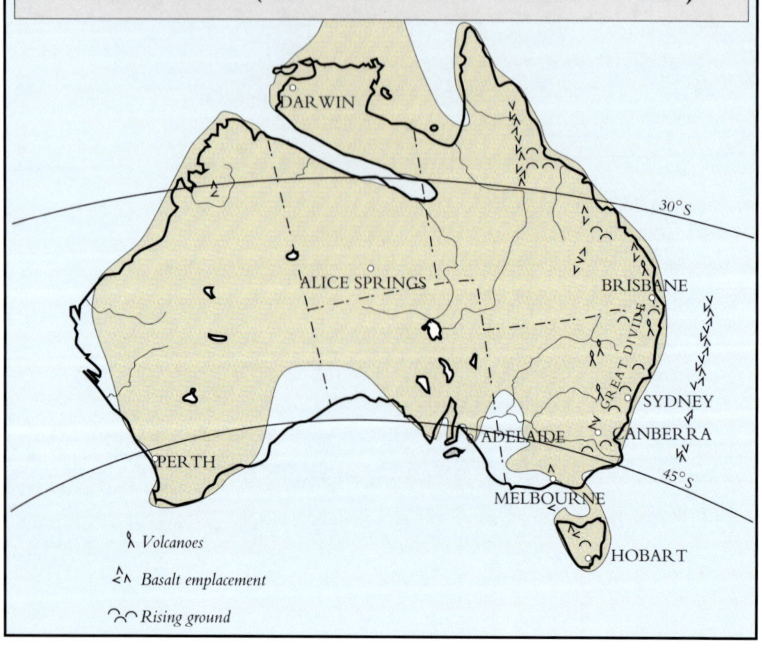

A synthesis of information on the changing landscapes, environment, climates and flora and fauna through the 18.5 million years of Miocene time in Australia follows. The Australian story has to be seen within a wider context. What was happening in Antarctica at this time is vital to an understanding of the changes that occurred to set Australia on a path towards becoming the driest and most fiery continent. Also, the consequences of the environmental changes of the Late Miocene and the Terminal Miocene Event in Africa have the most profound implications because the evolution of the Human species can be tied to these critical climatic changes.

By the Early Miocene, New Guinea, on the leading edge[165] of the Australian continental crust, had reached 20° South. New Guinea's northern province at this time comprised only the Sepik terrane, which had docked with it in the Late Oligocene at about 25 million years ago. The southern margin of Australia lay between 53° and 55° South, with the

Tasmanian southern margin about 58° South.

The temperature of surface waters in the Tasman Sea showed a 3°C rise over Oligocene temperatures. Sealevel was high and along Australia's southern margin the Eucla Basin, St Vincent and western Murray Basin and the Gambier Basin (the Otway Basin of the Mesozoic) were inundated. Small intramontane basins in the southern Mt Lofty ranges in South Australia were also flooded. (The Mt Lofty Ranges and those near Leigh Creek were higher in the Early Miocene than they are today.) At the same time an arm of the sea entered the Ord River region of northwestern[144] Australia and flowed across the east Kimberley district of Western Australia, into the Northern Territory and across the Barkly Tableland to near Roxborough Downs in western Queensland. Evidence for this marine incursion is found in the part-marine limestones of the Mt White Formation in the East Kimberleys, the Brunette Limestone on the Barkly Tableland and the Austral Downs Limestone in western Queensland. One of the foraminiferids in all these limestones is *Ammonia beccarii*, an open-ocean, near-shore species which can also live in brackish estuaries.

There was widespread carbonate deposition in the epicontinental seas and around the coastline during warm Early Miocene times, and some reef development took place in northern waters.

During the Early Miocene when sealevel was raised, rainfall, at least in the southern half of the continent, was high. There is widespread evidence of deep weathering and also of extensive swamps in which lignites (brown coal) accumulated. Further north the increased prevalence of dolomites indicates evaporative conditions. There were many shallow lakes and swamps in basins and palaeodrainage valleys. Some were alkaline, as evidenced by the argillaceous and calcareous mudstones that were formed in them. In Central Australia the Lake Eyre Sub-basin was sinking and contained large freshwater lakes. There were lakes in the Callabonna Sub-basin, and the lines of present-day salt lakes that curve round Lake Eyre were sites for Early Miocene freshwater lakes. The Etadunna Formation and Namba Formation were deposited in these lakes.

The arrangement of salt lakes and ancient dune ridges[113] round Lake Eyre has been interpreted by some people as evidence of a Pleistocene megalake — 'Lake Dieri' — an enormous lake that was supposed to have occupied an area ten times the size of present-day Lake Eyre. The small salt lakes were interpreted as old strandline features. There is no geological evidence that a Pleistocene 'Lake Dieri' ever existed.[97] Extensive geological mapping by the South Australian Geological Survey has failed to find any trace of the widely dispersed lake-bed sediments that would

SEA TEMPERATURE IN THE HIGH LATITUDE SOUTHERN OCEAN DURING THE CAINOZOIC

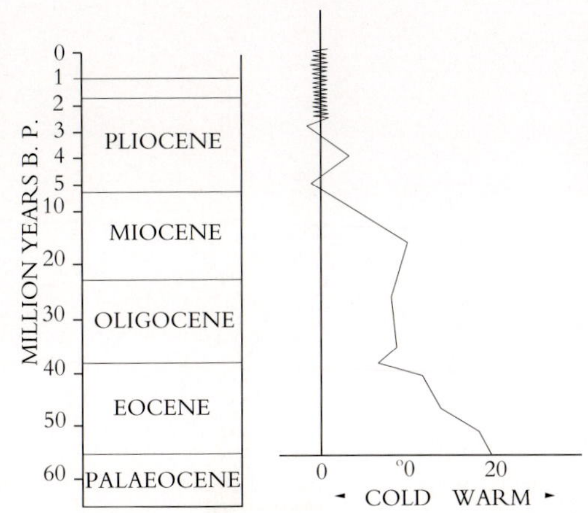

After Brain 166

Nothofagus gunnii a deciduous beech which is a glorious sight in its autumn colour. It grows in Tasmania's mountains where winters are very cold. The dead tree in the picture was a pencil pine, killed in a fire long ago and never replaced. Fire in these mountain areas is totally destructive.

DICK BURNS

THE MURRAY BASIN IN THE EARLY MIOCENE (24 TO 16 MILLION YEARS)

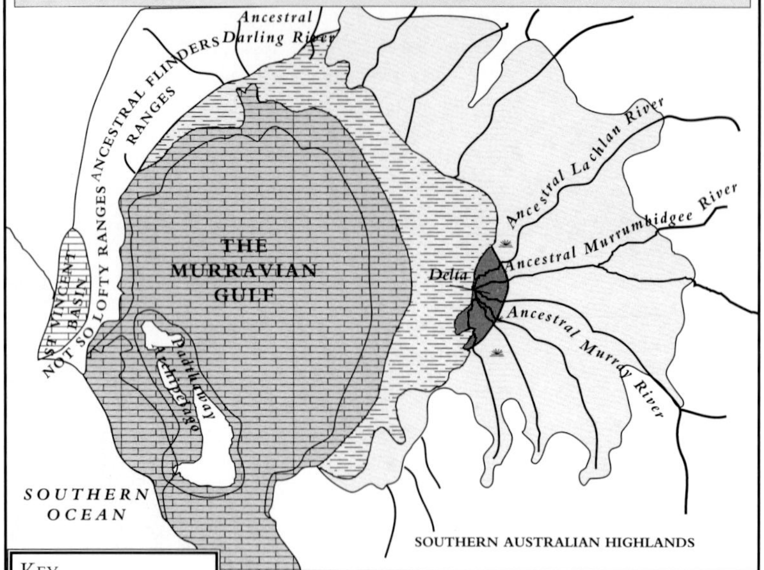

After Stephenson and Brown 147

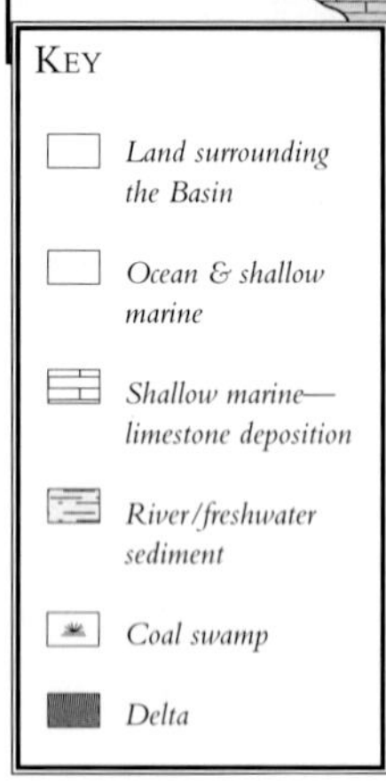

The Murray Basin was inundated during the high sealevel phase of the Early Miocene. An ancestral Riverine Plain in its eastern sector was traversed by rivers, including the ancestral Lachlan, Murrumbidgee and Murray, which formed a large delta at the edge of the Murravian Gulf.

indicate the previous existence of such a major body of water. However, on the basis of analysis of the source of the dunes, it has been concluded that Miocene Lake Eyre was considerably larger than its present size,[211] though not more than two or three times its present area. Its extension was into the Strzelecki Creek region when it was at maximum size.

(Had a lake the size of the mythical 'Lake Dieri' existed in the Pleistocene it would have had to be 150 to 175 metres deep in the middle, it would have overflowed through the region between Lake Eyre and Lake Torrens, and through the low area occupied by Cobham Lake and Salt Lake 80 kilometres south of Tibooburra in northern New South Wales. This can be deduced from contours in the landscape, and the scenario is patently incorrect. There is no evidence of its overflow into these regions. Many of the dunes that created the illusion of ancient shorelines have been found to be fluvial — made of river sands, not lake dunes.)

Lake Billa Kalina occupied the Billa Kalina Basin during the lake-full stage of the Miocene, its sediments forming the Watchie Sandstone along the shoreline and the dolomites of the Mirikata Formation deeper into the Basin. It was about 100 kilometres wide at its peak and its retreating shorelines as it dried up left arcuate ridges to its northeast in the desert.

In the Middle Miocene, as the seas retreated from the continent, the southern continental edge started to rise. During the Late Miocene the Eucla Basin platform was progressively uplifted and its sea-bottom limestone was exposed to form the flat Nullarbor Plain. Mild tectonic movements resulted in gentle warping and small-scale faulting in the St Vincent and Murray Basins, while the Lake Eyre and Torrens Sub-basins continued to sink. The onshore Gambier Basin underwent uplift from Late Miocene to Quaternary times.

Moderate uplift along the Great Divide in eastern Australia kept pace with erosion, and the relief since Oligocene times has probably remained about the same. The last stage of uplift of Victoria's southeastern highlands was post-Oligocene and less than 300 metres, and relief, drainage systems and position of divides were similar to today.[24] Basalt eruption occurred down the eastern side of the continent and into Tasmania throughout the Miocene.

During the last 20 million years Western Australia's continental shelf has been sinking rapidly and mild folding in the Carnarvon Basin has led to some land rising by 350 metres in the last 15 million years. In the western Kimberleys a spate of localised volcanism, with over a hundred separate eruption points, occurred during Miocene times. The diamond-bearing pipes that have brought wealth to Western Australia are part of this volcanic episode.

To Australia's north, and still part of the mainland,[65] New Guinea was evolving during the Miocene. Complex events in the collision zone with the Pacific Plate caused rapid uprising of mountains, and a second major terrane, the East Papua Terrane, docked about 15 million years ago. A further two terranes — the Irian Jaya Terrane of Western New Guinea, and the central Island Arc Terrane — became part of the landmass about 10 million years ago. (Emplacement of granite contributed to the elevation of the New Guinea mountains. Some of the granites are dated at only 3 million years, indicating the recent nature of some of the uplift.)

By 10 million years ago the leading edge of the Australian Plate had collided with the Sunda Arcs and the edge of the continent had moved northwards across the line where formerly subduction had occurred along the southern margin of Sundaland. As Australia continued to move northwards, while Sundaland moved westwards, the Sunda Arcs were bent. Today the leading edge of Australia's Plate is 800 kilometres beyond the inferred former position of the Sunda Arcs. The Banda Arcs represent a dragfold in the zone of interaction between the westward-moving Pacific Plate and the northward-moving continental edge of Australia. Ceram and Sulawesi have rotated 90° since the Miocene.

THE REFRIGERATION OF ANTARCTICA AND ITS EFFECT ON GLOBAL AND AUSTRALIAN CLIMATES

First glaciers had appeared on West Antarctica about 38 million years ago, during the Terminal Eocene Event. The temperatures in Antarctica during Oligocene times fluctuated but maintained overall cooler global climates than had characterised the Eocene. In the Early Miocene a warmer spell resulted in some ice melt on Antarctica, and caused higher sealevels. Then circum-Polar circulation became established and cooling of Antarctica began in earnest.

By the Middle Miocene there were substantial ice sheets on East Antarctica. An ice cap was established between 16.5 and 13 million years ago, with the greatest build-up of ice between 14.8 and 14 million years ago. By the Middle to Late Miocene the ice sheets in East Antarctica were discharging ice through gaps in the southern Transantarctic Mountains into an ice-free Ross Sea. Coastal uplands of Marie Byrd Land supported ice fields (for instance in the Jones Mountains)[84] but a full-bodied West Antarctica ice sheet was yet to be formed. With so much ice on Antarctica, atmospheric circulation intensified and the southern half of Australia would have come under the influence of westerly winds. Westerlies would have been felt further north than at present but parts of the centre, north and northwest probably experienced increasingly dry anticyclonic circulation.

Arid phases in the Lake Frome area are documented by the presence of dolomitic and other sedimentary rock types which formed when alkaline lakes dried up. However, in the far north there is no evidence of acute aridity. The mid-latitude arid belt of today's climatic zonation did not exist during the Miocene. In Western Australia the rivers that had run on the well-watered and forested Yilgarn Block and adjacent areas during the warm Early Miocene started to dry up in the Middle Miocene when

RAINFALL REGIMES IN ARID AND SEMI-ARID REGIONS: MIOCENE TO RECENT

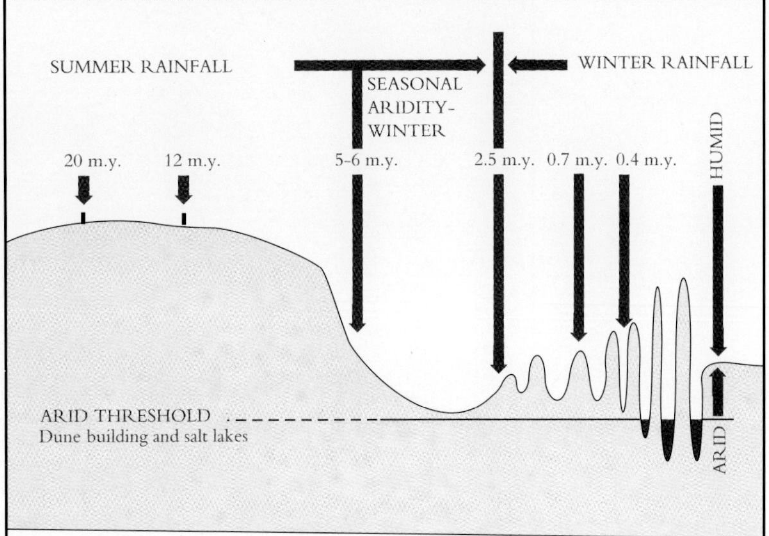

After Bowler 172

Global climate changed after the southern Ice Cap formed 15 million years ago and aridity became an increasing problem for Australia. In the southern half of the continent in regions which were becoming increasingly arid, summer rainfall patterns were replaced by winter rainfall regimes with pronounced dry seasons. From the time when the Northern Ice Cap formed (2.4 m.y. ago) widespread desert conditions (blown sand dunes and salt lakes) occurred in the semi-arid winter rainfall sector.

the East Antarctic ice cap expanded and caused global aridity and a drop in sealevel. Today, chains of salt lakes mark the course of these ancient drainages. The palaeodrainage patterns have been preserved because the region has been geologically stable and the change from a humid to an arid climate regime during the Late Miocene slowed down erosion and sedimentation.

The formation of the West Antarctica ice sheet post-dates the East by as much as 8 or 9 million years (which has significance in terms of the availability of refuges for plants). The Terminal Miocene Event of about 6 million years ago saw the full development of the West Antarctica ice cap, and the formation of the frozen Antarctic continent of today by ice-linking of east and west. This final establishment of an even larger volume of Antarctic ice than exists today had a profound effect on sealevel globally. The Event was short-lived, and a period of Antarctic ice regression followed.

Falling sealevel in the Terminal Miocene Event stranded the Mediterranean Sea and during several fluctuations within the Event it dried up and refilled, and vast amounts of salt were precipitated during the drying episodes. These today form deep deposits, which are estimated to contain a million cubic kilometres of halite and gypsum, below the Mediterranean. Removal of so much salt resulted in a decline by 6 per cent in the salt concentrations of global ocean waters. Wide continental shelves were exposed around all the

PALAEODRAINAGE OF THE CANNING BASIN SEEN BY REMOTE SENSING

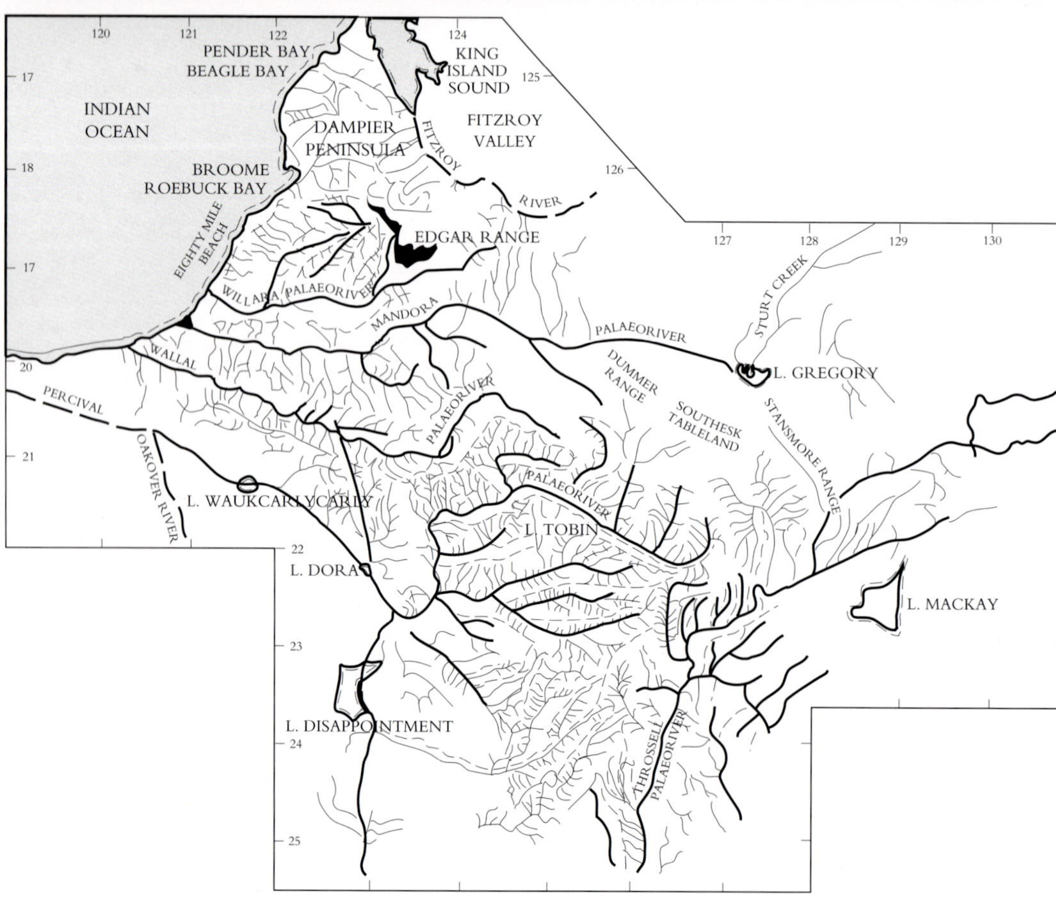

After Tapley 186

continents, including Australia. The final refilling of the Mediterranean occurred during the latest Miocene to Early Pliocene warmer period when ice melted and sealevel rose. (The Murray Basin was again invaded by the sea in this interval.)

The Mediterranean salinity crisis of the Terminal Miocene Event[167] actually contributed to deterioration of world climates. As a consequence of the removal of 6 per cent of the oceans' salt, seas became more alkaline and deep waters were under-saturated in calcium carbonate. Freshening of surface waters enhanced the formation of high latitude sea ice, because more ice was able to form at a higher temperature. More ice increased the albedo (reflection of heat) over Antarctica, promoting increase in growth of the ice cap. During the ice build-up towards the Terminal Miocene Event the Ross Ice Shelf extended 300-400 kilometres beyond its present limits, and so much sea ice must have had a dramatic effect on Tasmania and southern Australia.

After the Terminal Miocene Event,[169] a rise in temperature in the Early Pliocene saw sealevels rise.[169a] Surprisingly, in Antarctica[170] sufficient ice retreated for forests of *Nothofagus* to grow in the Transantarctic Mountains during this warmer interval. Abundant leaves and petrified twigs of a species of *Nothofagus* beech, closely related to a South American species, indicate that conditions must have been quite mild, with temperatures 15-20° warmer than today, in this 86°S latitude. The pollen record preserved in the *Nothofagus* leaf beds indicates that the forest contained other flowering plants, podocarps and mosses.

It has been suggested, in order to explain the local mildness when so much of Antarctica was still under an ice cap, that with decrease in ice and establishment of isostatic equilibrium a deep fjord probably ran from the Ross embayment deep into the Transantarctic Mountains, ameliorating climate. There in the Dominion Range the terminus of the ancestral Beardmore Glacier was flanked by glacial deposits on which the forest grew at an estimated elevation of about 600 metres. The glacier calved into the head of the Beardmore fjord. Very rapid uplift by a minimum of 1300 metres has occurred in the area during the last 3 million years, after the forest had gone.

After the Early Pliocene warming, the ice cap on Antarctica again expanded, extinguishing the continent's last traces of higher plant vegetation. (Discoveries of algae and microscopic lichens[159] that grow within the surface crust of rocks in Antarctica are fascinating reminders that Nature has a life form for every imaginable habitat. Mats of algae have long been known to grow in ice-free waters of Dry Valleys.)

Button daisies are among the showy ephemerals that grow rapidly in arid regions after rain. They flower and seed profusely and then die, leaving their seeds to wait for the next rain to carry on their presence in the region.

JIM FRAZIER

In the Late Pliocene, about 2.4 million years ago, the Northern Polar ice cap formed, and thereafter the ice age with its 17 fluctuations from glacial stages with ice build-up to interglacials which were warmer was in full swing.

The effect of Antarctica's refrigeration on global climate has been, and is, profound. Antarctica has temperatures always 10-12°C colder than the North Pole, a consequence of its isolation from other lands, and the temperature gradient that results affects global circulation patterns of oceans and the atmosphere. At times of ice build-up in the lead-up to the ice age, and during glacial stages of the last 2.4 million years when the ice age has been in progress, aridity was a feature of climates. In the Southern Hemisphere, strong winds characterised the arid times as well, increasing the effects of less rain and cooler temperatures.

In Australia, the drying of the continent accelerated during the Miocene but the final aridification would only take place during the ice age glacial stages.

In Western Australia, where many rivers on the ancient Yilgarn Plateau[186] and elsewhere have not flowed since Late Miocene times, night-thermal imagery (remote sensing from a satellite) has revealed palaeodrainages in the part

HOMINID EVOLUTION AND MIOCENE CLIMATIC CHANGE[166,167,168,191,192]

Southern Africa has a history of gradual drying with contraction of forest and establishment of sclerophyll vegetation from the time of the first glaciation of Antarctica in the Terminal Eocene Event. The changes were similar to the sequence of changes that occurred in Australia. Some areas of the continent, particularly on the western side, were already arid when the Terminal Miocene Event of six million years ago resulted in the strengthening of the cold Benguela Current along the West African coast. (The Current had developed first in the Middle Miocene.) The stronger Current led to increased aridity, and during subsequent cold fluctuations it penetrated further up the west coast than it does today, causing aridity to western Zaire and adjacent regions. With the continent already under arid regimes because of the great build-up of ice, the extra drying resulting from the cold current had an exaggerated effect on the tropical forest cover, reducing it to a dendritic mosaic greatly reduced in area and with a greatly increased forest fringe. In this interface between forest and savannah the humanoid apes started their evolution towards *Homo*.

Using new techniques with DNA analysis, scientists have shown that humans, chimpanzees and gorillas share a substantial common ancestry. (Recently a 13 million year old jaw has been found in Namibia and described as being a new candidate for 'common ancestor'. Previously East Africa was the only region where ancestral Miocene humanoid ape bones had been found, so the geographical range of these creatures was wide.) The point of divergence between them occurred between four and six million years ago. This suggests that the Terminal Miocene Event which changed the environment could well have been the force behind the evolutionary step towards becoming bipedal *Australopithecus afarensis* and *africanus*.

The next sharp climatic fluctuation with formation of the North Polar ice cap between 2.6 and 2.4 million years ago saw the forking of the ancestral line and separation of the *Homo* line from the robust apemen — *Australopithecus boisei* and *robustus*. These two forks represent different adaptations to the more open environments and the rapid swings in vegetation type that occurred in response to the glacial-interglacial fluctuations of the ice age. The robust apeman strategy was to increase the grinding area of cheek-teeth and power of jaws to cope with fibrous food, while the large body size was an advantage in changeable climate. There was no increase in brain size in proportion, and though robust *Australopithecus* survived for two million years, all were extinct by about a million years ago.

The *Homo* line did not change much in teeth and jaws or, originally, in size. Instead with increased development of manual dexterity came the evolution of a bigger brain. Development of technology, starting with stones and involving ever more use of the hands, promoted the enlargement of the brain and development of the capacity to reason and to solve problems. And with the coming of Man, the first species with these abilities, the troubles of the Earth were just beginning.

For the health of the living planet it would have been better if the ice age had not brought environmental change which forced humanoid apes out of the trees and demanded that they compete on the savannah!

This skull of an apeman was originally named Plesianthropus *when it was found at Sterkfontein in South Africa. It was affectionately known as 'Mrs Ples' being assumed to be female because the individual whose head it was was so small and gracile. It is now classified as a male skull of* Australopithecus.

JIM FRAZIER

HOMINID EVOLUTION

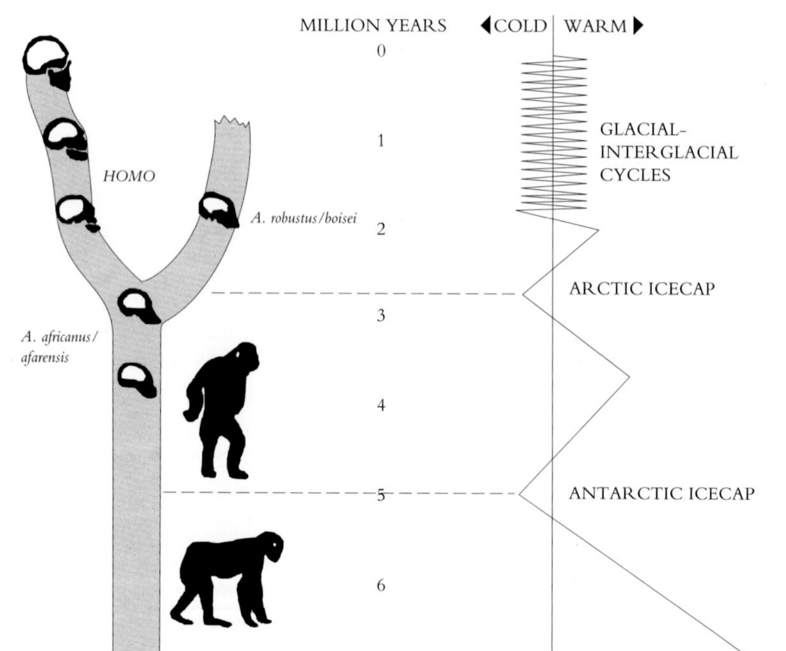

After Brain 168

A suggested correlation between global low-temperature events and critical points in hominid evolution is illustrated in this diagram.

of the Great Sandy Desert that occupies the onshore Canning Basin. At night the satellite picks up and photographs the thermal differences in the different rock and sediment types that form the region. The palaeodrainage channels with their groundwater-holding infill show up in thermal imagery because their dampness at depth renders them cooler than the surrounding terrain.

Three major palaeoriver systems are revealed in the Canning Basin, running westward towards the Indian Ocean — the Mandora to the north, the Wallal in the middle and the South Percival the most southerly. In addition it can be seen that the Fitzroy River once flowed across the base of the Dampier Peninsula on its southern edge. It then flowed on beyond Roebuck Bay to a shoreline 160 kilometres offshore from the present-day coastline at a time when the continental shelf was exposed during a low sea stand. (The Fitzroy River today runs into King Sound on the north side of the Dampier Peninsula.) Because the Fitzroy River runs through terrain upstream where diamond pipes occur in volcanic rocks (in the Lennard Shelf area and the Fitzroy Graben) the palaeoriver channel may prove to be a structure worth prospecting for alluvial diamond deposits.

Also revealed by the night thermal imagery are four linear channels radiating from the Edgar Ranges. They are either drainage from a domed structure or can be interpreted as major tributaries draining into a large east-west palaeoriver — the Willara.

The ancient river channels revealed in the Canning Basin are straight, indicating free-flowing rivers cutting down with considerable gradient to the sea. A much wetter climate than that of the present is inferred as catchments are about 1000 kilometres inland. It is thought that the lowered sealevel that exposed the continental shelf and resulted in down-cutting of the rivers and in the mouth of the palaeo-Fitzroy River being 160 kilometres beyond the modern coastline probably coincided with the Antarctic ice build-up around the Middle Miocene when the rivers were still running. By the Late Miocene when sealevels fell towards the Terminal Miocene Event the Fitzroy would have had diminished flow. The re-routing of the western end of the river was caused by local warping and the formation of a new drainage divide.

Waterholes in the Canning Basin palaeochannels were used by early explorers. The Canning Stockroute from Wiluna to Halls Creek depended on their water and on wells that tap the aquifers in the calcrete infill in the palaeodrainages. Saline lakes in the palaeochannels can also contain ephemeral freshwater supplies after rains.

The Terminal Miocene Event was short-lived and at the very end of the epoch sealevel rose with ice melt. The western Murray Basin was again invaded by the sea. The deep river valleys that had been excavated while sealevel was low were infilled with coarse sands and conglomerates of the Calivil Formation, forming deep leads. The Murray and Darling Rivers emptied into the Murravian Gulf (the sea in the Murray Basin) and the delta of the Late Miocene to Early Pliocene Murray River at the time of maximum transgression of the sea was located at the western margin of the modern Riverine Plain.

When the sea retreated in the Early Pliocene it left northwesterly trending beach ridges and inter-ridge deposits across a sandplain — the Parilla Sands. The estuary of the River migrated westwards and the Pliocene Murray River flowed on much the same course as it does today. It turned towards the sea when it reached the Morgan Fault where the high ground of the Mt Lofty Range stopped its westward flow then as now.

CHAPTER TEN

ADAPT OR MIGRATE

THE CHOICE FOR MIOCENE PLANTS AND ANIMALS

(23.7 to 5.3 million years ago)

Modernisation of the Australian vegetation proceeded during the Miocene. However, the fossil record is meagre and for a land as large as Australia it is difficult to obtain a good overall picture as a result. The climatic fluctuations within the epoch obviously led to sorting of elements of the vegetation and progression towards the communities that characterise some parts of the continent today. Some landforms that are now so characteristic of modern Australia — like the blown-sand deserts and dunefields and the gibber deserts with their highly specialised vegetation — had not yet appeared. Therefore much change had still to occur before modern equivalents were reached.

The vegetation formations that had characterised the Late Oligocene[173] persisted during the warm, wet Early Miocene. Temperate rainforests were still widespread over much of southeastern Australia, extending from Tasmania into Victoria, westward to coastal South Australia, and northward through western New South Wales and the Southern Tablelands, and into coastal Queensland. These forests often had a high proportion of *Nothofagus* of the *brassii* type and abundant gymnosperms and broadleaf LAURACEAE and MYRTACEAE.

There was variation within the forests related to latitude and local conditions. A vegetation gradient from the coast to the inland was from *Nothofagus* dominance to forests much richer in MYRTACEAE and *Araucaria* (hoop and bunya pines). A general south to north gradient resulted in more of the taxa that characterise modern warm, wet rainforest occurring in northern vegetation, and fewer in southern sites. The presence of *Nothofagus* with *brassii*-type pollen implies high rainfall of between 1500 and 1800 millimetres per year if the requirements of the Miocene species were similar to those of extant *brassii* species.

Some of the major elements of Australian open plains, *Acacia*, *Casuarina*, grasses, chenopods and ASTERACEAE (daisies) — were established in communities by the Early Miocene. As more forested land was replaced by open, drier vegetation types in areas that experienced seasonal aridity they became increasingly evident. *Eucalyptus* was present in drier forest and woodland but not a dominant component.[84]

Brown coal deposition continued in the Gippsland Basin throughout the Miocene. The Yallourn pollen floras show that gymnosperms were abundant — kauri, hoop pine, *Dacrydium*, *Phyllocladus* (celery-top pine), and *Podocarpus* (plum pine). *Nothofagus*, mostly of *brassii* pollen type, was very abundant in the older part of the sequence and the abundance decreased towards the Late Miocene,

Family LENTIBULARIACEAE, *these bladderworts growing in pure white sand at the site for the proposed space base on Cape York are water-margin plants. The probability of their pollen being preserved in lake sediments because of their choice of habitat is large even though they form only a small proportion of the local vegetation.*

JIM FRAZIER

indicating increasing dryness. Also present were *Casuarina*, Ericales, LAURACEAE, MYRTACEAE, OLACACEAE (wild olives), PROTEACEAE (including *Beauprea* and *Banksia*), restios, WINTERACEAE and SAPINDACEAE (*Cupaniae*). The closest living analogue to this assemblage is found in New Guinea forests that grow in the highlands at between 1200 and 2100 metres, where the rainfall is 1500 millimetres a year and the mean annual temperature is between 13° and 18°C. (It is tempting to equate the pollen of Ericales (a section of the the heath family ERICACEAE) with *Rhododendron* which today has one species in Australia and several in New Guinea, and to visualise these lovely plants in the ancient forests.)

In the Murray Basin the Miocene part of the Geera Clay[83] (laid down during marine inundations)[146] has a pollen record which shows the nature of the vegetation round the Murravian Gulf (the name given to the shallow sea that occupied much of the Basin until the Middle Miocene). The pollen record also has dinoflagellates which are useful in close-dating the sequence. *Nothofagus* is present throughout. MYRTACEAE is abundant — the closed forest genera *Syzygium* and *Acmena* (lillypillies) and *Tristania* (brush box, water gum) dominate but *Eucalyptus* is present. *Podocarpus*, *Dacrycarpus*, *Phyllocladus*, *Microcachrys* and *Dacrydium* are common, and

The hoop pine is a large and elegant conifer of the ARAUCARIACEAE *Today it is an inhabitant of drier types of rainforest in northeastern Australia. The genus* Araucaria *to which it belongs was widespread and abundant in the Early Tertiary.*

Araucaria reaches high frequency in some samples. CASUARINACEAE are always present and sedges and grasses may reach 10 per cent. The CYPERACEAE (sedges) show high diversity. *Acacia* pollen is present and also GYROSTEMONACEAE, and the first record of *Utricularia* (water plants — bladderworts, of

'Brassii' Nothofagus *disappeared in the Middle Miocene when the southern ice cap formed and a marked dry season became a feature of the climate. The progressive replacement of rainforest by sclerophyll together with the climatic changes led to greatly increased fire frequency. During fluctuations of climate when wetter conditions returned briefly,* Nothofagus *(* non-brassii*) returned, and fire frequency and abundance of Myrtaceae decreased sharply.*

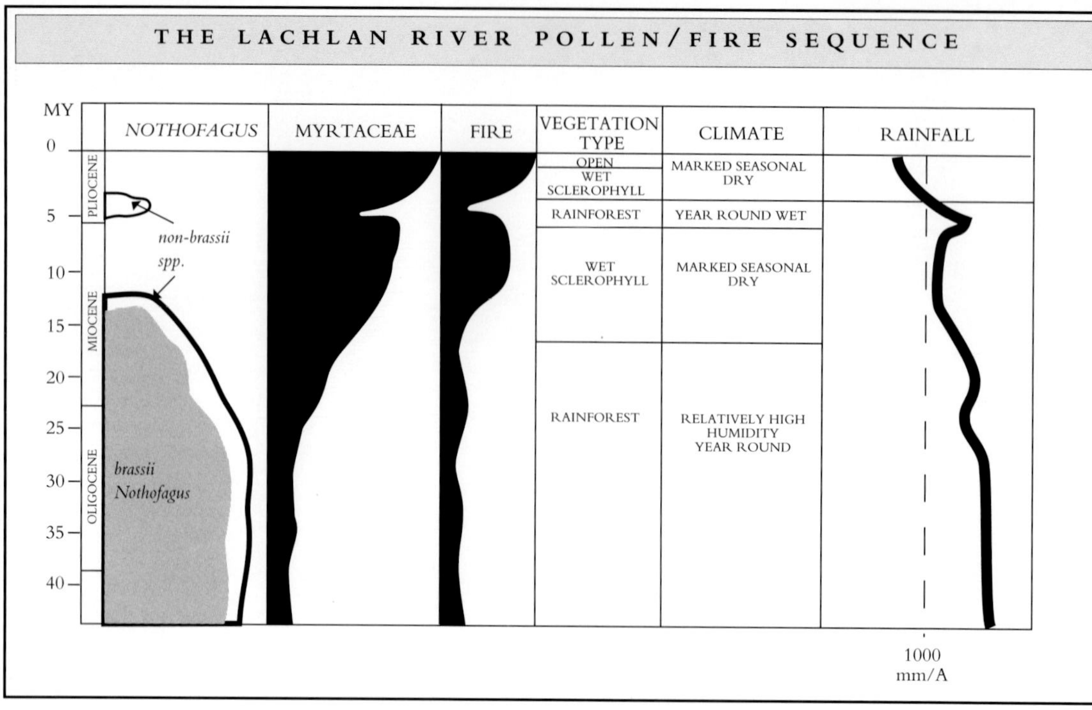

After Martin 145

the family LENTIBULARIACEAE) and *Gardenia* (RUBIACEAE) are from this Murray Basin sequence. Tree-ferns, *Ilex* (holly family), *Quintinia*, MALVACEAE (*Hibiscus* family), and CUNONIACEAE (*Weinmannia*) also occur.

The vegetation type was evergreen rainforest with a high proportion of MYRTACEAE and containing *Nothofagus*. Such forest requires year-round precipitation. Tidal flats and foreshore swamps supported the sedges, grasses and burr-rushes as well as chenopods (saltbush).

At Kiandra in New South Wales an Early Miocene flora[17] dated at between 18 and 22 million years contains abundant pollen[164] of *Nothofagus* of *brassii* type, *Podocarpus*, *Microcachrys*, *Dacrydium* and *Araucaria*. Leaves of LAURACEAE and MYRTACEAE are abundant and epiphyllous fungi and mosses growing on them indicate that locally conditions were very humid and warm.

Fossils from Chalk Mountain near Bugaldie in New South Wales,[171] in diatomite dated by basalts of the Warrumbungle Volcano as between 17 and 14 million years (Middle Miocene), contain leaves and gumnuts of *Eucalyptus*, and flowers of *Ceratopetalum* (New South Wales Christmas bush, CUNONIACEAE). Pollen analysis of the diatomite shows that the local forest was rich in gymnosperms and MYRTACEAE (including *Tristania*, *Angophora*, *Baeckia*, *Backhousia* and others). The presence of pollen of chenopods, AMARANTHACEAE, mistletoes, epacrids, restios, GOODENIACEAE, RUTACEAE (the *Boronia* family) and grasses shows that a large herbaceous element was present, implying open bush as well as forest.

The diatomite at Chalk Mountain resulted from the damming of streams in the highlands by basalt and the concentration of freshwater diatoms (single-celled algae) in the small lakes that resulted. Because the lakes were on and surrounded by new basalt and there was just about no erosion, the water remained clean and the pure white diatomite, which is composed of the siliceous shells of the organisms, was uncontaminated.

Towards the centre of the continent there was at least[173] localised development of rainforest close to Alice Springs, as evidenced by mid Miocene lignites several metres thick at Ti Tree; and near Lake Eyre and Lake Frome as indicated by the carbonaceous clays in the Etadunna and Namba Formations. The pollen spectra from these inland areas differ from those of the broad coastal belt in that they contain a high proportion of grass pollen in the *brassii* beech-podocarp assemblage. This suggests grasslands in the interfluvial areas and gallery forest along permanent watercourses. Such a situation suggests a seasonal climate, and the implication of times of aridity is reinforced by the geology of the Namba Formation. In this Formation dolomitic and lime-rich beds lie above the pollen horizon, indicating protracted arid phases superimposed on the warm-temperate climate. This situation is what would have been expected with the westerly winds affecting more of the continent in Miocene times than they do now.

In the Robe River district of Western Australia[139], below the pisolitic ferricrete which is commercially mined, rainforest pollen assemblages containing podocarps and some *Nothofagus* indicate that during the Miocene such forests were still present in the northern part of the State.

In mainland southeastern Australia a number of taxa which are today confined to lower latitude rainforests disappear in the Late Miocene. These include *Ilex*, *brassii*-type *Nothofagus*, *Austrobuxus* and *Dissiliaria* (EUPHORBIACEAE), *Cupaniae* (SAPINDACEAE) and others. A similar situation is found in central Queensland where the same forms are last seen in the Middle Miocene.

The decline of *brassii*-type beech, which is an indicator of non-seasonal rainforest, began in central areas of the continent and spread along the present rainfall gradient, eventually leaving New Guinea as the only part of the landmass in which it is now living. *Nothofagus* pollen appears for the first time in New Guinea during the Miocene. Although many rainforest taxa suffered reduction in their range, or regional extinction, there is no evidence that any taxon present in the Miocene became totally extinct. The processes of change were sifting and sorting according to the climatic preferences of components of the vegetation.

From the Middle Miocene onwards, when the continent started to dry out and vegetation types changed in many areas to drier sclerophyll, open woodland and grassland, fire became a more frequent factor in the environment. In the more fire-prone communities selection for fire tolerance must have started to modify the flora.

In the Late Miocene in the Murray Basin rainforest[145] was no longer widespread. It was replaced by wet sclerophyll forest.[194] The presence of large amounts of charcoal associated with the pollen in the pollen record indicates that these forests burned on a regular basis, presumably following more arid seasons.

Limited evidence from the Miocene to Pliocene transition[189] indicates dramatic change in the vegetation in southeastern Australia. This would be expected, bearing in mind the severity of the Terminal Miocene Event. In the Latrobe Valley there was complete replacement of rainforest by open woody and herbaceous vegetation. Because of the dryness of this phase there is very little pollen information and sequences in boreholes, which provide the detailed information on changing vegetation through time, exhibit a barren section just where one would most like to know what was happening. The Terminal Miocene Event was a foretaste of aridity and though a more benign interval was to follow much of the sorting and sifting and alteration towards ice age vegetation patterns had its acceleration then. The 'wide brown land' was established briefly and gave warning of what conditions would be like in the slide towards the ice age.

While the pollen record and a few occurrences of plant macrofossils give us some idea of the vegetation during the Miocene, there is a fossil record of animals that provides information on the creatures that lived in Australia during this time of change. Abundant vertebrate fossils provide faunal evidence which enables us to populate the gallery forests and rivers and lakes of Early to Middle Miocene times with a host of creatures.

The Pinpa fauna from the Namba Formation southeast of Lake Frome[178] contains fossils of the crocodiles and platypus, turtles, lungfish and ray-finned fish which swam in the waters of the lakes and rivers. Water birds were abundant, including flamingoes, pelicans, ducks, geese, rails and grebes. The presence of flamingoes lends strength to the argument that some lakes were already saline, the nitre- and salt-rich waters of lakes in rift valleys and arid lands being the preferred niche of extant flamingoes. Marsupials included early koalas and wombats; possums, including gliders and ringtails; and carnivorous types related to thylacines (Tasmanian tigers). A high percentage of the animals were arboreal, inhabiting the gallery forests of the times.

The Ericmas and Tarkarooloo faunas from the same region[177] contain an early platypus — *Obdurodon*, dolphins[178] and four different cuscus/possum relatives. As well they contain primitive diprotodontids, palorchestids (*Ngapakaldia*) and macropodids (kangaroo family). These browsers and grazers probably inhabited the forest fringe and grasslands.

The finding of bones of dolphins in dry salt pans in the dead heart of the continent makes one realise the amount of change that

A carpet of everlastings form ephemeral ground cover among jam Acacia *in the Coalseam Reserve, Irwin River district, Western Australia. The jam* Acacia *is so named because its leaves smell like raspberry jam.*
M.E.W.

has occurred through time. The dolphins must have travelled up rivers from the sea and entered the inland lakes while the Lake Frome Sub-basin still drained to the sea. Contemporaneous (Middle Miocene) faunas in the Lake Eyre Sub-basin (the Ngapakaldi fauna from the Etadunna Formation) have no dolphins as that basin was inward-draining with no connection to the ocean.

A younger fauna which also occurs within the Lake Eyre Sub-basin, the Kutjamarpu fauna from the Wipajiri Formation, has a wider range of terrestrial taxa, including bandicoots, marsupial mice, and native cats (dasyurids) as well as a koala and several different possums, a diprotodontid and a primitive kangaroo. Fish, turtles and water birds are abundant, and an emu and a large running bird — a dromornithid — are also present. Increase in the proportion of animals that live on the ground to those that live in the canopy indicates the drying out of the area and the substitution of grassland and open scrub for forest.

Further north, at Alcoota and Bullock Creek in the Northern Territory, eight and ten million years old respectively, Late Miocene sites have yielded a fauna different from others in the Centre because of the preponderance of large terrestrial birds and mammals. Bones of one crocodile, ducks and flamingoes are the water component, and diprotodontids, a palorchestid, primitive kangaroos, dromornithids and an emu form the large animal component. *Neohelos*, a diprotodontid the size of a large pig, was very abundant at Bullock Creek.

Apart from the water birds at Middle Miocene sites, eagles, cranes, shore birds, gulls, curlews, nightjars and frogmouths show the similarity of Miocene and modern bird fauna. In addition some bird fossils have been found in the Lake Frome area which do not belong to any modern family and may represent remnants of an earlier bird fauna. The large dromornithid flightless birds, some larger than ostriches, characterised the Miocene. Seven species of four genera are known and their abundance may be the result of the appearance of grasslands and open country which offered them opportunity in new niches, just as it suited the emu and led to the Pliocene radiation of kangaroos with the long hopping stride that is an adaptation to open plains. The large running birds may have been the first large grazers of open grassland.

CHAPTER ELEVEN

COUNTDOWN TO AN ICE AGE

THE EARLY TO MIDDLE PLIOCENE

(5.3 to 2.4 million years ago)

During the Miocene to Pliocene transition, after the Terminal Miocene Event (which is not well named as it occurred nearly a million years before the end of the epoch) rapid Antarctic ice melt occurred with global warming. This interval saw the final refilling of the Mediterranean Sea.[169] It also saw the return of *Nothofagus* forest to Antarctica as evidenced by the leaf and wood fossils found there in the Sirius Formation of approximately three million years ago.

The Antarctic continental record for the Pliocene[191] is hard to discover because of the present size of the ice cap. There may have been several deglaciation events when conditions were warmer than they are today and high sea stands (up to 60 metres higher than today) resulted. Two such high sealevel episodes, the 4.5 million year peak of the Miocene to Pliocene transition phase and another less well dated at round about three million years ago, indicate that fluctuations from cold to warmer (and from arid to wetter for Australia) were becoming more rapid and more serious. The presence of warm-water diatoms in strata that supply evidence of ice retreat on Antarctica, as well as the *Nothofagus* leaves and wood and pollen of other taxa, confirm the deglaciations and the considerable temperature rises involved.

The warm, wet Early Pliocene was an interruption in the general cooling and drying trend that had begun in the Middle Miocene. At the warmest time between 4.5 and 4 million years ago the West Antarctic ice sheet may have disappeared and the Pensacola and Wilkes Basins became open sea. The significant cooling that recommenced during the Middle Pliocene at 3.4 to 3.2 million years ago was mainly abyssal water cooling and its effects were mainly felt in the Southern Hemisphere.

An Early Pliocene global tectonic event which affected circulation patterns and contributed to establishment of ice age regimes was the elevation of the Panama Isthmus between North and South America about 5 million years ago. This interrupted the circulation of warm equatorial waters.

Leaves of Antarctic Beech, *Nothofagus*, from the Sirius Formation in the Transantarctic Mountains. Forests grew in parts of Antactica during a warm interval in the Pliocene about 3 million years ago when global temperatures were a few degrees higher than they are today. Conditions in east Antarctica then were probably like those in southern Patagonia now.

PHOTO: BOB HILL.

(At much the same time, sea-floor spreading in the Gulf of Aden between Arabia and Somalia started, after about 25 million years of continental extension by crustal thinning, and the continental margins are now about 90 kilometres apart. This rifting, however, has had little effect on ocean circulation except fairly locally.)

Australia's northern margin was colliding with the Timor region during the Pliocene so that at times of lowered sealevel and exposure of continental shelves migration between Southeast Asia and Australia was easy.[177] Rodents made their way into the continent about 5 million years ago (or possibly considerably earlier) along this route. The collision of the recessed northwestern part of the Australian continental margin at about 3.5 million years ago involved the thrusting of a five kilometre thick set of thrust slices from the basement of the Banda Arc over Timor in the dramatic buckling-up of the region that resulted. There has been 4000 metres of uplift in New Guinea since the Middle Miocene and 6000 metres in Timor as a result of interaction between the edge of the Australian Plate and the Southeast Asia region.

Sea-floor spreading started in the Woodlark Basin near the eastern tip of New Guinea about 3.5 million years ago. This most recent rifting represents the latest of the rifting events which had started in the northwest of Australia with the departure of Argoland more than 132 million years ago. Rifts had spread progressively round Australia delineating its margins, ending with this continuing episode in the northeast.

As the sea retreated from the Murravian Gulf a strandplain of mixed river and marine sediments was left behind.

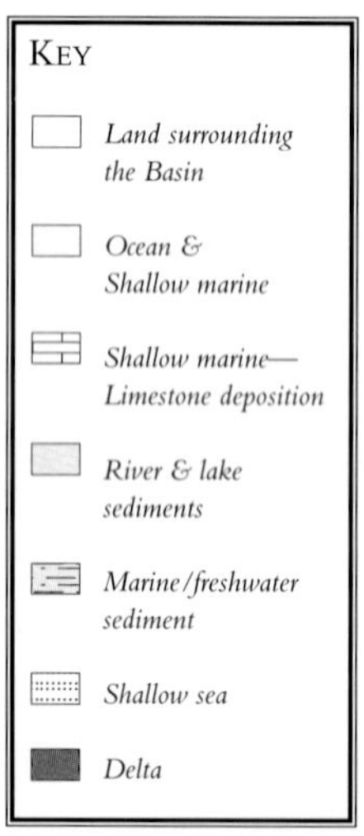

THE MURRAY BASIN IN THE PLIOCENE

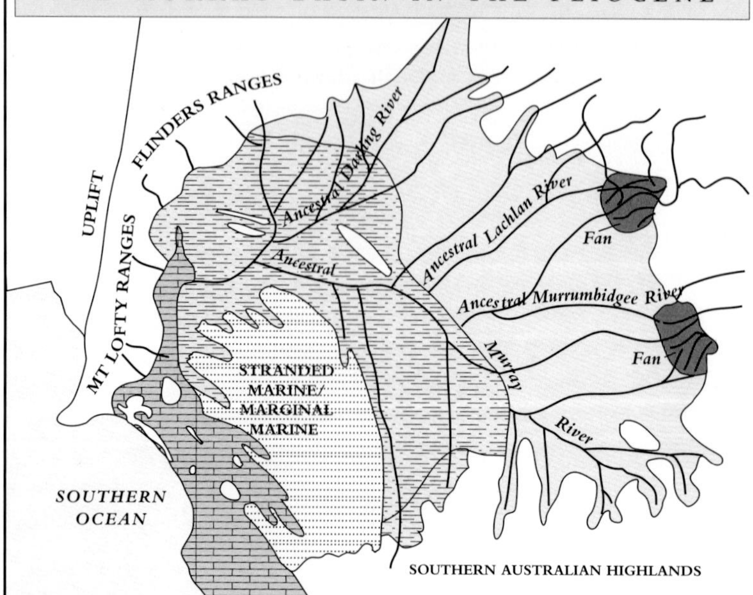

After Stephenson & Brown 147

Deep weathering took place in the Early Pliocene during the wet and warm interval and episodes of ferricrete and silcrete formation resulted when aridity followed later in the epoch. The extensive Dundas Surface was formed in southeastern Australia[141] and there were some smaller areas of duricrusting in the Eromanga Basin. In the Gulf of Carpentaria there was extensive ferruginisation (enrichment of sediments with iron) in the Karumba Basin, equating to the Dundas Surface. The iron-enriched and silica-cemented Karoonda Surface which extends from Lake Frome through the western Murray Basin is a further indicator of deep weathering which occurred under high rainfall regimes.

With the high global sealevels at the start of the Pliocene[172] the western Murray Basin again became an inland sea and rainforest with *Nothofagus* (but not of *brassii* type)[174] returned briefly to the Basin. This '*Nothofagus* phase' forms a well-defined stratigraphic horizon, marking the warmer, wetter phase of the Early Pliocene, and can be picked up in pollen spectra from the major river systems on the western slopes of the Eastern Highlands.[194] The high proportion of chenopods (saltbush, bluebush) and daisies that is found in the western Murray Basin reflects the abundance of these plants in the salt marsh and sand habitats which surrounded the retreating inland sea. (The CHENOPODIACEAE which today characterise the acutely arid and saline desert and semi-desert areas of Australia are a worldwide family which inhabits seashore, salt marsh and sand areas that are not necessarily in arid zones. The adaptation of the family to such conditions made them the obvious candidates to colonise the vast arid and saline areas that appeared intermittently during arid glacial times and permanently when true aridification established Australia's dead heart in Pleistocene times.)

Early Pliocene vegetation on the Eyre Peninsula in South Australia[173] was apparently not forest but open woodland and grassland. Abundant myrtaceous pollen, some of it *Eucalyptus*, with *Casuarina*, considerable amounts of grass, chenopod and daisy pollen characterise the assemblage, and the presence of podocarp conifers suggests that forest remnants occupied gullies and river banks. Rare grains of *brassii*-type *Nothofagus* may have blown from distant mountains where pockets of forest remained.

In western New South Wales the forests were rich in MYRTACEAE, once again frequently dominated by genera other than *Eucalyptus*, and *Casuarina*. There was no *Nothofagus* in these assemblages.

At Lake George in New South Wales in cores of lake bed sediments[84] the last phases of rainforest are seen alternating with intervals of open *Casuarina* woodland with groundcover of grasses and daisies.

At Lake Tay in Western Australia (between Ravensthorpe and Norseman)[180] a five million year old flora contains many taxa that today are part of the Kwongan (sandplain) flora of seasonally arid southern parts of the State. PROTEACEAE including *Banksia*, *Dryandra*, *Isopogon*, *Xylomelum*, *Grevillea*, *Hakea* and *Sterlingia* (all sclerophyll type); CASUARINACEAE; GYROSTEMONACEAE; MYRTACEAE; sedges, grasses, restios and chenopods dominate the assemblage. Rare associated rainforest elements and *Nothofagus* show that remnant forest still persisted in the area. The southern part of Western Australia was already isolated from eastern parts by the Nullarbor, which had been an arid area approximating a desert since the Middle Miocene. A distinct Western Australian flora[172] was developing in the southwest of the continent in response to the isolation, the laterisation of the land surface in many parts and sandplain conditions in others, and probably in response to fire. The summer rain-winter drought climate that seems to have been the successor to the wet interval at the start of the Pliocene would have served to increase the endemism in the flora, and when mediterranean style climate with winter rain and dry summers followed about 2.6 million years ago the evolution of a yet more specifically Western Australian flora was the result.

Unpublished reports of pollen records from northern Western Australia show that remnant forest with *Nothofagus* still persisted in pockets there in Early Pliocene times.

Pollen records obtained from cores of sea-bed sediments off the Queensland coast show that during the Pliocene open savannah with a high frequency of daisies and chenopods was the vegetation type over large areas. This vegetation type would only become widespread in the southeastern part of the continent much later, in Pleistocene times.

The animals which lived in Early to Middle Pliocene Australia[178] are known from a number of fossil faunas. The most productive in terms of numbers of fossils found is the Rackhams Roost local fauna from Riversleigh in Queensland. There in a limestone cave formation an amazing array of bats which occupied the caves have left their bones: ghost bats, leaf-nosed bats, plainface, sheath-tail and others, mainly insect-eating. Rats and mice (immigrants from Southeast Asia), tiny to middle-sized carnivorous marsupials, ringtail possums and rat kangaroos are there in plenty, and odd bones of kangaroos and big diprotodontids add to the mammal inventory. Lizards (dragons, geckoes and monitors) and snakes represent the reptiles. Fossils of frogs and fish which swam in the pools in the cave system are also there in abundance. The nature of the deposit, which accumulated in caves, results in it having a poor record of the larger animals of the Pliocene. Gigantism was a character of many Pliocene marsupials and large, weird and wonderful creatures were roaming the countryside.

The appearance of the first true grazing kangaroos was a Pliocene phenomenon. The evolution of the kangaroo and its long hopping stride paralleled the development of open grasslands and grass-chenopod-shrubland open plains to which its locomotion and grazing were adapted. (Miocene faunas had primitive kangaroos whose relationship to modern-style kangaroos is not fully known.)

The Middle Pliocene Bluff Downs fauna of northeast Queensland has grazing kangaroos and large diprotodontids. The Chinchilla fauna of the same age has similar grazers and also waterbirds and crocodiles. The vegetation in that part of southern Queensland was already open savannah, a vegetation type which did not appear in southeastern Australia till Pleistocene times.

The Lake Palankarinna fauna from the Mampuwordu Sands east of Lake Eyre is about four million years old, Early Pliocene. It contains fish, crocodiles, diprotodontids, a bandicoot, kangaroos, dromornithids and emus. The bandicoot is of interest because it is of the same group as the bilby, which today occupies the arid centre of the continent.

About 2.8 million years ago global ice volume began a cyclic[191] but steady increase that culminated in an abrupt peak at 2.4 million years ago when the northern ice cap had been established and was of considerable size. The

Kangaroos with their long hopping stride evolved in parallel with grasslands towards the end of the Tertiary.

Jim Frazier

Right: Stirlingia latifolia, *a living representative of a genus whicch was already part of the sandplain flora in southern Western Australia 5 million years ago, and which has a lineage going back to the Late Cretaceous in the rift valley between Australia and Antarctica.*

K. W. Dixon

Far right: Casuarina *woodland, a semi-arid vegetation type, was much more widespread in the continent in the past.*

M.E.W.

'Pleistocene' ice age starts here, a million years before the official start of Pleistocene time. Up to about 2.6 million years ago, global sealevel rises and falls associated with changes in ice volume had been driven by Antarctica — a uni-Polar situation. From the time when a substantial northern ice cap formed the control of sealevel fluctuations became increasingly bi-Polar.

The dramatic ice increase at the peak of northern ice at 2.4 million years ago had profound environmental effects worldwide. The truism about environmental change being the driving force of evolution, with natural selection the agent, is exemplified in the separation of the *Homo* line from the 'robust' Australopithecines at this time when drastic environmental change resulted from the aridity associated with the build-up of ice.

The effects of North-Polar glaciation were felt and can be recognised worldwide. In East Africa micromammals, bovids (grazers related to buffalo etc.) and pollen spectra show the expansion of arid and open grassland close to 2.5 million years ago. In the Transvaal in South Africa changes in the grazing animal populations suggest a fundamental shift from relatively mesic and bush-covered environments to more arid and open savannah at the same time. In sub-Saharan lands grassland areas expanded. Cores of sea-bottom sediments taken in the South Atlantic for pollen analysis show that the African continent as a whole became significantly drier at 2.5 million years ago.

The same environmental changes with increased dryness[192] from 2.5 million years ago were seen in north Central Europe where tundra-type vegetation appeared for the first time. Extensive iceberg rafting of glacial sediments began in the North Atlantic at the same time. In central China loess deposition (wind-blown sediments) dated between 2.3 and 2.4 million years ago marks the change from warm and humid environments to harsh continental steppe. Pollen records from Colombia show the lowering of the forest line in response to dryness at 2.5 million years ago. In Australia, as we shall see, fundamental changes occurred at this time as well, making it the arbitrary starting point for the Drying phase of Australia's evolution in this book.

The introduction of bi-Polar ice caused a general decline in mean deep-water temperatures. Antarctic glaciation became

Above: Wreath Lechenaultia, *one of Western Australia's endemics, grows on the hot, dry, duricrusted soil. It is surprising to see such delicate flowers in such a situation. The adaptation of the plant to those specific conditions is achieved by an underground storage and deep root system.*
M.E.W.

Left: Kangaroo paws, like this species Anigozanthos manglesii *which is the State flower of Western Australia, are only found in the wild in that State.*
M.E.W.

increasingly Polar and from then to the present fluctuations have been in the extent of the ice shelves that ring the continent of Antarctica. The volume of southern sea-ice has been determined largely by fluctuations in the northern icecap. Changes in the Antarctic cryosphere (amount of glaciation)[191] prior to 2.4 million years were sufficient to influence Southern Hemisphere climates at mid and low

THE PLIOCENE ICE CAP EVOLUTION AT 4.5, 3 AND 2.6 MILLION YEARS

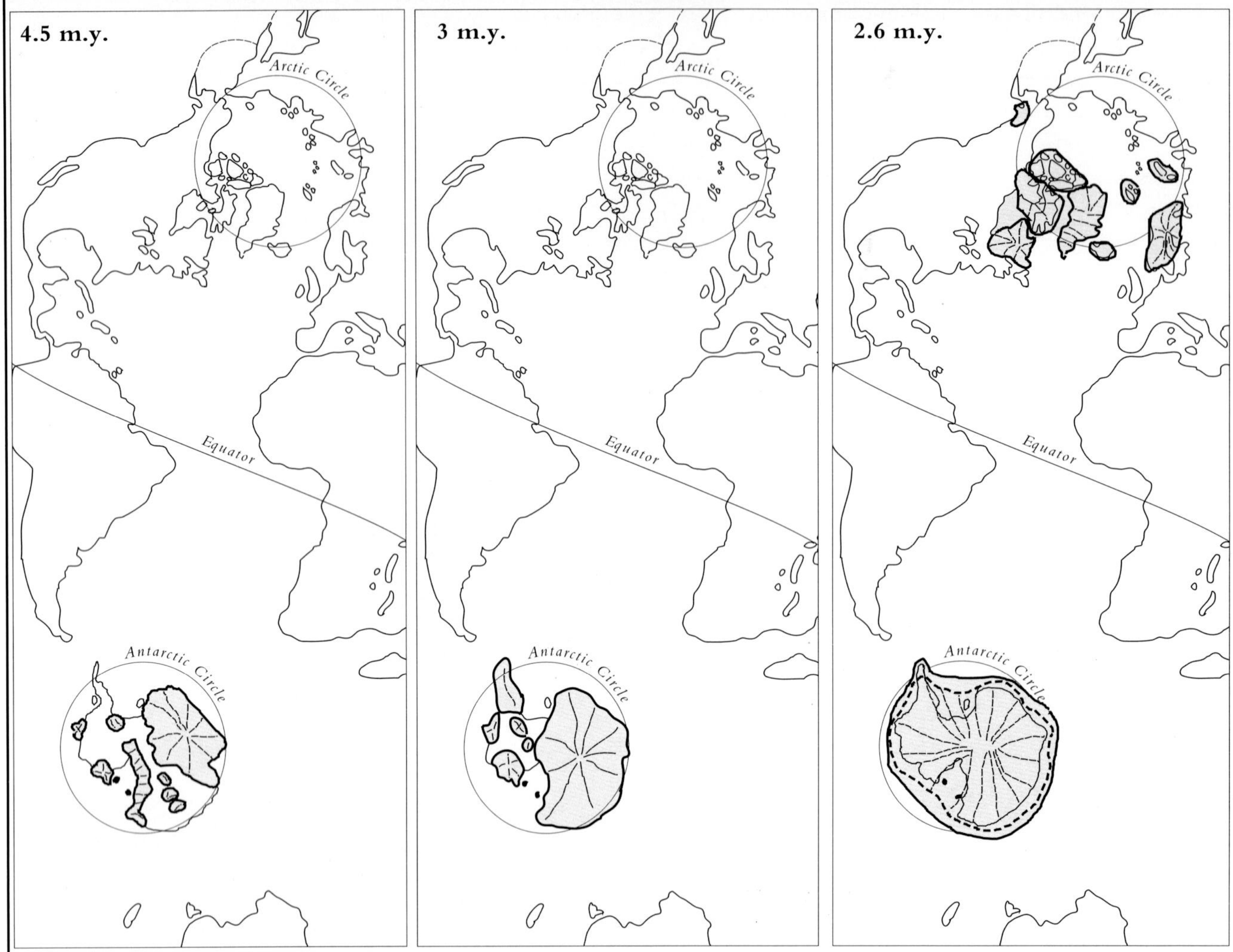

After Prentice and Denton 191

latitudes, but under bi-Polar control the influence of increased global ice volume intensified the effects that increase of ice in Antarctica alone had produced. The Polar anticyclone and strength of westerlies was greatly increased, air temperature was reduced, and climatic zones shifted to become similar to those of today.

After the 2.4 million year peak of northern ice, fluctuations within Pliocene time continued. Between 2.4 and 2.1 million years there was a steady decrease in ice volume. The volume stabilised at 2.1 million years ago and small fluctuations for the rest of the epoch were around this constant mean with a periodicity of 40,000 years. This rapid small fluctuating continued into Pleistocene time (which starts at 1.6 million years ago) and up to 900,000 years ago when ice volume increased dramatically to a new maximum which was achieved by 700,000 years ago. Since then major fluctuations have had a periodicity of 100,000 years.

The present Polar glaciation pattern which was established at 2.4 million years ago[24] seems to be controlled by features of the Earth's orbit around the sun — the Milankovitch Hypothesis. This hypothesis involves 100,000 years of increasing ice cap volume with associated fall in sealevel (which reached 130 metres lower than the present at the peak of the last two glacial stages), followed by a largely North-Polar melting of ice. A stable 5000 to 10,000 year interval follows and the cycle then begins again. Seventeen such cycles have been detected in the last 1.8 million years. The Earth's present-day situation within this cycle is in the 5000 to 10,000 year stable sealevel phase.

Before proceeding to the Drying section of this book, which takes up the story at 2.4 million years ago — only 800,000 years before the end of the Tertiary Period — it might be useful to review the changes that had occurred in vegetation and climate to this point.

In the Early Tertiary (Palaeocene and Eocene) rich and diverse rainforest containing many taxa that are now confined to northeastern Queensland, New Guinea and New Caledonia covered much of the well-watered continent. Composition of the forests varied and types ranged from those classified by reference to modern types as simple notophyll vine forest (SNVF) to complex mesophyll vine forest (CMVF). There was no 'pan-Australian forest dominated by *Nothofagus*' as is often stated in early literature. *Nothofagus* was a minor component in some forests, absent in others.

During the Oligocene the climate became more seasonal and cooler overall and *Nothofagus* became a more important component of forests. Species diversity was lower in the forests which were of temperate rainforest types. The Early Miocene was warm and wet with temperate forests with a high proportion of *Nothofagus*, podocarps, LAURACEAE and MYRTACEAE. From the Middle Miocene climate was more seasonal and cooler and the vegetation shows changes from closed forest to wet sclerophyll forest and from dry forest to open woodland with grassland. Eucalypts, *Acacia*, chenopods, *Casuarina*, grasses and daisies became prominent.

The Terminal Miocene Event saw very arid times, decrease in forest overall and increase in dry-land vegetation. Transition to the Pliocene saw a brief resurgence of forest in southeastern regions and other suitable areas, with *Nothofagus* returning to some forest communities. Grasslands and open forest and woody shrublands were widely developed. With increasing fluctuation of climate towards the Middle Pliocene the mosaic of different vegetation types, expanding and contracting, was adjusting to the approaching ice age. The climatic sifting of the flora that had accelerated in the Miocene became pronounced. Plants either adapted to the changing conditions, or became confined to refugia where the conditions that they preferred continued to exist, or migrated into newly emerging suitable areas. Many of the original Gondwanan taxa of the Early Tertiary forests could survive only in certain limited refugia in Australia and many migrated into New Guinea where rising mountains were creating suitable environments.

PART THREE

DRYING

(FROM 2.4 MILLION YEARS AGO TO TODAY)

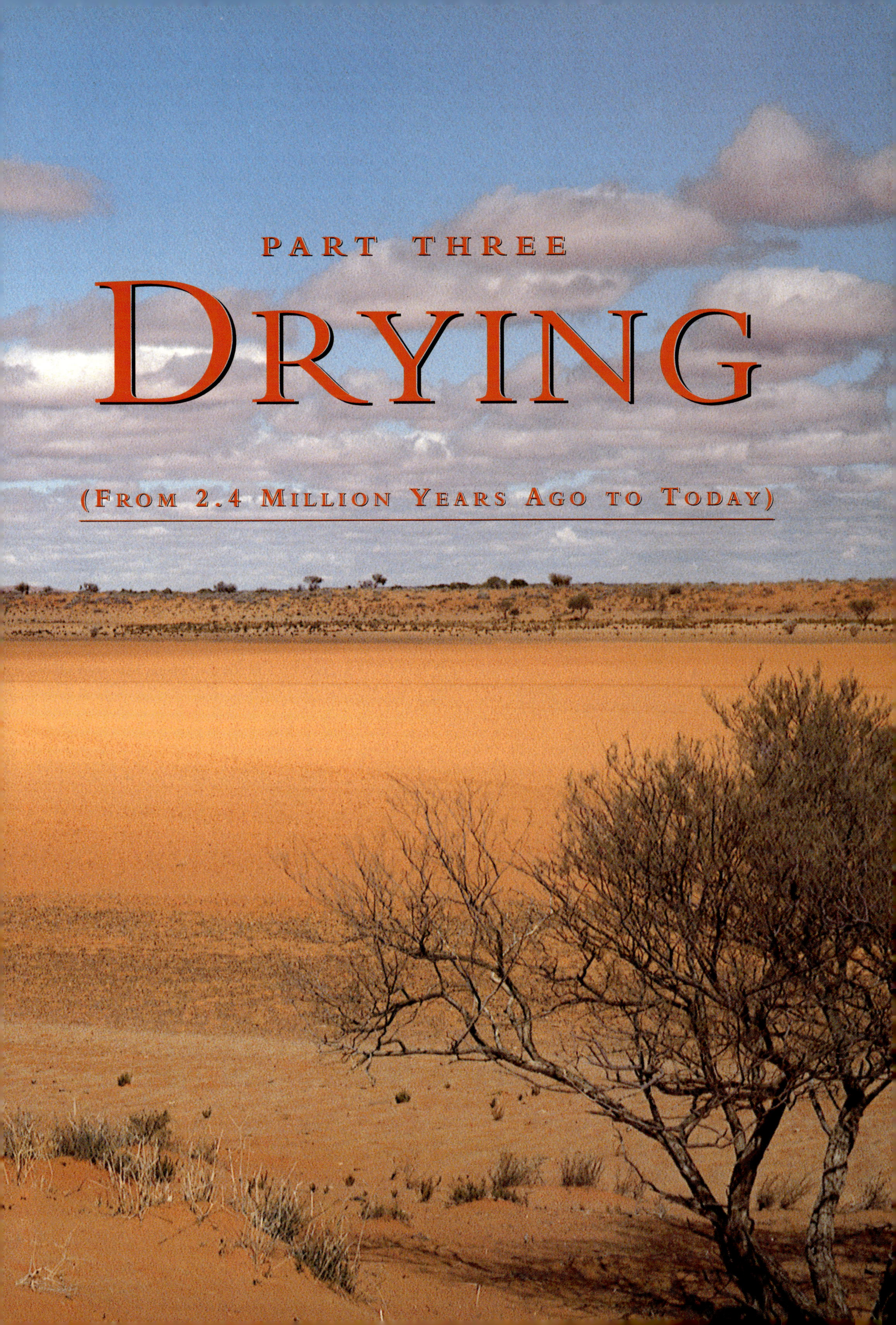

INTRODUCTION

Ice Age Australia

By 2.4 million years ago, when a very arid phase related to the development of the northern ice cap affected climates worldwide, Australia was already a land with only remnants of the ancestral closed forest that it had inherited from Gondwana. These remnants had maintained their presence in the areas that remained suitable or had migrated into newly emerging suitable areas during the Tertiary as the continent dried out.[189] The present-day distribution of the forests and what little information the fossil record supplies of the migration rates of such complete ecosystems leads to the conclusion that the position of refugia and core forests within them has not changed during the Quaternary. Response to the rapid climatic changes that are encompassed by the Pleistocene and Holocene has been contraction or expansion of the boundaries of the forests within the areas.

A similar conclusion has been reached in New Zealand where once it was thought that glaciation had meant a wipe-out of forest over large areas and eventual re-colonising from quite distant sites. Now it is believed that small refugia were widespread within the regions that were largely deforested and that the forest spread back from these nuclei within the local area after the worst of the glacial stages had passed.

Overall in Australia forest and woodland communities became much restricted in their distribution during glacial stages and were replaced by grasslands. Over most of the continent this was not due to conditions being sub-alpine. Temperature variations were not so vastly different in a land that lies in middle latitudes with hardly any high ground. The great expansion of grassland and change to ecosystems that looked like ones responding to extreme cold was as much a response to seasonal aridity, high winds and increased evaporation, and to the fire frequency that accompanied the drying-out stages towards the glacial peaks.

Many dryland vegetation types were established by 2.4 million years ago. In pockets where extremely arid conditions existed because of topography, soil type and salinity, many arid-adapted plants had already been sifted from less drought-influenced communities to fill the increasing variety of niches that the drying continent presented. The processes of aridification that were progressive during glacial fluctuations, and which were temporarily halted in some areas and lessened in others during interglacials, were now on a time-scale that operated in thousands, not millions, of years. The boundaries of the mosaic of vegetation types that covered the continent expanded and contracted in response to the climatic fluctuations.

With the change in regimes caused by bi-Polar control of world weather patterns, winter rain-summer dry climate came to southern Western Australia and to the southern sector of the continent. The zonation over the rest of Australia was becoming more like it is now, with the northern part within the tropical monsoonal belt, middle latitudes subject to aridity, and warm-temperate conditions to the south.

DRAINAGE DIVIDES AND BASINS ACCUMULATING SEDIMENT

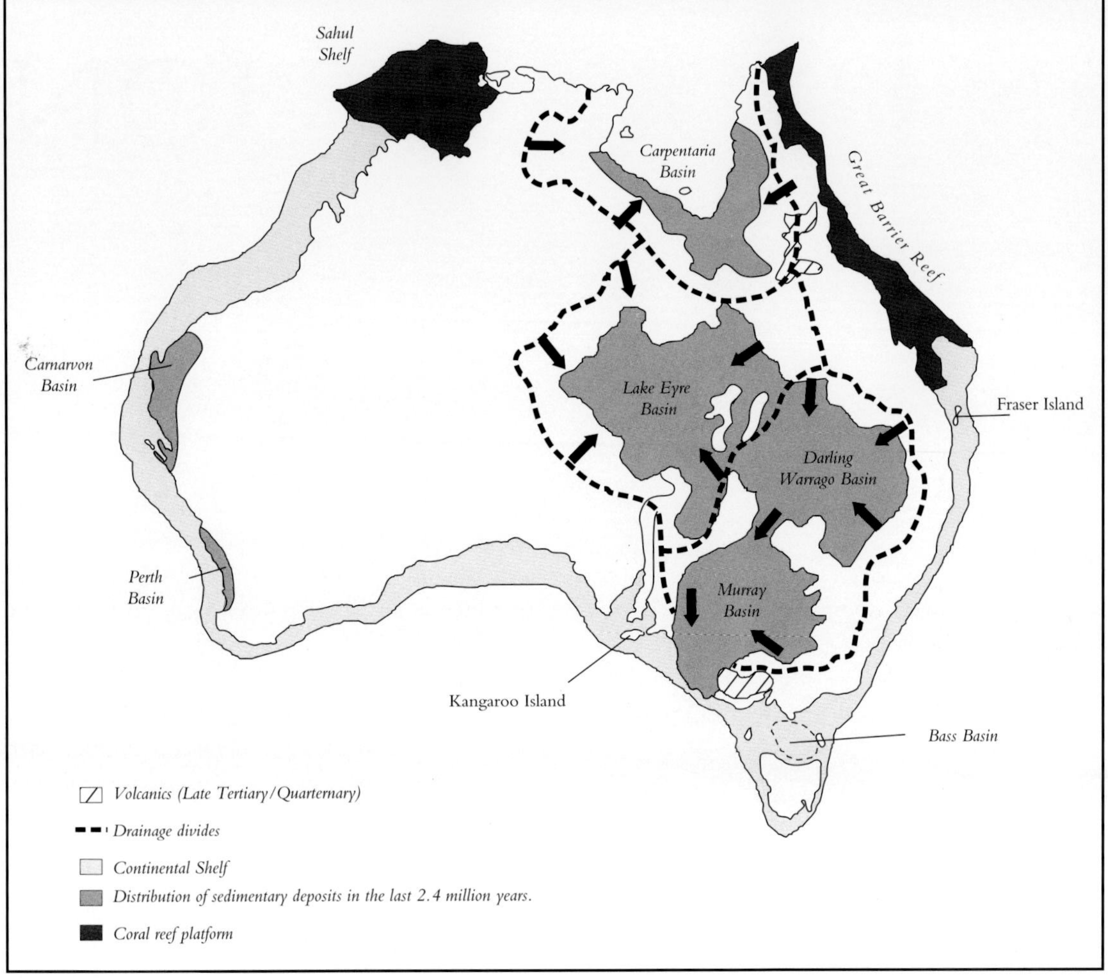

Only Tasmania and the highlands of the southeastern corner of the mainland now have cool-temperate climate. Glacial stages meant not only greatly increased aridity and windiness but also reduced incidence of tropical cyclones in the north, more wind-blown sand, cooler sea and air temperatures, and snow and ice in Tasmania and round Mt Kosciusko on the mainland.

The severity of glacial stages varied so that sealevel falls were to different depths depending on the amount of increase of polar ice. Similarly the degree of warmth and wetness of interglacials also varied, some being warmer and wetter than the present and resulting in the spread of mesic vegetation into larger areas than it occupies now, others being less benign.

The repeated fluctuations from intensely dry to wetter have, as it were by a process of attrition, turned more and more of the continent permanently arid. Fire became an increasing factor as more of the land dried out. A pattern was established in which it seems that Australia was doomed from the start of the slide towards the ice age some 15 million years ago to become ever more arid and ever more fiery. It is possible to see this process in action as the story of the last 2.4 million years of the continent's history unfolds. How the intervention of humans into the scheme of things greatly accelerated the natural processes and now threatens to turn the Island Continent into a *desert* island also becomes obvious.

Ephemeral grasses and daisies cover a wide clay-pan swale between dunes in far western New South Wales. Flocks of galahs congregate to eat the grass seed.

Jim Frazier

Spinifex, red dune sand and scattered Acacia *in the red centre of the wide brown land.*

Jim Frazier

CHAPTER TWELVE

How Australia Changed in an Increasingly Icehouse World

Late Pliocene to Early Pleistocene

(2.4 million to 700,000 years ago)

Mallee woodland in the Murray Basin.

M.E.W.

The Pliocene is the last epoch of the Tertiary Period. The Quaternary Period, which encompasses the last 1.6 million years, comprises the Pleistocene and the Holocene epochs. The Holocene is the last 10,000 years — our present interglacial.

During the Quaternary our present-day landscapes, climatic patterns and types of variability have become established and the biota has adapted to the new environments. In terms of tectonism Australia has been a stable land and only minor adjustments have occurred in the last few million years. One region affected by uprising of a block of terrain in this interval was the Murray Basin.

During the Pliocene, uplift of the Pinnaroo Block[147] in the southwestern part of the Murray Basin was taking place. This altered drainage in the Basin and from 3.2 million years ago started to dam the River Murray and the ancestral Darling, Lachlan and Murrumbidgee which flowed into it. By 2.8 million years ago a huge freshwater lake — Lake Bungunnia — had formed. It covered an area of about 68,000 square kilometres at the time of its greatest extent. It extended through the Murray Valley, inundating a large area of southeastern South Australia, southwestern New South Wales and northwestern Victoria. The Lake developed on the Parilla Sands Formation with its areas of deeply weathered and lateritised Karoonda Surface. (The Parilla Sands were laid down in the Late Miocene-Early Pliocene marine invasion of the Murray Basin.)

Lake Bungunnia formed when the Pinnaroo Block was elevated, blocking the Basin's drainage to the ocean.

THE MURRAY BASIN IN THE LATEST PLIOCENE TO EARLY PLEISTOCENE (2.5 MILLION TO 700,000 YEARS AGO)

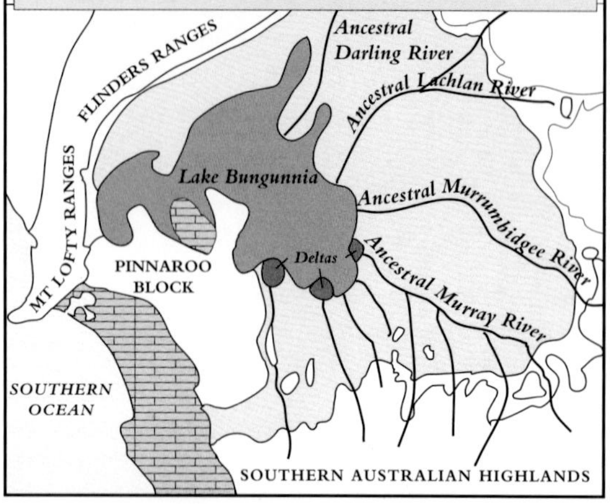

The southern end of the Murray River was formed when Lake Bungunnia drained through a breach in the southwestern corner of the Basin.

THE MURRAY BASIN IN THE LATE PLEISTOCENE (600,000 TO 100,000 YEARS AGO)

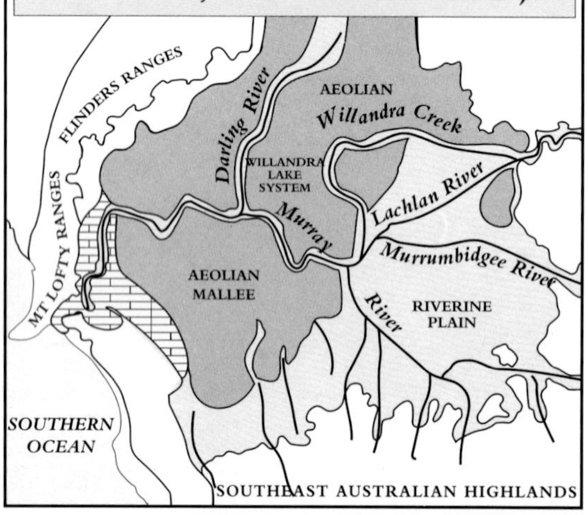

After Stephenson & Brown 147

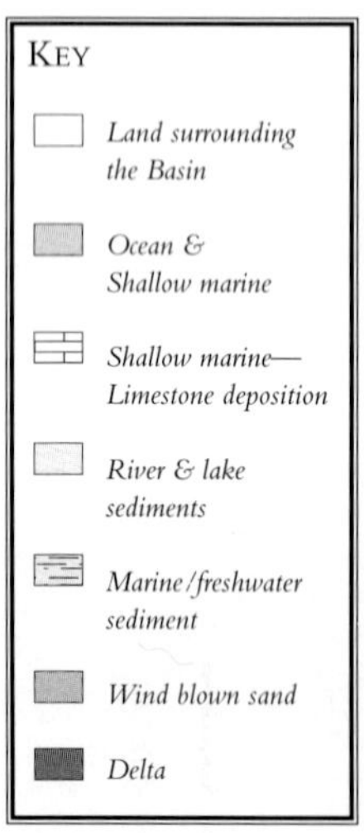

Wide alluvial plains extended to the north and east of Lake Bungunnia. Today's Riverine Plain, whose amazingly flat and featureless expanse will be remembered by anyone who has travelled from Hay to Balranald, must be very little different from those Pliocene alluvial plains. To the south a shallow marine shelf occupied the edge of the continent between the Pinnaroo Block and the present-day coastline and marine limestone was deposited on this shelf.

The date at which Lake Bungunnia started to accumulate sediments and the time span that it covers (3.2 million to at least 700,000 years) are known because the lake sediments, the Blanchetown Clay Formation, can be dated by magnetic-pole reversals which are recorded in them.

The sediments also tell the story of the fluctuations in water level and hence reflect climatic conditions. From 3.2 to 3 million years ago water level fluctuated with four shallow events. From 3 to 2.8 million the water was moderately deep, the climate humid. From 2.8 to 2.6 sandy sediments tell of a relatively dry phase with a slight fluctuation to wetter at 2.53 million years. Between 2.5 and 2.4 million years ago the lake was dry. This was the time when the northern ice cap formed and intensely dry conditions affected much of the globe. Between 2.4 and 1.9 million years the lake was filling and then showed minor fluctuations until 1.7 million years ago. It remained a vast expanse of water with minor fluctuations until 700,000 years ago when it started to drain, through a breach in the raised southern margin. The southern end of the modern Murray River formed when Lake Bungunnia drained.

As the lake retracted it left behind the scattered ponds of Menindee, Tandou and other minor lakes along the Darling and Murray Rivers. Lake Tyrrell, a large salt lake of The Mallee, developed after one of the southern arms of Lake Bungunnia had retracted and it provides evidence that arid conditions prevailed after Lake Bungunnia had drained. Its saline Tyrrell Beds with abundant gypsum overlie the Blanchetown Clay. They formed at the time when the volume of polar ice had stabilised at a new high level.

The Blanchetown Clay Formation laid down in Lake Bungunnia underlies The Mallee with its surface array of salt lakes and blown-sand dunes. It has been deduced from a study of the sedimentary succession in the part of the Murray Basin affected by Lake Bungunnia that the arid-zone landforms, dunes, palaeosols and salt lakes of The Mallee all formed within the last 700,000 years, and probably within the last 4-500,000 years. Thus the evolution of its unique vegetation type dominated by mallee eucalypts has taken place in about half a million years.

Another Plio-Pleistocene lake[208], in a setting very different from Lake Bungunnia in the Murray Basin is Lake Buchanan which is situated within the Great Divide in central Queensland about 350 kilometres west of Mackay. Today it is a salt lake, usually dry, or with a pool in its southeastern corner, and differs from the salt lakes of the arid centre of Australia in that it has no chemically precipitated carbonate or gypsum. A thin halite crust forms in the southeastern part when that is dry. When full it is about 23 kilometres long and 7 kilometres wide.

Its sediments can be dated because they show the magnetic pole reversals that also date the Lake Bungunnia sediments and those at Lake George and elsewhere. Beach ridges from ancient high water stands occur on the western side of the lake, indicating easterly winds. The lake sediments show repeated iron-stained layers which represent old soils formed during times when the lake was dry and its bed was vegetated. The sediments were laid down from 1.67 million years ago and testify to numerous wet and dry phases. At least four long wet phases occurred in the last 730,000 years. The record is incomplete because during times when the lake was dry its bottom sediments were deflated (blown away).

A change to arid-type deposition with gypsum and salts and windblown silt at about 2.4 million years ago, corresponding to the dry stage of Lake Bungunnia, can be detected as an horizon in sedimentary sequences in many parts of the continent. Throughout large areas of southern and inland Australia the deeply weathered profiles of Early Pliocene age which are acidic and often cemented with iron (duricrusted) are frequently abruptly overlain by calcareous sediments and soils which are alkaline. These fine-grained sediments and soils can only be explained by wind transport.

In Central Australia where shallow lakes had been sites for sedimentation throughout the Miocene and into the Early Pliocene (the Etadunna and Namba Formations and their equivalents) a clay layer often caps the lake sediments. The fine, wind-blown material forming the clay is evidence of the arid and windy times when the lakes were drying, their waters saline and gypsum-rich and their lake-floor sediments exposed and being deflated. A change in sedimentation regime is seen in the Upper Tirari Formation where a gypsum-cemented layer of sediment occurs above the Etadunna Formation. The development of alkaline soils and extensive aeolian (wind-blown) deposits on a large scale in Late Pliocene times marked the start of desertification. As far as can be determined this large-scale change occurred in central Australia also around 2.4 million years ago.

The effect of rising and falling sealevels[143], coinciding with interglacials and glacials, on Australian coasts was profound. During high sea stands the edge of the continent was inundated and marine sediments, including limestone, were deposited. Dune strand lines and shell beds further inshore than the modern coastline bear witness to higher sea stands. At times when sealevel was low and continental shelves were exposed, rivers had to cut deep channels to reach the sea. When the sea rose again the deep-cut valleys were inundated and the deep river beds were silted up. Drowned river valley systems form the well-known coastal features of Sydney Harbour, Broken Bay and others right round Australia's margins. The final sculpturing of the coasts occurred during the ice age.

At times when continental shelves were exposed, which were the windy and dry glacial times, the sand from the continental shelf was blown inland. Dunes formed and in time were often blown into sandsheets.

A large-scale example of the sort of continental margin development that resulted

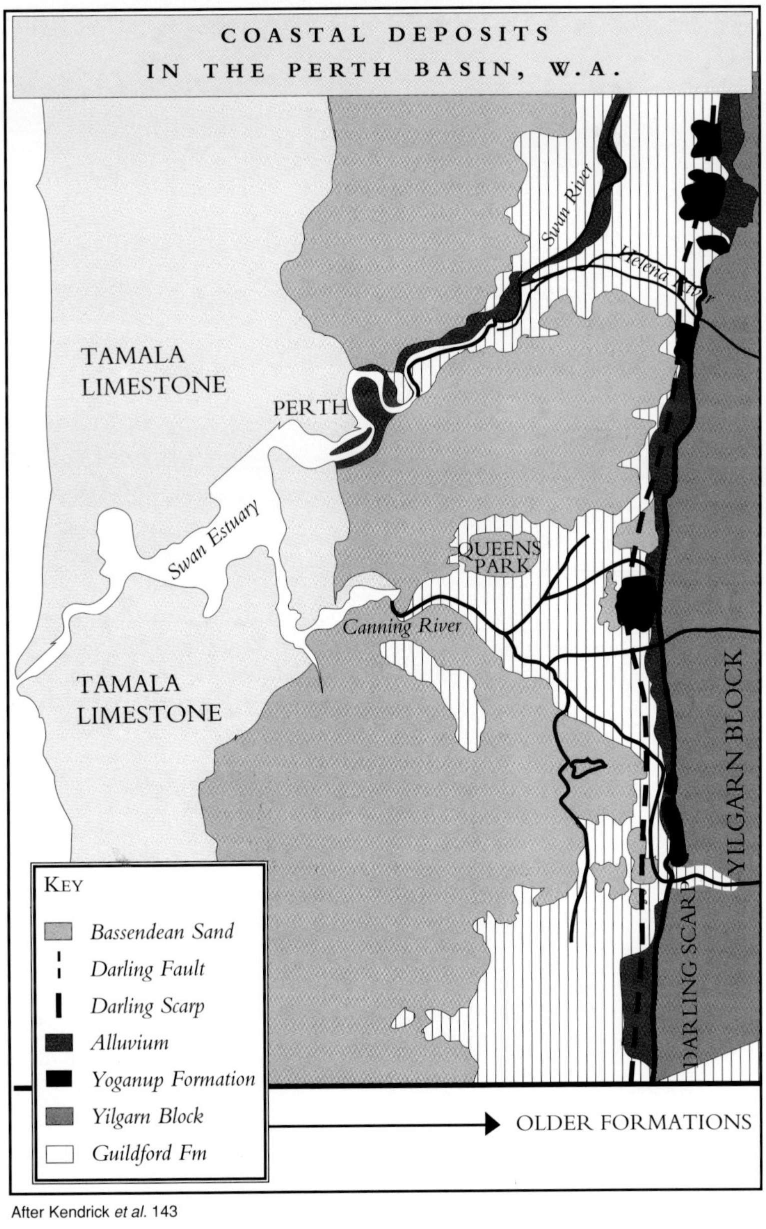

After Kendrick *et al.* 143

from ice age fluctuations in sealevel is the Western Australian coastal strip which constitutes the onshore section of the Perth Basin. This coast has been relatively stable and evidence of events which span the Late Pliocene to Pleistocene interval has largely been preserved. In the central sector of the Basin, in the vicinity of Perth, Pliocene to Quaternary marine sequences form deposits which lie parallel to the present coastline and are bounded to the east by the sharp line of the Darling Fault. This Fault separates the Perth Basin from the ancient Yilgarn Block, whose edge forms the 'ranges' of the Darling Scarp.

Far right: Kwongan heathland on the sandplains near the coast in the Perth Basin has grass trees in profusion among the low-growing shrubs and brightly flowering woody herbs.

M.E.W.

Above: Coastal limestone at Burns Beach near Perth.

M.E.W.

Right: Granite of the ancient Yilgarn Block with eucalypt woodland on Mt Latham.

M.E.W.

The marine sequences increase both in age and elevation towards the scarp, reflecting different heights of sealevel at different stages in the Pleistocene. However, there has been a certain amount of uplift of the Basin (and of the Yilgarn Plateau) during the Pleistocene and this has amplified the levels at which the deposits now lie. As the continental shelf margin has been subsiding throughout, the general gradient from west to east has been maintained, emphasising the sequential deposition of the sediments.

The Yoganup Formation, Ascot Formation and Bassendean Sand represent a complex of

LIVING REPRESENTATIVES OF THE REGATTA POINT FOSSIL FLORA

Far left: Dicksonia antarctica *tree-ferns at Liffey Falls.*

Jim Frazier

Left: The leatherwood, Eucryphia, *is famous for its honey as well as for the beauty of its flowers. The family* EUCRYPHIACEAE *of which* Eucryphia *is the only genus, is a Gondwanan one. Two species occur in Chile, two in Tasmania and one in New South Wales. All the species are rainforest or rainforest-margin plants.*

Jeanette Closs

Left: Nothofagus gunnii *the deciduous beech, and* Athrotaxis selaginoides *the King Billy Pine at Cradle Mountain.*

Dick Burns

sediments deposited at ancient shorelines, on inner shelves, and as dune sands. In the central and eastern part of the Perth Basin the Pliocene-Pleistocene units which constitute the Ascot Formation are entirely sub-surface and form one of the important 'water-mound' aquifers on which Perth depends for its water supply. These sediments are quartz sands.

Heavy-mineral sands in fossil shoreline deposits[218] have been mined on the Swan Coastal Plain since 1956; on the Scott Coastal Plain near Augusta at the southern tip of Western Australia; south of Shark Bay in the Carnarvon Basin; and in tidal rivers in the north of the State. The first deposit to be mined was at Koombana Bay near Bunbury, and it was mined out in ten years. Others are still being mined today, particularly those near Eneabba. The minerals in the placer deposits (deposits in which heavy ore minerals that had been eroded from their original bedrock have been carried by water and concentrated in an area) contain ilmenite, zircon, monazite, rutile and leucoxene (a whitish mineral resulting from the decomposition of ilmenite). The weathering of the ancient granite rocks of the Yilgarn Block provided the sediments that contain the minerals being mined in the Perth Basin. They were carried by rivers and deposited on or near the coast. With subsequent alternating high and low sealevels they were sorted according to weight and grain size and formed strandline deposits during interglacial highs.

The coastal Tamala Limestone originated as offshore calcareous sands which were brought ashore when sealevels were high and later blown into dunes. The warm Leeuwin Current was strong at warm times during the Pleistocene leading to abundant marine organisms which produced the calcareous material in the sands. Warm-water corals like *Acropora* spread down the coast at such times extending their range far beyond modern limits.

The dunes may have been up to 300 metres high and were originally mobile. (Modern calcareous-sand mobile dunes are still a feature of parts of the coast.) Over time rainwater seeped through the sand dunes. Being slightly acidic it dissolved some of the calcium and carried it down to the watertable. The nature of the climate that prevailed throughout the Pleistocene — a cool wet winter and hot dry summer mediterranean climate — meant that when the dune dried out in the dry summer the dissolved lime at the watertable was precipitated, cementing a layer of the sand and stabilising the dune. Vegetation then played an increasing part in the stabilisation process and a soil developed on the dune's surface. Soils and their vegetation cover resulted in the groundwater becoming more acidic, dissolving more calcium, cementing and solidifying more sand into limestone. (This was a deep weathering process which resulted in calcrete formation. Duricrusting took place near the surface at the junction between soil and dune where the most acidic reactions occurred and a hard caprock type of limestone resulted.)

The Tamala Limestone, which formed from mid Pleistocene times[209] on, forms a wide band adjacent to the coast in the Perth Basin extending south from Shark Bay more or less continuously. It also occurs in patches on the southern coast of Western Australia as far as Albany.

The Pinnacle Desert, where upstanding pillars of Tamala Limestone stand in a sand desert, has been created by groundwater solution of the once-continuous limestone formation. As the limestone dissolved, eaten away by acidic groundwater, sand free of calcium was left filling the gaps between columns of limestone. When the area was deflated the sand blew away but the solid pillars were left standing and have been preserved additionally by their duricrust capping. Many have acquired a mushroom form through further erosion of the soft limestone below the indurated cap rock.

An interesting phenomenon in this already amazing Pinnacle Desert is the presence of rhizoliths — fossilised roots. The roots that grew down through the soil into the dunes created acid groundwater in their immediate vicinity. This dissolved lime and resulted in its re-precipitation round the root surface, encasing it in a lime crust. With the deflation that removed sand cover many of the roots stand up out of the ground.

The Bassendean Sand represents an arid depositional phase. At dry times when sealevel was low and continental shelves were exposed, quartz sand was blown inland. The wonderful sandplain flora of Western Australia is attuned to the low-nutrient, effectively arid, often glaringly white, sandy soils of the Perth Basin.

The fossil record of plants in the Late Pliocene

and Early Pleistocene is fragmentary. The period was obviously one when change was occurring rapidly. The sorting and sifting of the flora was taking place as forest decreased, dry sclerophyll vegetation was becoming widespread and large regions of central Australia were becoming acutely arid.

Tasmania began to feel the impact of the ice age and quite extensive glaciation occurred during the Ross Glaciation of Late Pliocene or Early Pleistocene times. The exact dating of this Tasmanian episode is not known, but may coincide with the short, sharp drop in global temperatures at 900,000 years ago which was felt in Africa to the extent that it caused the extinction of the 'robust' Australopithicines and the spread of *Homo erectus* into Eurasia.

The continuation of the cooling trend saw a new high reached by polar ice at 700,000 years ago. This is approximately the time given for the Linda Glacial Stage in Tasmania by the magnetic-pole reversal which also helps to date Lake Bungunnia[162] and Lake Buchanan, fixing the glaciation date as about 730,000 years ago. Little is known of this glacial stage because the glacial deposits are extremely weathered and they are overlain by those of later glacial events.

A flora from Regatta Point on Macquarie Harbour[201] in western Tasmania has both macrofossils and a pollen record which indicates a Late Pliocene to Pleistocene age. Two vegetation types were present in the area — a rainforest community of a type differing from the forest which grew in Oligocene times at Pioneer by decrease in leaf size and lower diversity, and a sclerophyll community much like that which grows in the area today — *Eucalyptus*, *Casuarina*, *Banksia*, epacrids and tea trees. The rainforest contained *Nothofagus cunninghamii*, *Eucryphia* (leatherwood), *Atherosperma moschatum* (sassafras), *Quintinia* (possum wood, now extinct in Tasmania), *Acacia*, *Lagarostrobos franklinii* (Huon pine), *Athrotaxis selaginioides* (King Billy pine) and *A. cupressoides* (pencil pine), *Podocarpus*, *Tasmannia* (WINTERACEAE) and the tree-fern *Dicksonia antarctica*.

This list of temperate rainforest plants could as well be a list of modern plants from Tasmanian beech forests, apart from the *Quintinia* which has now disappeared (but which hung on here at Regatta Point far longer than it did in southern parts of the mainland).

At Hazelwood in the Latrobe Valley in Victoria[189] a pollen flora from the Pliocene to Pleistocene transition has a great abundance of daisies and grasses with *Acacia* and *Casuarina*. This may indicate cool or dry conditions or both. The interval is preceded in the pollen core by an assemblage that is probably pre-northern ice cap in age. It has high values for forest taxa including abundant *Eucalyptus*, and a rainforest component which includes *Nothofagus* (*brassii* and *menziesii*), Huon pine, CUNONIACEAE and ELAEOCARPACEAE.

At Butchers Creek on the Atherton Tableland in northern Queensland, where today there is still rainforest, pollen floras of the Pliocene to Pleistocene transition indicate that the area had complex rainforest without any taxa which indicate drier conditions. *Nothofagus* was still present, though it has since become extinct in the area.

At Lake George, in part of the pollen record[189] which is of earliest Pleistocene age, high levels of daisies and grasses are seen replacing the *Nothofagus brassii* type forests. The climate was probably moderately seasonal there but with reliable rainfall.

In general, by the end of the Tertiary Period rainforests had contracted to the eastern fringe of the continent, essentially within their present range. Boundaries of rainforest communities would again expand and contract with the climate fluctuations of the Quaternary Period.

The fauna of the Late Pliocene to Pleistocene interval[178] from the Murray-Darling Basin when Lake Bungunnia occupied part of the region is known from fossils in the Blanchetown Clay and its equivalents. The fossil sites are near Lake Victoria on the Murray River. The oldest fauna is from Moorna Sands, river sediments that lie just below or just within the Blanchetown Clay. This Fishermans Cliff local fauna contains marsupial dasyurids (native cats), wombats, potoroos, wallabies and kangaroos, and (placental) rodents. The rodents are of two groups, one of modern genera that today occur in arid regions, and the other of an extinct kind that is not closely related. The marsupials include *Lagorchestes,* a hare-wallaby, which today is a genus of arid Australia. Thus, already arid-adaptation was a feature of fauna in the Late Pliocene.

Another fauna, higher up in the Blanchetown Clay — the Bone Gulch local fauna — has fish, turtles, *Thylacoleo* the marsupial lion, wombats and indeterminate macropodids (wallabies and kangaroos).

Far right: Nothofagus cunninghamii, *myrtle beech, forest on the Gordon River today is an analogue of the Regatta Point fossil flora.*

Dick Burns

Right: An ancient, twisted tree of Athrotaxis cupressoides, *the pencil pine, growing on a lunette at Great Lake on Tasmania's Central Plateau.*

M.E.W.

Diprotodon bone fragments also occur in this fauna.

Further north, in the Menindee Lakes area, the diprotodontid *Euowenia* and the giant kangaroo *Protemnodon* are found in boreholes, and a lungfish tooth has been found at Lake Tandou. Lungfish do not occur in southern Australia after this time and their restriction to their present range in the Mary and Burnett Rivers in Queensland took place after the time of the drying out of Lake Bungunnia (700,000 years ago) and well before the last major phase of aridity between 18,000 and 15,000 years ago.

In the Lake Eyre Basin in the Late Pliocene to Early Pleistocene interval, the Kanunka fauna from stream channel deposits, the Katipiri Sands, contains a diverse assemblage of animals. Lungfish, chelid turtles, crocodiles, cormorants, flamingoes and other water birds, hare-wallabies and nailtail wallabies are among the smaller animals. The large diprotodontids *Diprotodon* and *Zygomaturus*, and *Megalania* the giant goanna are the megabeasts in the assemblage. (The Katipiri sediments are palaeodrainage deposits of the ancestral Cooper and Warburton Rivers which merge into wind-blown dune sediments of latest Pleistocene and Holocene age in which fossils of creatures of the last glacial maximum occur.)

The flamingoes of the Kanunka fauna were of two kinds — one that is now confined to Africa, and the other more widespread in tropical and temperate regions. The reason for the extinction of flamingoes in Australia is obscure but may be related to competition with other aquatic species like filter-feeding ducks which can feed in and adapt to a wider range of water conditions. The fluctuating water levels and repeated drying up of shallow lakes in glacial stages of later Pleistocene times must have made it difficult for birds that were adapted to very specific niches. The presence of flamingoes in central Australia from Miocene times until Late Pleistocene is an indicator of the presence of large bodies of shallow water there.

The Malkuni fauna, also from Katapiri Sands, from the lower Cooper River contains lungfish, crayfish, turtles and crocodiles, and a diverse bird fauna which includes grebes, pelicans, cormorants, rails, spoonbills, ducks, herons and a crane. The mammals present are modern *Sarcophilus* (Tasmanian devil), brushtail possums, and bettongs. Extinct forms include *Sthenurus* (browsing kangaroos), a giant wombat (*Phascolonus*), *Protemnodon* (a giant wallaby), *Procoptodon* (giant browsing kangaroo) and the diprotodontids *Diprotodon* and *Zygomaturus*. The presence of lungfish and crocodiles indicates reliable continuation of water in the area. As soon as intermittent drying occurred these animals disappeared from local faunas.

CHAPTER THIRTEEN

BIOTAS ON THE ICE AGE SEE-SAW

MIDDLE PLEISTOCENE TO THE START OF LAST GLACIAL CYCLE

(700,000 to 130,000 years ago)

Ice ages interrupt the usually warm state of the Earth every 300 million years. The Quaternary, which comprises the Pleistocene and Holocene epochs, with its ice age regimes, represents such an interruption. The conditions which were experienced by the biota were an extraordinary episode of Earth history. A repetition of conditions last experienced in the Late Carboniferous to Permian ice age of 300 million years ago has moulded the flora and fauna and altered landscapes on a grand scale in our island continent. The result is modern Australia — the driest vegetated continent with a unique flora and fauna. The production of the largest arid zone within the territory of a single country, of desert areas as large as those of North Africa and the Middle East added together, was a cumulative process which itself experienced many interruptions. The see-sawing of climate has sculptured the land and fine-tuned the flora and fauna.

A gorge in the North Flinders Ranges on Moolawatana Station where a permanent waterhole provides a refuge centre for animals.

M.E.W.

With the increase in polar ice to a peak at 700,000 years ago, the 100,000 year cycle of climatic change was established and this regime remains active up to today. Warm interglacials take up only 10-15 per cent of the cycles, 85 per cent of the time is encompassed by intervals which are cooler and drier, and glacials which are much cooler, much drier and much windier.

By early in the Pleistocene seasonal aridity and widespread droughts were well established in inland Australia, as indicated by high levels of grass, daisy, *Casuarina* and chenopod pollen. Eucalypt-dominated communities occupied less harsh but still seasonally dry environments. The tropical north, including the Arafura Plain which connected New Guinea to the mainland, bore semi-deciduous rainforest and non-eucalypt savannah. Closed forests were confined to parts of the eastern ranges from New Guinea down to Tasmania.

The tempo of climatic fluctuation from 700,000 years ago onwards hastened the sifting and sorting of flora and fauna into environments that suited them. Those plants and animals that were very specialised in their requirements had problems and the Quaternary was a time when many extinctions occurred. However, generalist species were able to take advantage of the changing times, migrating or dispersing to keep up with the changes.

Each time a glacial stage arrived many complete ecosystems as well as localised populations were wiped out and when conditions eventually improved areas had to be re-populated by individuals which had survived in refugia or by the expanding of boundaries of adjacent vegetation.

The fragmentation of populations that occurred due to the inability of individuals and their associates to survive over their complete range resulted in small pockets, where climate, soil and conditions remained suitable, acting like islands. One result was speciation, with new forms arising, some of which were suited to the new niches that were repeatedly appearing. The great radiation of *Acacia* with its 640 species was a result of the changing environments into which the original few species were thrust during changing Pleistocene times, isolation of small populations and the genetic ability of the genus to adapt. Another consequence was the establishment of disjunct distribution patterns for taxa which had been unable to recolonise their original range.

The importance of refugia in the scheme of things in Australia is clearly seen in its prehistoric past. In our modern continent in which the boom and bust regimes of drought and flood create stress for populations, and human activity has changed landscapes, refugia are still of the utmost importance on both the long-range and short time scale. Unless pockets remain where species diversity can be maintained in hard times, there is no way that regeneration of depleted areas can occur. This obviously implies that during times of maximum stress in the past such refuges have existed in sufficient numbers and have been located within core areas of all the major vegetation provinces.

Migration routes must also have existed allowing connection across inhospitable regions, and alluvial tracts along rivers must have been significant in this regard. Species associated with gallery forest, better soils, swamps and permanent water could spread along valleys with alluvial soil. The Darling River system, which provides an inland connection between well-watered areas in the northeast and the southern part of the continent, probably acted as a corridor during the fickle changes of the Quaternary Period. Western Australia had no such north-south migration track. Its inland rivers had ceased to flow in Miocene times and the southwestern region in particular was isolated and responded by developing an individual flora with a very high proportion of endemics.

Pleistocene history from 700,000 years ago started literally[223] with a bang in Tasmania.[224] A meteorite slammed into the side of Mt Darwin south of Queenstown in the western part of the island. Luckily it was only a small body, unlike the huge bolide that collided with West Kampuchea at the same time. The Darwin Crater that it made is not a landscape feature which stands out like Gosses Bluff, but from the point of view of supplying information about Pleistocene vegetation and climate it has been very useful. The crater, in a down-faulted valley in ancient rock strata, was originally about 200 metres deep and 1000 metres in diameter and it became a lake. The sediments that accumulated in it and ultimately filled it, merging the structure into the landscape so that it took geological detective work to discover it, preserved a pollen record of the vegetation.

It was the strewnfield of Darwin glass, the product of fusion of silicate rocks under the enormous forces generated by the impact, that led to the discovery of the crater. Glass is found over an area of about 400 square kilometres and interest in it was generated because of its use by Tasmanian Aboriginals in making artefacts. They traded it and carried it far and wide in Tasmania. The amount of glass formed by the impact was probably not more than 100 tonnes, while the amount produced by the Kampuchean bolide's impact and strewn over a vast area (about 4 per cent of the Earth's surface) is estimated as at least a million tonnes. These microtectites are the same age as Darwin glass and their strewnfield covers Tasmania, mainland Australia, South China, Thailand, the Philippines and parts of the Indian and Pacific Oceans.

Tasmania experienced glaciation of its high mountains[162] between the older-than-700,000 year Linda Glaciation and the last glacial cycle. Dates of the glacial stages involved are uncertain.[233] The Governor Glacial Stage is older than the Henty[234] which occurred round about 200,000 years ago.[235] There were many ice advances and retreats within the glacial stages.

The start of continental dune and sandsheet

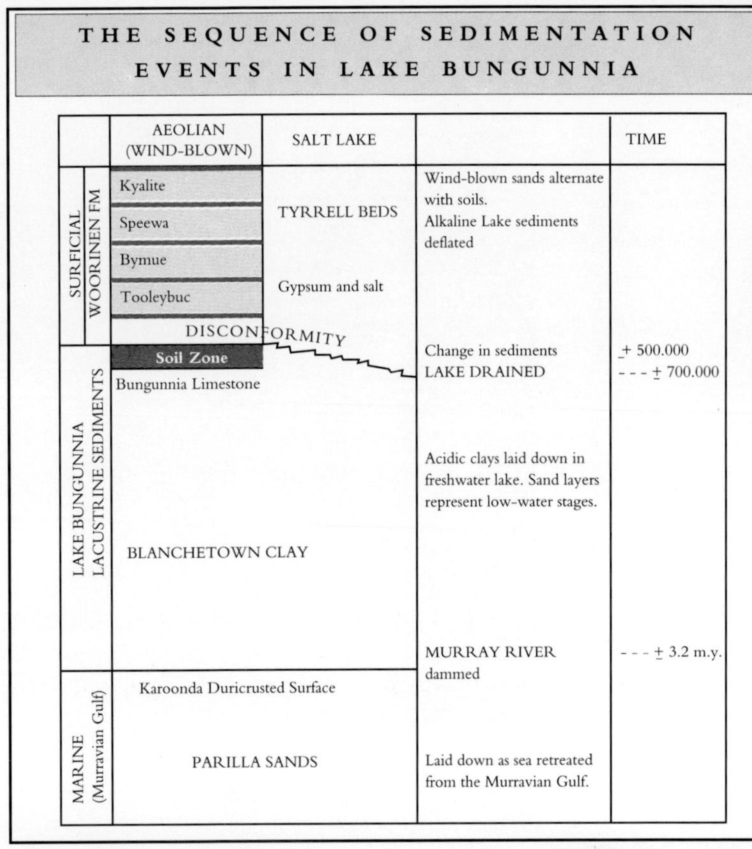

After An Zhisheng, Bowler *et al.* 193

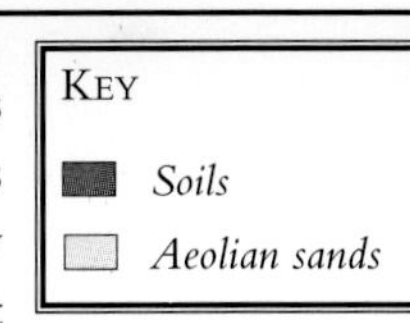

development, creating sandridge and sandplain deserts of any large extent and widespread distribution, began with the windy and arid glacial stages some time after the 700,000 dateline that saw amplification of climatic fluctuations. Because of the replication of cycles and the repeated reworking of dune sands and wind-blown sediments, it is not yet possible to determine more than a handful of dates for the early establishment of these desert features. New techniques of age determination of sediments using thermoluminescence may eventually clarify the picture - if such work is ever carried out over the vast area that is occupied by aeolian (wind-blown) sediments today. The situation becomes much clearer in the latest glacial cycle when the age of some of the dunefields is known and there has not been another cycle to cover up evidence of the processes of dune formation.

Just as Lake Bungunnia provided valuable evidence in its Blanchetown Clay Formation[193] of the climatic fluctuations[185] in the Late Pliocene to Early Pleistocene time slot, so later deposition in its vicinity provides evidence of the start of dune formation. The Woorinen Beds, correlated to the saline beds of the Tyrrell Formation (which show the coming of salt lake arid conditions to Lake Tyrrell), are an east-west linear dune sequence that is now buried under the dunes produced in The Mallee in the last glacial stage. Four episodes of dune formation are preserved in the Woorinen Beds. They can be seen in the railway cutting at Nyah West and elsewhere near Woorinen. The most probable date for the dune sequence is believed to be from 500,000 years, making them, until recently, some of the oldest dunes so far detected below the blanket of recent dunes and sands that make ancient dune history so hard to decipher.

Recent research in Lake Amadeus in Central Australia[217] has found that saline and aeolian deposition started there at about 900,000 years ago (unlike the post-700,000 year and most likely 500,000 year commencement of aridity seen in the Tyrrell Beds). Parts of Central Australia may therefore have been experiencing characteristic arid zone conditions leading to salt lakes, gypsum deposits and blown dunes for almost a million years. This latest evidence comes from sedimentary cores taken in a drilling programme on Auger Island in Lake Amadeus.

Thermoluminescent dating of the floodplain sediments[225] of the Channel Country, whose tangle of river channels now carries water during flood times to Lake Eyre, has revealed vast sandsheets indicating greatly increased river activity during interglacials at 250,000 to 200,000 and 120,000 to 100,000 years ago. The sand-dominated phases are interpreted as having occurred in semi-arid climatic regimes when sandsheets resulted from floodouts over the area, in a manner similar to that which occurs today. Above the sandsheets, mud deposition layers indicate a much drier climatic regime from 85,000 years ago onwards, right up to the present. Rivers no longer carried a heavy sand load. This suggests that the interglacial in which we live today is so far not as humid as were previous interglacials. In other words, the Centre is drier now than it ever was during previous interglacial times.

It seems to be generally agreed that from 300,000 years ago[185] large areas of Australia were arid with blown sand dunes[172] and sandsheets forming during the acutely arid glacial stages.[219] It has to be remembered, in the context of adaptation of the biota to aridity, that the intervals between glacial stages which represent about 70,000 years of each 100,000 year cycle had climatic regimes that included seasonal aridity and probably frequent droughts over a great proportion of the continent. Thus adaptation to dryness, and in some ecosystems

Right: Kyabra Waterhole is a permanent large stretch of water on Kyabra Creek, a channel on the floodplain of the Cooper River. Its water and surroundings supply a refuge between rains in this arid region. Below: River red gums in Bratchina Gorge in the Flinders Ranges in South Australia. The sheltered gorges in these rocky mountains where permanent water in pools is life-supporting when the surrounding terrain is waterless are refugia which enable animals and plants to survive in hard times. From such refugia they can recolonise surrounding parts which have become almost lifeless during severe droughts.

M.E.W.

to frequent fires, was a continuing process for the biota, interrupted by benign 'holidays' of wetter times which were too short to undo the evolutionary changes in physiology, anatomy and programmed behavioural patterns that stood the organism in good stead during arid and fiery times.

The changes in vegetation which occurred on the Southern Tablelands[189] near Canberra throughout the 700,000 to 130,000 years interval are recorded in a pollen sequence obtained from cores of sediments deposited in Lake George. A very detailed sequence is known for the last 350,000 years and the fluctuations in vegetation type that it shows have been extended back through the whole interval. In two interglacials, from 347,000 to 297,000 and 251,000 to 195,000 years ago, forests were dominated by CASUARINACEAE with cool-temperate rainforest species frequently represented. The intervening glacials show high levels of herbaceous taxa, with grasses and daisies predominating, and with a minor component of sub-alpine types.

Temperatures during glacials must have been between 8° and 10°C lower than at present, and rainfall was substantially reduced. This vegetation pattern that characterised the interval from 700,000 years ago up to about 130,000 years ago is then disrupted. Suddenly a great increase in carbon particles in the core, indicating the frequency and severity of fires, occurs, and there is a great increase in *Eucalyptus* pollen. The late Dr Singh and his co-workers on the Lake George project saw this carbon increase as representing the time when Aboriginal people came to the area and increased the fire frequency. Their interpretation was not widely agreed upon because the evidence so far suggested that the time of arrival of people in the continent was not more than possibly 60,000 years. However, now evidence from sea-bed sediments off the north Queensland coast has confirmed a huge increase in carbon at about 140,000 years ago, and once again it has been suggested that this is evidence of human intervention in the scheme of things. No other explanation seems to satisfy the facts.

Whatever the truth is about the reason for increased fire frequency, the result on the vegetation was clearly evident. The major consequence of increased burning in the Late Pleistocene and Holocene has been to decrease diversity in floras and to accelerate the processes of selection of plants suited to local conditions. Fire-tolerance became a prerequisite for much of the Australian flora.

At Lake George the fire-sensitive sclerophyll communities dominated by *Casuarina*, together with temperate rainforest patches, had survived under regimes of infrequent burning determined by the periodicity of lightning strikes and the rate of litter accumulation. During glacials the carbon had disappeared from cores because with the open vegetation there was insufficient accumulation of fuel to sustain fires. The increase in fire frequency led to the replacement of fire-sensitive sclerophyll and cool-temperate forest with fire-promoting eucalypts. These could maintain high fire frequencies even into the next glacial stage because of their inherent

Rocky gorges in Central Australia provide microclimates where cycads, native cypress pines, white-stemmed eucalypts and many other plant species find sanctuary from the hot and arid surrounding terrain. These refuge areas are vital to survival of animals and maintaining of diversity.

M.E.W.

THE LAKE GEORGE POLLEN SEQUENCE: MIOCENE TO HOLOCENE

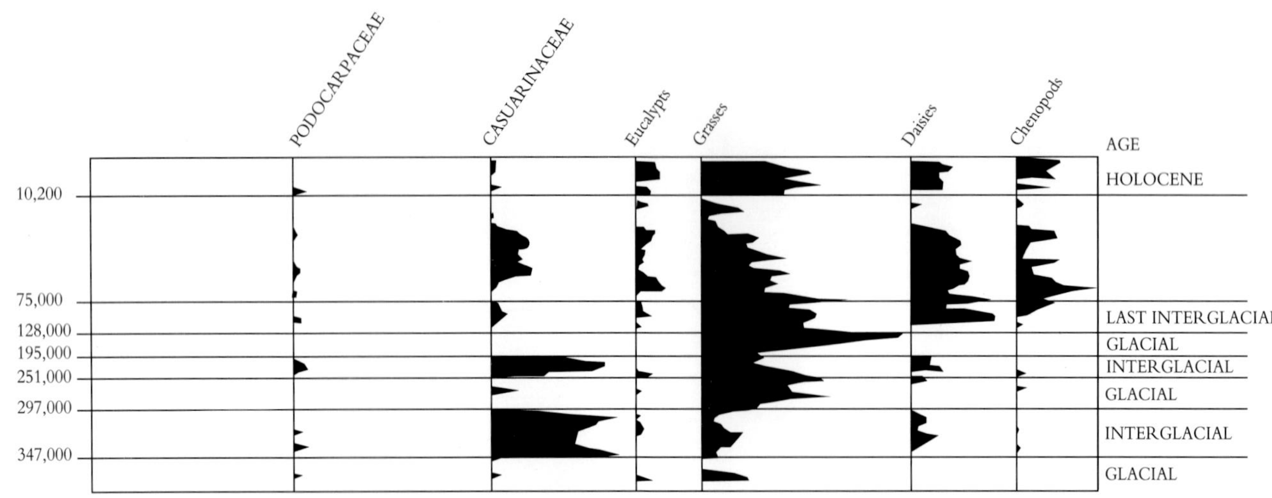

After Kershaw 189

flammability and their production of flammable litter.

Between 130,000 and 75,000 years ago only minor fluctuations occurred between open forest and sclerophyll woodland at Lake George. Only during the last interstadial, when temperatures were a little lower than at present but rainfall was higher, did *Casuarina* and cool-temperate taxa make a small comeback. These taxa are now regionally extinct. The increase in chenopod saltbush along with the eucalypts may have been promoted by increased soil salinity due to hydrological changes under a less stable vegetation cover.

Lynchs Crater in northeast Queensland has provided a pollen core[189] that starts about 180,000 years ago. It indicates that in the segment up to the start of the last interglacial about 130,000 years ago the forest contained abundant podocarp conifers and CUNONIACEAE; lesser amounts of *Elaeocarpus*, MYRTACEAE, and *Casuarina*; and very low frequency of grasses, daisies, chenopods and *Eucalyptus*. The glacial stage saw a change to high levels of grass pollen, an increase in *Eucalyptus* and CASUARINACEAE, and low levels for CUNONIACEAE, MYRTACEAE and PODOCARPACEAE. Araucarian conifers showed an increase indicating that a much drier forest type had replaced rainforest and the vegetation was open-type with abundant grass.

The Darwin Crater pollen sequence[162] for this time leading up to the last interglacial showed the same fluctuations between forest and grassland as seen in the Lake George core — but in this Tasmanian example the climate was cooler and the crater lay close to the snows on the west coast mountains during glacials. Thus the alternation was between cool-temperate rainforest and alpine-style grassland.

At Lake Callabonna and at Billeroo Creek[178] southeast of Lake Frome a fauna of megabeasts, with a suggested age of about 120,000 years based on the pollen flora, contains *Diprotodon*, *Phascolonus* the giant wombat, *Sthenurus* a browsing kangaroo and *Genyornis* a giant running bird. The large flightless birds progressively lost diversity during the Pleistocene, so that by the end of the epoch only *Genyornis newtonii* was left. (A recent advance in technology for dating cave deposits has resulted in eggshell of the now-extinct *Genyornis* from a site[370] in Wentworth being determined as between 60- and 70,000 years old. The eggshell was in an Aboriginal midden deposit, so it also dates the presence of people there at that time. This is the earliest positively dated time for an Aboriginal presence so far in Australia.)

CHAPTER FOURTEEN

On the Slide Toward the Last Glacial Maximum

(130,000 to 18,000 years ago)

Australia had already become a land not unlike the one which we know before the penultimate glacial maximum lowered sealevels around it, creating wide areas of continental shelf and linking it to Southeast Asia. It was the flattest, driest land where drought was part of the scheme of things over most of the continent, where soils were leached and weathered and had had no renewing of nutrients by volcanism except along the Great Divide and in a few areas in the eastern part of the southern margin. Leached soils are not good moisture holders so plants had become accustomed to aridity even when climatic conditions were not extreme. There had not been (nor would there be) any revitalising of soil cover by the passage of ice over the land during glacials as was the case in parts of the Northern Hemisphere.

Glacial stages for Australia had meant increased dryness, more wind and colder, hard times for the biota. Its vegetation had sorted itself out to inhabit the different zones and fill the multitude of different niches that such a large landmass presented, and its animal population had undergone adaptation to their environments. A mosaic of different vegetation types with their animal inhabitants had moved or expanded or contracted, in response to the fluctuations within previous 100,000 year cycles.

The fluctuations of climate that were to come in this last cycle were more extreme — or so it would appear from the evidence we can find to recreate the picture. Maybe the fact that the changes that occurred are the most recent and have not been obscured by later cycles allows us to see them more clearly in the record of evolution of landscapes and of the biota. Our interest in and empathy with the changes that occurred in this last lap of the journey towards the present day are heightened by the fact that people were part of the biota. They experienced the good times when lakes were full and rivers ran, the hungry dry times when survival depended on skill and acquired wisdom to survive, and the quite appalling conditions of the last glacial maximum in Tasmania when some of them lived near the edge of the ice cap on the Central Plateau.

The chronology of these times is better established because dating techniques give more precise results for this time span. We live in and can visit the environments which were formed by this slice of the past, and we can recognise the plants and animals of our modern land among the biota of those times.

There is evidence of higher than present-day sealevels during the 128,000 to 124,000 years ago interval. Low areas adjacent to the coastlines of the continent were inundated. An

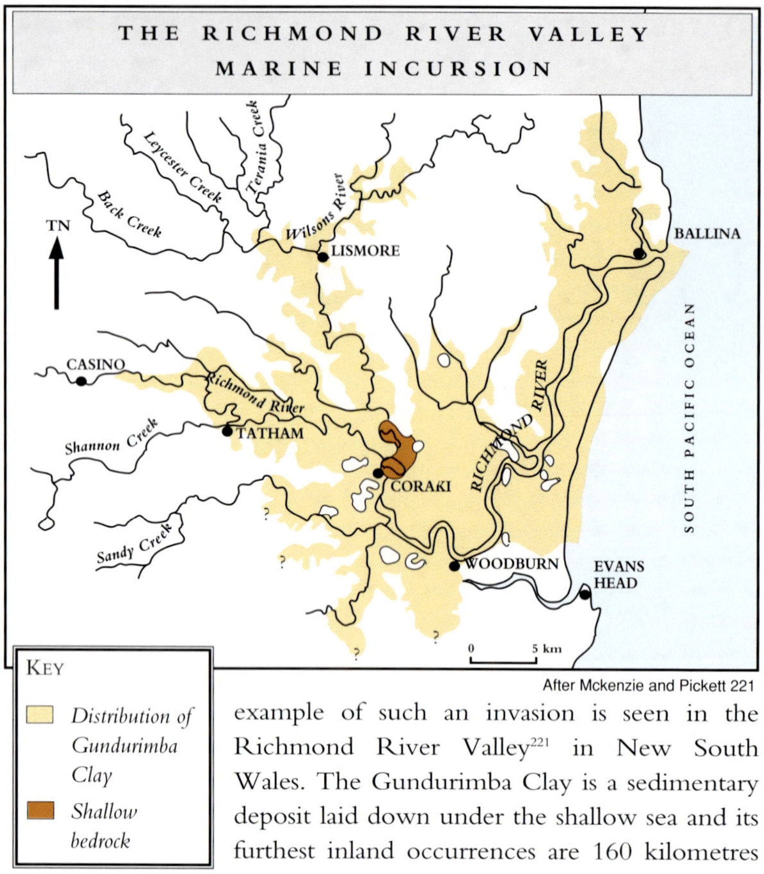

After Mckenzie and Pickett 221

example of such an invasion is seen in the Richmond River Valley[221] in New South Wales. The Gundurimba Clay is a sedimentary deposit laid down under the shallow sea and its furthest inland occurrences are 160 kilometres from the present mouth of the Richmond River at Boatharbour (upstream of Lismore), and extend up another valley to a point four kilometres east of Casino. The area occupied by the Pleistocene sea was irregularly shaped, with arms extending up tributary valleys and with islands of higher ground remaining emergent. The Clay contains marine Ostracodes and extends beyond the modern coastline where it lies on the continental shelf. It is overlain on the coastal fringe by dune sands which extend seawards as sand bars.

Environmental interpretations based on the Ostracode species in the Clay indicate a relatively protected bay with an immediate hinterland of generally low relief. The embayment was open to the sea and this implies that the Inner Barrier Beach ridges that characterise the area today were deposited later when sealevel fell and sands from the continental shelf were redistributed. This dune-sand barrier is composed of the Woodburn Sand. Ostracodes and a Late Pleistocene coral fauna from Evans Head[227], outside the limits of the Clay, show that the environment there was more exposed to the open sea than it is now because the sort of corals present are of a type that require open marine circulation. The Barrier beach ridges had not formed there either at the time when the coral was alive. Water temperatures were probably slightly warmer in the area than they are today.

North of the Richmond River mouth the coast of southeastern Queensland is marked by striking accumulations of quartz sand which form parabolic dunes of Pleistocene age along 350 kilometres of coastline. These dunes are orientated northwest to southeast lying obliquely to the present coast and rising locally to 200 metres. Five barrier islands lie off the coast, part of the greatest coastal sand deposit in the world — North and South Stradbroke, Moreton, Bribie and Fraser Islands. At Stradbroke Island the sands extend to 60 metres below sealevel and the dunes are fixed and vegetated. They contain peat and sandrock layers and the sand is strongly iron-pigmented in many places.

The Great Sandy Region encompasses the Cooloola Sands and Fraser Island and the strait between Fraser Island and the mainland. The sand originated from erosion of the land along the eastern edge of the continent and was carried up the coast by along-shore currents and deposited on the continental shelf and rocky ledges ready to be carried ashore to form dunes. Fraser Island has been forming throughout the Pleistocene and probably its oldest dunes (on the east of the island) started to form from times of lowered sealevel in the Late Pliocene when continental shelves and their sands were exposed. It is the largest sand island in the world and its development resulted from its small area of rock blocking the path of the sand-bearing ocean currents and causing them to drop their sand in the lee of the rocky ridge.

It is fortunate for the Great Barrier Reef that north of Fraser Island the Capricorn Basin, a deep-sea incursion towards the Australian coastline at the Tropic of Capricorn, acts as a trap and stops sand from building up on the continental shelf any further northwards. The shelf would not support reef growth if it was subject to sand build-up.

In general, quartz sands form dunes on the eastern and northern margins[222] of the Australian continent while the southern, southwestern and northwestern margins are characterised by calcareous sands. These lime-rich sands form coastal limestone and only the youngest remain as mobile dune sands.

After the episode with raised sealevels, the last cycle began at 120,000 years ago with environments similar to those of today.[141] Sealevel was approximately the same, so coastal

Fraser Island—the largest sand island in the world

Top: Indian Head, a rocky outcrop at the tip of the Island.

Above: The coloured sands of Fraser Island, compacted and eroded into fantastic shapes, are a feature of the island.

Above right: Lines of shifting dunes advance into the Island from the east.

Right: A big sand blow gradually engulfing the vegetation.

Glen Carruthers

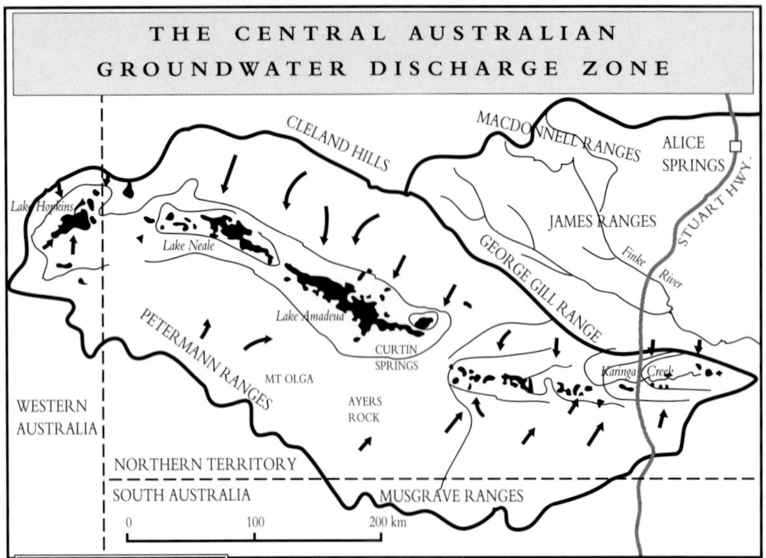

After Arakel and Wakelin-King 241

KEY
Salt lakes
Main drainage divides

features and shorelines were similar. The interior of the continent was dry, with stabilised dunes that had been produced in previous arid and windy intervals that equated with glacial stages. Only minor fluctuations in climate seem to have occurred during the first 40,000 years of the cycle.

Increased fire frequency was promoting the spread of eucalypts and their diversification.[253] The great variation shown within species and species groups in *Eucalyptus* supports a relatively recent radiation and massive extension of range.

Evidence from Central Australia about periods of aridity[240] after the first 40,000 years of the cycle comes from the chains[241] of playa lakes that extend from Lake Hopkins in Western Australia[237], through Lake Neale, Lake Amadeus and numerous small playas to the Finke River in the Northern Territory. These salt lakes are remains of a palaeodrainage system that once flowed to the ancestral Finke River and thus to Lake Eyre. Groundwater calcrete deposits, frequently silicified, and gypsum and glauberite deposits in this palaeodrainage can be dated. They represent intensely arid phases.

Near Curtin Springs a relatively long arid interval is recorded between 75,000 and 35,000 years ago. Sluggish groundwater formed a valley floor deposit of silicified calcrete below the watertable. (This is important as an aquifer in the region today.) A shorter arid episode from 16,000 to 8000 years ago was initiated by the last glacial maximum.

From 27,000 to 22,000 years ago there is evidence of increased rainfall. Vadose calcrete, formed by the seepage of water through the layers above the watertable, characterises this interval. (This interval corresponds to the lake-full stage at Lake George.) Lakes in the Centre were full during this wetter interval. Over all the arid regions of the continent watertables were higher and dry times became seasonal rather than constant, but droughts were probably still part of the pattern. However, high evaporation, aeolian sands, salinity, clay substrate in swales and gibber plains, and other conditions that had become established still imposed limitation on the vegetation. The absence of migration corridors from areas that supported more mesic vegetation could not have allowed much change in the vegetation patterns over most of what was already a 'dead heart' of the continent. The processes of aridification had so altered the landscapes that simply adding more water over a few thousand years could not return vegetation types suited to non-arid regions to them. Margins of the different sorts of communities would have fluctuated, however.

Lake Woods near Newcastle Waters in the Northern Territory[244] is a documented example of a lake that expanded during the wetter times in what is now a semi-arid region. It was very much larger and six metres deeper about 22,000 years ago.

The start of the 75,000[141] to 35,000 arid interval in Central Australia corresponds to the 75,000 to 64,000 cold interval seen in the Lake George pollen core as a treeless time when grassland surrounded the usually dry lake and fire frequency dropped because of the decreased ability of the vegetation to carry fires. The 64,000 to 22,000 interval at Lake George was a warmer and better-watered period when eucalypt woodland predominated and the lake was overflowing between 26,000 and 23,000 years ago. Stone tools made by Ngunawal Aboriginals occur in the oldest aeolian sands at Lake George dated at between 23,000 and 17,000 years. These sands indicate the coming of arid times (and a dry lake bed) to the area.

The constant aridity that affected the Centre between 75,000 and 35,000 years ago was not experienced in the Murray Basin either. There from about 60,000 years ago the catchments on the southern half of the Great Divide supplying water to the Murrumbidgee and Murray and the southeastern rivers were well-watered and by 55,000 years ago all the lakes in the Basin were full. The Willandra Lakes, fed by a tributary of the Lachlan River, covered a total of about 1000 square kilometres.

Lake Mungo is one of these lakes and is of particular interest because of its archaeological record. At about 36,000 years ago there was an arid period of unknown but relatively short duration and the lakes shrank, became increasingly saline and most dried up. Lunette dunes were blown on the eastern side of the dry lakes, a great deal of salt was carried from saltpans into the surrounding landscape where it caused salination of soil and later contributed to a very saline watertable. Then a return to wetter conditions allowed the lakes to refill and to remain at moderate levels until about 25,000 years ago. After that drier conditions prevailed and the catchments began to experience much cooler temperatures.

Lunettes at Lake Tandou and Lake Menindee contain faunas[178] probably older than 30,000 years which contain modern arid zone species as well as the extinct *Diprotodon*, the giant kangaroo *Protemnodon*, the browsing kangaroo *Sthenurus*, *Thylacoleo* the marsupial lion, and *Procoptodon* the giant browsing kangaroo. The modern arid zone forms are hare-wallabies, nailtail wallabies, bettongs, pig-footed bandicoots and bilbies. Similar faunas at Lake Menura and Lake Victoria also contain koalas, indicating that these creatures lived in the riverine forest in the seasonally arid environment just as they do in parts of Queensland today.

Further evidence of good rainfall and heightened river activity[267] during the interval between 60,000 and 36,000 years ago[248] comes from sites on the Nepean River near Sydney. A large alluvial terrace deposit was laid down by a braided river system between 47,000 and 40,000 years ago. Interest in the age of the deposit was sparked by the discovery of Aboriginal stone artefacts in the river gravels of this deposit (the Cranebrook Terrace). Many other rivers in the continent show a period of increased activity which corresponds to this wetter part of the last cycle.

In Tasmania at Rocky Cape and near the mouth of the Welcome River[232B] in the northwestern corner of the island large alluvial gravel deposits dated at between 44,000 and 22,000 years show greatly increased river activity.

In the extreme north of Australia the lowered sealevel[246] associated with ice build-up in the last 40,000 years[230] resulted in exposure of the floor of the Gulf of Carpentaria. A drop of 70 metres in sealevel turns the Gulf into land. (A similar situation existed in regards to Bass Strait. There the Bassian Plain connected Tasmania to the mainland when sealevel dropped by 65 metres.) From about 35,000 until about 12,000 years ago a very large freshwater to brackish lake — Lake Carpentaria — occupied the centre of the Gulf. Its area was about 30,000 square kilometres and its maximum depth about 10 metres. Sediments laid down in this lake have been cored and it appears that the lake surrounds supported a savannah with vegetation like that on today's black soil plains to the south of the Gulf. Sedges and swamp and water plants predominated and grassland with MYRTACEAE and some *Callitris* and *Casuarina* were present. An interesting point about the pollen record in this area, which was in direct land connection with New Guinea, is that there is no evidence of invasion of any foreign taxa into the Australian flora there during this 35,000 to 12,000 year interval.

The time of high lake levels in southeastern Australia[226] between 50,000 and 30,000 years ago is known as the Mungo lacustral phase. It was followed by a period of major water deficit.

LAKE BASINS IN THE MALLEE OF WESTERN NEW SOUTH WALES

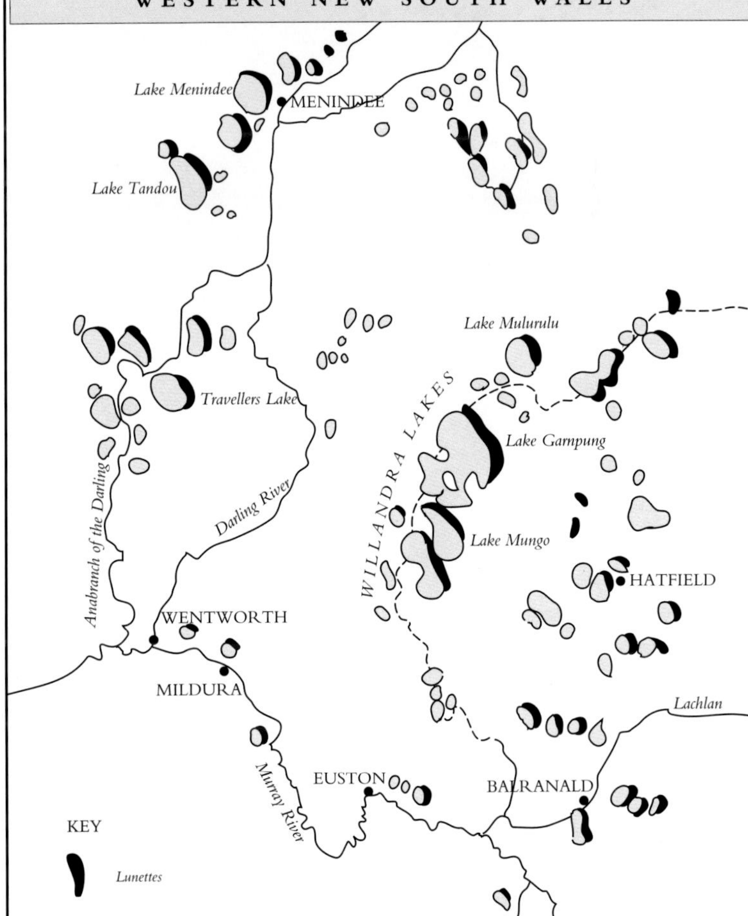

After Bowler 207

EUCALYPTS FOR ALL ENVIRONMENTS

Right: White stemmed gums in arid grassland.

M.E.W.

Far right: A wandoo forest in Western Australia. Eucalyptus wandoo *is a beautiful tree with silver and grey bark.*

M.E.W.

Bottom far right: Eucalypts are often festooned with mistletoe which saps their strength during times of water deficit. The parasite takes up and uses the host's water at times when the tree has done all it can to shut down its operations until conditions improve. The mistletoe seems to continue its operation regardless throughout the times of stress.

M.E.W.

Bottom right: White-stemmed gums in arid grassland.

M.E.W.

On the mainland the drying trend escalated between 25,000 and 14,000 years ago and maximum aeolian activity leading to the formation of continental dunes was between 18,000 and 15,000 years ago, coinciding with the maximum extent of Northern Hemisphere ice sheets (the glacial maximum). The behaviour of rivers on the Riverine Plain in the eastern part of the Murray Basin shows the climatic changes of this period from lake-full to lake-empty: During the Mungo lacustral phase the river systems crossing the Plain were wide and meandering with dunes on their northern and northeastern margins. (The Tallygaroopna Stage of river evolution.) In the drying phase of increasing aeolian activity towards the glacial maximum the rivers narrowed, the amplitude of their meanders was less, and sandsheets spread from the marginal dunes. (The Kotupna Stage.) In the interglacial after 14,000 years the rivers took on their modern form with highly sinuous channels. (The modern Goulburn-Murray Stage.) The same patterns occurred in the ancestral Darling River, which is today also characterised by narrow, sinuous channels.

The pattern of change in these rivers reflected the total water budget of the times. The big meander stage corresponded to the high groundwater/lake-full stage, smaller meanders to the decrease in water and lower watertable, and the final stage represents the

Above top: Old-man saltbush and dune canegrass grow on the red sand round Lake Menindee.

M.E.W.

Above: Trunks of large trees that died when the water level in Lake Menindee was artificially raised stand on the beach and create a drowned forest.

M.E.W.

Across the gibber plains to the distant Flinders Ranges — one of the emptiest landscapes in the continent.

M.E.W.

LAKE LEVELS AND RIVER CONFORMATION IN SOUTHEASTERN AUSTRALIA

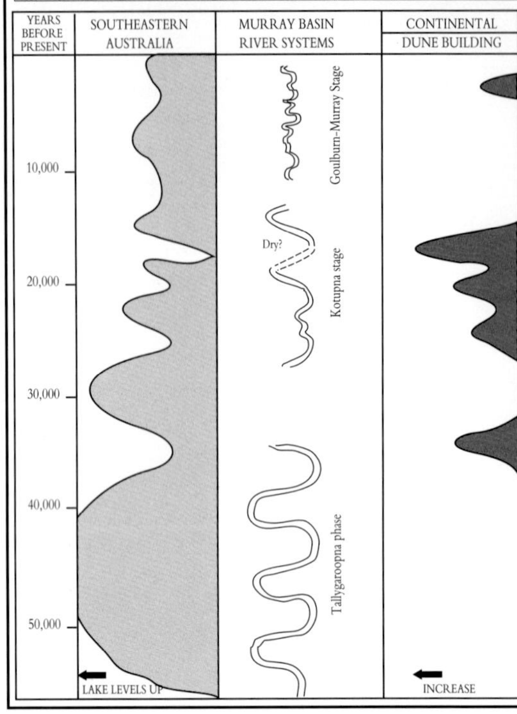

After De Deckker 226

status after recovery from the intense aridity of the glacial maximum.

In the highlands of the southeastern sector of the mainland and in Tasmania from 30,000 years onwards to the time of the glacial maximum, glacial conditions were on the increase. The snowline was depressed by more than 1000 metres at the glacial maximum compared with today. The Margaret Glaciation[183] in Tasmania saw the West Coast mountains glaciated and an ice cap on the Central Plateau. By 20,000 years ago ice cover was continuous from the Plateau to the West Coast Ranges. It formed an ice cap at least 700 metres thick on the Plateau covering an area of about 6000 square kilometres. (This estimation of the area is conservative and omits fringe areas which would bring the figure up to 7000 square kilometres.)

Two caves in the Florentine and Shannon River valleys[265] were occupied by Aboriginals during the Margaret Glaciation. Bluff Cave in the Florentine Valley shows occupation between 31,000 and 11,500 years ago, and cave ORS7 in the Shannon Valley at the edge of the Plateau was occupied between 31,000 and 2500 years ago. Emu eggs and bones of wallabies and other creatures eaten by the Aboriginals occur in the caves with great numbers of stone artefacts. There is no evidence of megabeasts in the cave deposits, either because they were not eaten or because they were already extinct in Tasmania by 31,000 years ago. (They became extinct on the Mainland 15,000 years ago.) The mean temperature during the glacial maximum was at least 6.5°C colder than today.

Mt Kosciusko on the mainland was glaciated, but the amount of territory involved was small — perhaps no more than 25 square kilometres. The periglacial area, however, was quite large affecting all the mountainous country in the southeastern corner with frost and winter snowfalls. The freeze-thaw oscillations led to breakup of rock and increased sediment in upland rivers, which often spilled out as alluvial fans on the plains.

During the 25,000 to 20,000 year interval Antarctic sea ice expanded to its greatest extent. South of 55° South latitude open water probably existed for only three months of the year.

Globally sealevels were more than 100 metres lower than today and very large areas of continental shelf were exposed round continents. As an example of what this meant in terms of changed coastline, about 20 kilometres of coastal plain lay between today's coast and the ocean off Sydney, and the Queensland Plateau where the Great Barrier Reef is today was a coastal lowland. The dry and windy conditions that accompanied the build-up of global ice caused the most intense aridification of the Australian continent that had ever occurred in the 2.4 million years of the Pleistocene Ice Age.

Lakes dried up, many never to fill again.[185] At Lake Eyre and Lake Frome huge gypsum dunes were produced, a phenomenon not seen before or since, implying very unusual conditions. Regionally high groundwaters were required prior to desiccation to mobilise salts and provide a concentration of sulphate in surface waters. On drying the salts crystallised and gypsum formed with halite at the interface between sediment and water (not as it does today, sub-surface). In dry, windy phases the crystals were blown across the lake to form the enormous dunes.

In addition to the increased wind strength which characterised glacial times, southern Australia is believed to have experienced a shift in the seasonal pattern of prevailing westerly winds. The orientation of clay lunettes shows that wind direction was modified due to the summer anticyclone's effect being felt 5°

History of Lake Mungo and the other Willandra Lakes

An example of lake-level changes and lunette-building activities

Between 55,000 and 30,000 years ago the lakes of the Murray Basin[207] were brim-full except for a brief interval at about 36,000 years ago. During the period westerly winds and associated wave action produced sandy white beaches on the eastern sides of lakes. During dry summers foreshore dunes were formed from sand blown from the beaches. On Lake Mungo these dunes form the Mungo Unit. The dunes built up to a height at which they became stable, when they were colonised by shrubs and trees growing on soils that formed on their surface.

When the dry times came at about 36,000 years ago, salinity of the lake water increased and clay zones within the Mungo Unit indicate times when saline clay pellets were blown onto the dunes. Salts were blown in clouds of dust across the landscape east of all the drying lakes causing widespread salination of soils and groundwater.

Lake levels rose again after that drying episode and remained fairly high until about 22,000 years ago when aridity increased dramatically towards the glacial maximum. The Mungo Unit merges progressively into the Zanci Unit of salt-clay-rich sediments during this time. When all the lakes dried completely the volumes of salty dust and sand forming clay-rich lunettes and travelling across the land became enormous because not only were the Willandra Lakes dry, but all the other major lakes and the multitude of minor salt lakes and salt pans were also being deflated. The Zanci Unit in Lake Mungo represents this time of terrible dryness and salt. Mobile white sand caps the sequence, comprising dune sands reworked during more recent times.

The Murray Basin landscape during the glacial maximum would have been a saline wasteland. The salt which was spread and recycled in the Basin remains in groundwater there today.

The development of deep sumps of extremely salt groundwater below the drying salinas was another phenomenon of the times. (The mechanism by which this occurs is obscure.)

The Willandra Lakes did not refill after the glacial maximum except for a brief event in the highest small basins in the lake chain. The Lachlan River changed its course and the Willandra Creek channel on which the lakes lie no longer carries any water except as overflow in floods.

Today the eroding dunes form the Walls of China at Lake Mungo. The lake bed is a saltbush plain where kangaroos and sheep graze. The history of the dry and wet phases of the lake can be read in the layers of clean quartz sand which alternate with bands of clay-rich sediment in the dunes. The pre-lake-full Gol Gol Unit underlies the Mungo and Zanci Units, representing older dune sands from a previous phase.

Aboriginal people lived around the Willandra Lakes, particularly Lake Mungo, camping on the beaches, fishing and eating the freshwater mussels that abounded. Shell middens are testimony to the supply of shellfish in those times. There were fish feasts when hundreds of fish were devoured and their bones piled at a single season's campfire.

The oldest Aboriginal human remains so far found in Australia are preserved in the Mungo Unit of dune sands. They were dated by radiocarbon dating techniques as about 30,000 years old, and artefacts up to 40,000 years old were found in lower layers of the Unit. However it is now considered that these dates are too young — and the source of the information that changes the estimates is somewhat surprising.

Aboriginal hearths and ovens have been exposed by erosion of the dunes.[276] The stones that were used in the fireplaces have undergone heating beyond the temperature at which their crystal structure changed and they were coded with the direction of the magnetic field at the time of heating. (When volcanic rocks containing magnetite cool from molten magma to form rock the crystals of magnetite within them align themselves with the Polar axis. If they are reheated, either in a volcanic event or in this case by humans, the crystals re-align themselves according to the north-south axis. Palaeomagnetism of this sort can be read with the technology that is now available.) It was found that two magnetic field shifts could be identified in the hearth stones.

The Earth acts like a giant magnet and has a magnetic field which changes its strength and direction in what are known as magnetic excursions and reversals. The polarity of the Earth's magnetic field reverses at intervals, north becoming south and south becoming north. Sometimes the polarity starts to change and the strength of the field fluctuates as if a reversal were about to take place and then things return to normal. During the Lake Mungo lake-full stage there were two fluctuations in which the magnetic field was stronger than usual at between 31,000 and 28,000 years ago and much weaker at 26,000 years ago. This 'Lake Mungo Geomagnetic Excursion' lasted about 3000 years and involved a deviation of 120° from the axial dipole field at 30,000 years ago.

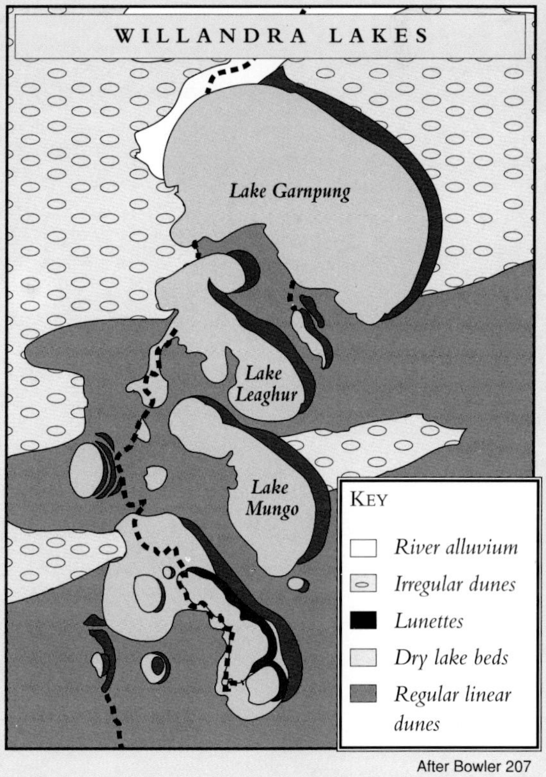

After Bowler 207

Radiocarbon dating involves calculating the rate of decay of carbon isotopes under, among other things, normal geomagnetic strength. The rate of decay is accelerated when more cosmic rays reach the Earth when the magnetic field is weak. On the basis of the abnormality of the magnetic field during the time when the Mungo human remains were undergoing carbon isotope decay it has been reasoned that the radiocarbon dates obtained are too young by between 3000 and 5000 years, making the age of the oldest Mungo skull between 35,000 and 40,000 years.

The China Wall at Lake Mungo is eroding and humps and pinnacles of more resistant composition remain.

M.E.W.

The sands that formed the different layers in the dunes at Lake Mungo are of different colours. The pure white sand of the surface layers overlies pinky-orange sand.

M.E.W.

further north than today. This together with the greater continentality that resulted from the exposure of wide continental shelves would have affected climate. Outbreaks of hot dry air from the interior of the continent were probably more frequent during glacial stage summers than they are now, bringing aridity closer to the humid coastal fringe.

Evidence from Kangaroo Island and from northwestern Tasmania support the extension of aridity further south. On Kangaroo Island the Seton rock shelter provides information on the presence of open grassland with bettongs and hare-wallabies between 20,000 and 10,000 years ago where today there is dense heathland and woodland. In northwestern Tasmania the presence of big grey kangaroos and open dry grasslands tells the same story.

The Nullarbor was drier than it is today during the 20,000 to 15,000 year interval.[247] The last remnants of mallee vegetation had moved onto its fringes by 20,000 years ago, and woodland was never to return. The Koonalda and Madura Cave deposits and those of a cave near Eucla and the Norina rock-shelter confirm the increased aridity of this time span.

In the southern corner of Western Australia in the Cape Leeuwin-Cape Naturaliste region the Mammoth Cave has a faunal record from 37,000 years onwards and the Devils Lair has one covering the 35,000 to 5000 year interval. A very diverse fauna is preserved with arid-land species like the bettong and the burrowing rat-kangaroo.

In northern Australia the southeast trade winds would have intensified from 20,000 years ago onward to the glacial maximum but they would not have changed their direction. It can also be seen from the alignment and structure of dunes in the arid centre of the continent that wind directions there were not significantly different from those of today.

The most visible expression of the ultimate drying that coincided with the maximum extent of global ice in the 18,000 to 15,000 years interval are today's dune deserts.

CHAPTER FIFTEEN

A LAND OF DRIFTING SAND

The high velocity winds related to the intensified pressure systems[211-215] which operated during the glacial maximum reactivated the dune sands which blanketed much of the arid zone as a result of previous glacial stages. Vegetation which had stabilised the dunes during wetter times could not provide enough cover when drought made survival precarious and when, in many areas, fires were more frequent and devastating in the dry and windy times.

Sands in dunes on the down-wind side of lakes and rivers, which had built up during the lake-full phase of only a few thousand years before, were blown across the landscape, forming new dunes and sandsheets. When the salt lake beds were waterless and exposed, their salts and sediments were blown down-wind, first forming clay-rich deposits in which the gypsum and salt were concentrated, and then being dispersed by the wind to form layers in dunes that were forming. Fine dust and salts from the dry saltwater discharge areas, which are such a conspicuous feature of the inward-draining basins that constitute the centre of the continent, spread a mantle across wide areas, making a significant contribution to salinity of groundwater. Where the fine salty dust accumulated in quantity it formed a red silty clay which is known as parna. (Equivalent wind-borne material distributed in Europe, America and China during the glacial maximum formed loess deposits.)

Dunefields in this last glacial stage covered a vast area of the continent from near the Gulf of Carpentaria in the north to northern Tasmania in the south. The dunefields of the Strzelecki, Simpson and Great Sandy Deserts and The Mallee lie within sedimentary basins where the

Mobile red dunes are spreading onto the Strzelecki floodplain, engulfing whole eucalypt trees as they advance. In some cases only the top branches (still living) of trees are visible.

M.E.W.

vast quantities of sand were readily available for their construction. The other dunefields lie in topographically more complicated areas. The Tanami dunes spread over areas of low relief between low ranges, duricrusted tablelands, and adjacent and intervening floodplains. The northern Gibson dunefield is an extension of the Great Sandy dunefield, but in the southern part the landscape is one of low tablelands and ridges with dunes lying in the lower areas between them. The Great Victoria dunefield is similar to

THE ANTICYCLONIC SWIRL OF DUNE RIDGES, AND DISTRIBUTION OF LUNETTES

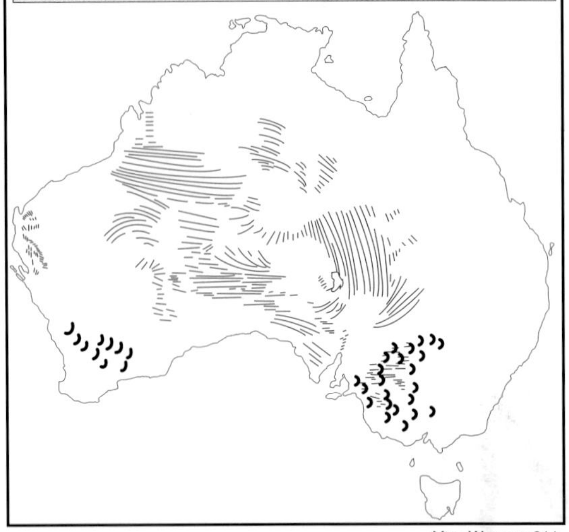

After Wasson 211

AUSTRALIA'S DUNEFIELDS

After Wasson 214

the southern Gibson. The western Gibson dunefield extends onto the Yilgarn Plateau where dunes occupy the valleys between the duricrust-topped breakaways and the outcrops of rocky terrain.

The exposed continental shelves were also sites for dunefields and sands moved from them onto the edge of the continent.

The Bassian Plain connected Tasmania to the mainland and remnants of the dunes that formed on it are today seen on the Bass Strait islands that were the higher parts and all that remained of the Plain when sealevel rose during deglaciation. Flinders Island is one of the mountainous remnants of the Plain in eastern Bass Strait. The lake beds and river margins from which the sediments were blown to form its dunes are now below the waters of the Strait.

Groups of Aboriginals lived on the coastal fringes[266] of the Bassian Plain more than 20,000 years ago[204] and throughout the glacial maximum. As the sealevel rose after 18,000 years ago they moved with the changing shoreline, and its increasing length as the Bass Basin in the centre of the Plain was inundated from the west was to their advantage. They were largely obtaining their food from the sea and the coastal fringe and they had a much increased length of coast to inhabit. With the entry of the sea into the Basin between the mainland and King Island, the King Promontory extended from Tasmania's northwestern tip, while Victoria and the northeastern tip of Tasmania were still connected. When this connection was breached at about 13,000 years ago the Bass Plain Aboriginals became Tasmanian. Sealevel continued to rise creating the separate islands on the Furneaux promontory. A population was isolated on Furneaux Island and archaeological investigations there show that for several thousand years people lived and hunted there. Then they died out and were not replaced. Flinders Island also had a prehistoric Aboriginal population but for how long is not known as chronology was blurred when in the last century the Tasmanian government of the time transported all the remaining Aboriginals on mainland Tasmania to the Island. They rapidly died out or were blackbirded by whalers.

Cave Bay Cave on Hunter Island off Tasmania's northwestern tip has a history of human occupation between 23,000 and 19,000 years ago when the cave was inland and used as a temporary shelter for hunters of brush wallabies, bandicoots and tiger cats. Bones of these creatures form 80 per cent of the total bone count in the cave deposits. The cave was used intermittently between 15,500 and 7000 years ago, during which time it evolved into a coastal encampment. King Island has also yielded Aboriginal artefacts, as yet undated, from its dune sands.

Sand from the continental shelf of the Bassian Plain[239] also blew onto the present-day Victorian coastal plain. It forms prominent parallel limestone (calcarenite) ridges, which trend west-northwest and east-southeast and are separated by interdune flats. The ridges extend throughout the Mt Gambier coastal plain, and are known as the Bridgewater Formation. Within the calcified dune sands on the Bellarine Peninsula structures of organic origin are found. They are the calcified remains of tree trunks or

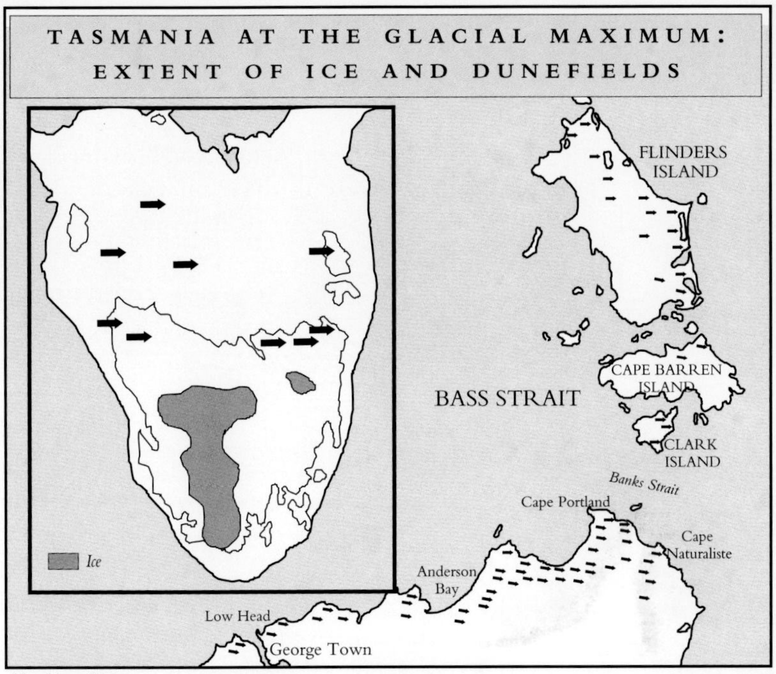

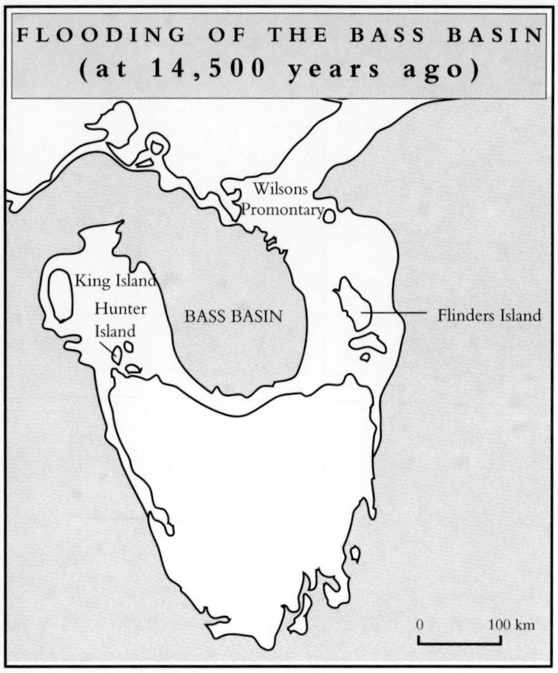

After Hope 204

roots of tea-trees that used to grow in this sandy region more than 20,000 years ago.

It has been found that some of the calcified boles of trees which occur standing erect *in situ* can be used to discover the direction of the prevailing wind at the time when they were alive. This is because it is rare to find a tree whose trunk is completely circular in cross-section when it is growing in an open and windy place. Trees as they grow elongate their trunks in the direction of the prevailing wind, thus the long diameter of their oval cross-section indicates wind direction, and measurements of its azimuth can be plotted on a polar grid. It was found in the case of those near Queenscliff, Point Lonsdale and Barwon Heads that the prevailing palaeowind direction was mainly southerly. The structure of the dunes also indicated that the prevailing wind was from that direction.

Kangaroo Island, which today lies 14.5 kilometres[232] from the coast of mainland South Australia[204], was joined to the mainland during the latter part of the last glacial stage. It was a low range of hills on the exposed continental shelf. An investigation of the shelf around it indicates that the Murray River flowed in a southwesterly direction across the shelf for 70 kilometres beyond its present mouth, passing within 10 kilometres of the eastern end of the present island to enter the sea 20 kilometres further south. This major estuary was a favourite Aboriginal site and Kangaroo Island has an archaeological history with the Kartan stone industry concentrated there. Remnants of linear dunes with an orientation consistent with the major mainland dunefields occur near White Lagoon, showing that during glacial times the region suffered the same aeolian activity. Underneath the dunes are remnants of lunettes which indicate that prior to the glacial maximum the area of the lake was six times greater than its present day maximum extent.

At the height of the dune-building activity, which coincided with the glacial maximum, large dunefields advanced eastwards across the Murray Basin forming the Big Desert and the Little Desert. The dunefields of the Simpson and the Strzelecki Deserts were greatly extended beyond their previous limits. Desert dunes encroached into western New South Wales and northern Victoria.

The orientation of dunes on the continent follows an anticyclonic wind direction pattern. Longitudinal dunes are the most characteristic form, extending parallel to the wind direction. The dune trend is constant for any given part of a desert area. Those in the southern Strzelecki Desert trend northeast. In accordance with the anticyclonic swirl those in the Simpson, Sturt, Stony and Tirari Deserts back to 10° West. Some linear dunes extend for up to 300 kilometres. Their lengthwise extension in a particular direction is due to the strong prevailing wind; their width and height are determined by the action of cross winds. In some areas dunes are closely spaced with up to 15 per kilometre, in others there may be only three or less per kilometre.

Lunette dunes, developed on the

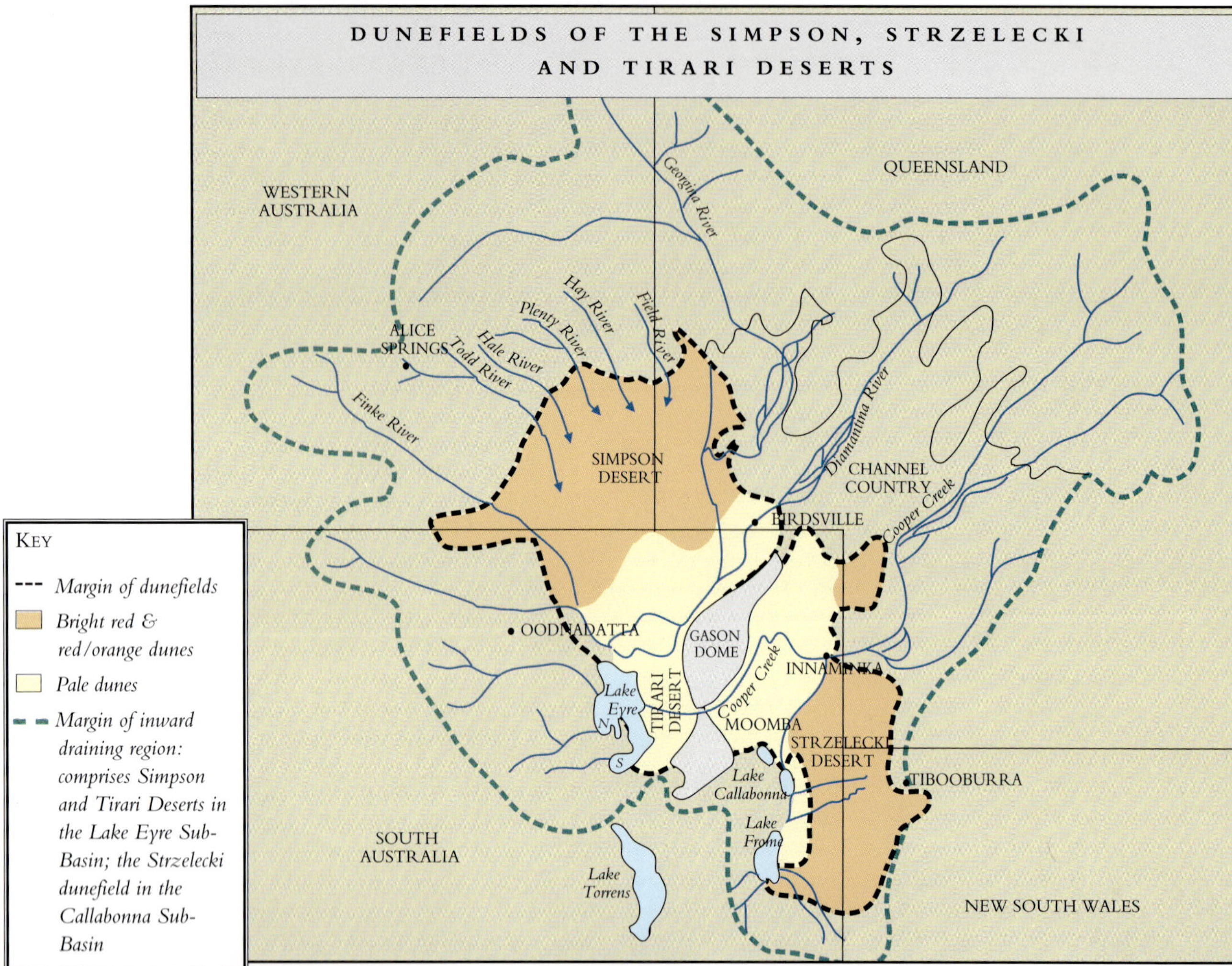

After Wasson 211, 212

Above right: Spinifex growing on the red sand in the Strzelecki Desert forms rings by growing outward from the centre, which dies off as the plant ages.
Above : A wide clay hardpan between widely spaced dunes grows little except sedges and holds pools of water after rain.
M.E.W.

downwind side of lakes, are transverse instead of longitudinal. Like classic barchan dunes they lie at right angles to the wind direction and their ends are curved downwind in two horns as a result of the wind sweeping round, as well as over, them. The lunette dunes of Lake Mungo and the other Willandra Lakes have been studied because of the record of Aboriginal history that is preserved in them, and they also record the stages from lake-full to the present day.

The dunefields of the Simpson and Strzelecki Deserts serve as an example of fields composed of longitudinal dunes. They lie within the well-defined Cainozoic Lake Eyre and Callabonna Sub-basins which connect through a gap in the Gason Range — the Kopperamanna Gap between the Gason Dome and the Corryanna Dome. Both basins contain Eocene Eyre Formation overlain by Miocene limestones, dolomites and clays of the Etadunna Formation (Lake Eyre Sub-basin) and Namba

Top left: Crotalaria eremaea *grows on the red dunes of the Strzelecki Desert.*

Top right:A sea of grass, now bleached white and seeding, grew in great haste on this swale between two widely spaced red dunes in the Strzelecki Desert. The largely ephemeral grasses and forbs that appear on the swales after rain have to complete their life cycles as quickly as possible so as to leave a seed bank ready for the next time it rains.

Above left: A mobile pale dune with dune canegrass growing on it.

Left: A pale orange dune near Tibooburra is being eroded where its plant cover of saltbush and mulga has been destroyed by grazing.

M.E.W.

DUNES OF THE COOPER FLOODFLATS NORTH OF MOOMBA

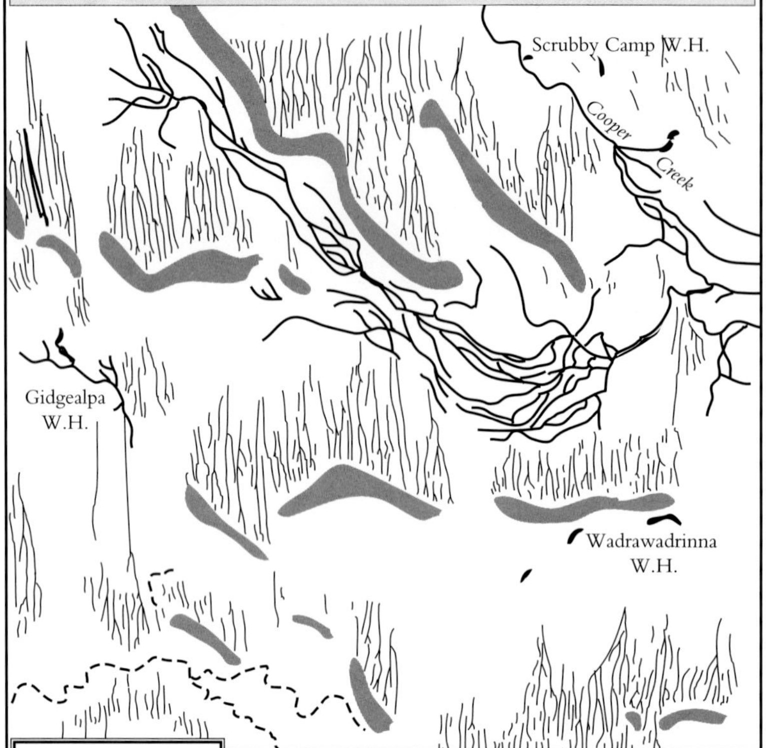

After Wasson 211

Formation (Callabonna Sub-basin). Tectonics in the area during the Tertiary uplifted the edges of the basins and raised the Gason Range between then, while downwarping occurred in the Kallakoopah Pans in the Eyre Sub-basin and around Moomba in the Callabonna Sub-basin. The Ettadunna and Namba formations are overlain by sediments spread by river and wind activity, and the dunefields that form today's spectacular landscapes overlie all this.

The colour of dunes varies from the glowing red-brown that is one's image of the 'Red Centre' through paler orange and cream to glaringly white. The brilliant red grains of the longitudinal (seif-type) dunes owe their colour to the oxidation of iron on the surface of grains, a process which has taken place during their transport from their source. The dunes have sinuous crests and in any area are of constant distance apart. Parallel dunes, however, converge occasionally forming tuning-fork or Y junctions. Why this happens is not altogether clear. Some people suggest that the wind funnels between rows of dunes and sets up the same sort of Venturi effect that causes two boats under sail running before the wind to be drawn together when they come within a certain distance; others blame the convergence on the operation of strong cross winds. The irregular spacing of convergences makes the phenomenon hard to explain.

In all cases the linear dunes merge upwind with sand accumulations on the lee side of playa lakes, flood plains or salt pans. These source-bordering deposits are often dramatic landscape features and may stand like ramparts up to 50 metres high.

Research on the dunes of the Simpson and Strzelecki Deserts has shown that colour of the sands increases with the distance of the dune from the source of the sediments, and that sediments have many sources. The reddest sands have the smallest percentage of clay and are 98 per cent quartz.

In the Strzelecki Desert dunes east of Strzelecki Creek are formed from redistributed Miocene lake sediments, lake-shore dunes and beach deposits. West of the Creek they are formed from muddy alluvium (clay rich) which was being deposited at the same time as the dunes were being built. In the southern Simpson Desert the sediments come from the deflation of pans controlled by groundwater levels and salinity. Clay pellets result from deflation of saline lake-bed sediments. In the northern Simpson Desert the dune sands are alluvial (spread by rivers) and are quartz sands with very little clay. Fine sands tend to occur near the modern channels of rivers and are pale coloured; the coarser, redder sands are further afield. Dunes within a region are composed of sands derived from that region, not transported from distant sources.

The form and nature of dunes varies according to their colour (which depends on their content). The red-brown dunes are almost always the simple longitudinal type. They rest directly on alluvium or bedrock and gibbers. The interdunal swales in the dunefield all lie at about the same elevation, are free of dune sand and are hard-pans or clay-pans.

The pale-brown longitudinal dunes most commonly occur in groups, with each longitudinal feature lying on a swell of sand. Aeolian sand forms the basis of the swales and both dune and swale contain clay pellets and support calcareous soils. Adjacent to Cooper Creek in the Strzelecki dunefield the swells with their longitudinal dunes extend downwind from transverse dunes of the river floodout. The pale dunes in the Simpson dunefield are downwind of pans, salt lakes and some river channels. In both dunefields the pale-brown compound dunes occupy the topographically lowest areas, near modern streams.

NITRE BUSH — *NITRARIA BILLARDIERI*[272-4]

Nitraria (family ZYGOPHYLLACEAE) is a plant with one of the most unusual disjunct distribution patterns. Its only occurrence in the Southern Hemisphere is in Australia. It occurs in Africa in the north in a strip bordering the Mediterranean Sea, and in the northwest in Morocco and the Western Sahara. A narrow strip along the eastern end of the Mediterranean connects the African province to the major regions in which the genus occurs in Eurasia.

Logically such a distribution ought to be from a Gondwanan ancestor, bearing in mind that Arabia and Iran were part of Gondwana. If the genus had its origins in Eurasia it would be easy to explain its presence in Africa, but not in Australia. Long-distance dispersal seems most unlikely. Distances involved are so great that transport of seeds in the gut of migrating birds is not a possibility. The time taken for migrant birds from central Europe and Asia to reach Australia is such that even the most costive bird would have voided the seeds long before it made landfall in Australia. *Nitraria* does have seeds with tough outer coats and the fruits that contain the seeds are small drupes with a fleshy envelope which is edible.

While the distribution pattern of the genus remains a mystery, the most likely explanation seems that it was once widespread but possibly not common over Gondwana, and, if old enough, it may have been cosmopolitan, occupying pockets of saline sand. It may have become a rare plant restricted to widely scattered pockets. Only when the Pleistocene brought the arid steppe to central Eurasia and the salt-sand desert and arid lands to Australia did it have its preferred conditions to spread over continuous large territories.

Australia's nitre bush (also called dillon), *Nitraria billardieri*, is a long-lived perennial shrub growing to a height of 0.5 to 2 metres on clays, loams and sands which are generally saline. It is an opportunist which takes advantage of reduced competition when the chenopod shrubland of saltbush and bluebush is overgrazed. In this capacity it has spread along all the major stock routes through arid southern Australia. In very arid areas it stabilises dune sands and provides periodic feed for domestic stock in droughts.

In Western Australia it is restricted to coastal habitats and occurs on offshore islands. It grows on the coastal fringe of the Nullarbor on the Roe Plains and spreads inland from Spencer Gulf to Lake Eyre. It occurs south of the Tropic of Capricorn and up to the Great Divide in the southeastern half of the continent. Major concentrations are in the Lake Eyre region and on the Riverine Plain in New South Wales where the plant associations are dominated by *Nitraria*. The two regions are strikingly different. In the former the soils are sands, in the latter heavy-textured clays. The common denominator is the salt content, which in both was derived from deflation of salinas in glacial stages of the ice age.

WORLD DISTRIBUTION OF NITRE BUSH

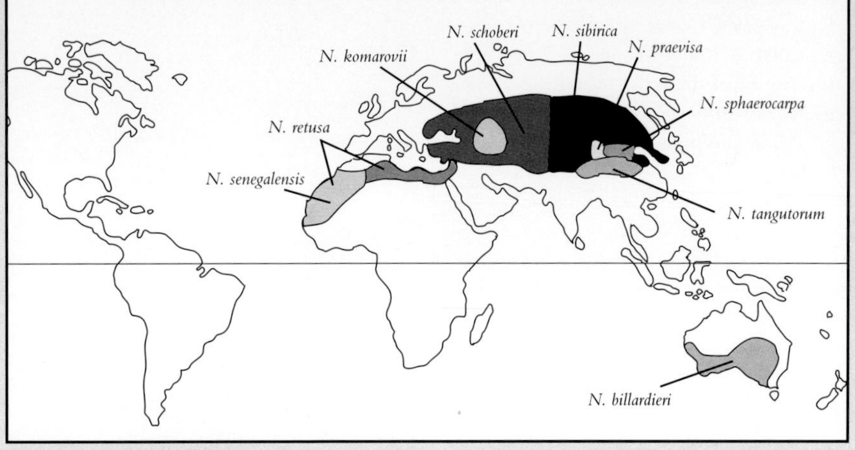

After Noble and Whalley 273

Nitre bush has been an active coloniser of unsuitable country that was cleared for wheat-growing last century. When the wheat failed, because it had been planted in the belief that the few good years of abnormally high rainfall were the norm, the soil blew away and nitre bush has been the only plant able to stabilise the dunes that formed during the droughts.

The old wheat research station at Mannahill between Peterborough, South Australia, and Broken Hill, New South Wales, now is a sea of nitre bush hummocks. Saltbush was originally ploughed under to grow the wheat — and man-made desert results (and spreads) in this land where 'civilisation follows the plough', the maxim in 19th century Europe, should have been 'desert follows the plough'. Too often even today the same disaster is repeated — as in the mallee lands of Western Australia and in all the marginal arid lands which are capable of growing one or two crops beyond the limits of a sustainable wheat belt in good years. When drought inevitably follows the vegetation does not return because the soil has blown away.

Emus play a special role in the distribution of nitre bush on the Riverine Plain. In December to March the fruits are ripe on the plants and emus congregate in large numbers to devour them. The hard seeds are not digested and when they are deposited in emu droppings they are able to germinate ten times more rapidly than seeds which have not been eaten. Also, five times more germinate than is the case with seeds which lie on the clay soil around the bushes. (The nature of the clay determines that seeds are seldom covered over and able to absorb moisture.)

The increased viability of seeds that have passed through emus results from the scarifying of the seed cover which has taken place in the gizzard of the birds, and the fact that digestive juices remove the salt concentrations in the outer layers which inhibit germination. This inhibiting until salt level in the seed coat declines is Nature's way of assuring that the seedling will only germinate when there has been sufficient rain to soak the seed, and that there will be sufficient water available to allow it to become established.

The advantage conferred by the ability to germinate fast is of particular significance to the nitre bush. Its root system in the seedlings and the rate of growth are such that it is not able to compete well with other plants if not given a head start. It spreads at the expense of saltbush when overgrazed land is being re-vegetated because of the help it gets from emus — an all-Australian adaptation to specifically Australian problems of a land of boom or bust! (Other animals eat the seed but passage through mammal guts does not confer the same advantages.)

In the sands in the Lake Eyre region seeds germinate when they are covered by dune sand — and in the intensely arid conditions competition with other plants is limited.

The leaves of nitre bush concentrate salt and they are shed in the summer. In this way the levels of salt within the plant are controlled. It has been suggested that nitre bush would be useful in desalinating soils, but this would require that the fallen leaves were collected and removed before their salt was recycled into groundwater. In any event they are useful as re-colonisers of saline scalds resulting from elevation of watertables in areas where the trees have been removed for cropping. They would also be useful for stabilising mine dumps and marginal dunes along roads which otherwise are enveloped when dunes advance.

Above top: Nitre bush, growing on white dune sand protects the area on which it grows from erosion and a sea of hummocks results. The nitre bush provides shelter for rabbits and reptiles which burrow under it.

M.E.W.

Right: The scene near the Montecollina Bore where nitre bush hummocks create a weird landscape.

M.E.W.

Emus are active spreaders of nitre bush. They eat its small fruits and the hard seeds inside them pass through the birds' digestive systems with an enhanced ability to germinate.

JIM FRAZIER

Dune colour can also be related to the geomorphology of the regions concerned. Pale dunes have received their sediments from the large rivers like the Diamantina and Cooper Creek which drain the Jurassic and Cretaceous rocks of southwestern Queensland. These rocks are commonly fine-grained sandstones and mudstones. The red dunes in the Strzelecki dunefield are in the area which receives river sediments from Cretaceous rocks of the Grey Range in the east, while in the south (east of Lake Frome) channels that used to enter the area drained the crystalline rocks of the southern Barrier Range and Olary Arch and the arenaceous rocks of the northern Barrier Ranges (parts of the ancient craton). In the Simpson Desert the Hale, Hay and Plenty palaeorivers brought sediments from ancient rocks of the Arunta Block in addition to Mesozoic sediments from the sandstones and mudstones of the Eromanga Basin. In other words, the difference in dune colour frequently relates to the different source of the sediments that they contain.

In the Strzelecki dunefield today many areas are capped by very recent white sand which has been reworked from the pale brown dunes beneath it. Near Moomba the Montecollina Bore, which is an artesian bore with high salt content, flows into a pool and overflows in a drain through the landscape of much-eroded white sand hummocks that overlie the dune sands. The edges of the drain are colonised by samphires that grow in the salty rind which covers the mud. Erosion of the white sand sheet has been retarded by nitre bushes that preserve the area of sand on which they grow, while the

An extensive sea of hummocks extends over vast areas of the Strzelecki floodplain. Each hummock is the result of the partial protection from erosion of a small area covered by nitre bushes. These salt-tolerant plants have deep and extensive root systems which stabilise the sand.

The Montecollina Bore flows into a tank and a wide drain, approaching a wide river in size close to the tank, runs off into the nitre bush hummocks of the Strzelecki floodplain. The water is salt and the edges of the drain are samphire-fringed.

M.E.W.

intervening areas are blown away. The result is an endless landscape of white hummocks, capped by nitre bushes.

Satellite imagery of the Simpson and Strzelecki Deserts gives the impression of great uniformity within the dunefields. Detailed analysis of the source of sediments and the way in which the dunes evolved in different areas indicates that the picture is much more complex.

The dunes in the Simpson Desert are still advancing — spreading out over the desert plains. Today the area occupied by the Simpson Desert is 130,000 square kilometres which is half of the total dune-desert area of the continent. Brilliant red dunes can be seen advancing onto the edge of the Strzelecki floodplain engulfing whole *Eucalyptus* trees in the process. Only the crown of some trees remains visible.

Today 40 per cent of the continent is covered by dunes.

This beautiful desert lily, Crinum flavum *lives in areas which become swampy after rain. It has a large underground bulb which enables it to survive during the very arid times between rains. When rain comes, the lily rushes its leaves and flowers up and large seeds are formed in haste. These have a root and shoot emerging while they are still attached to the plant and are able to grow as soon as they fall to the ground, so that when there is enough water about in a good season new plants can become established before dry desert conditions take over again.*

M.E.W.

CHAPTER SIXTEEN

THE WIDEST, WIDE BROWN LAND

(18,000 to 15,000 years ago: The last glacial maximum)

It is lucky for modern Australians that the glacial maximum had passed 15,000 years before Captain Cook came to Botany Bay. It is doubtful whether any European explorer would have thought the continent worth settling then. Sealevel was about 130 metres lower[229] than now and the coastline was very different with wide coastal plains on the continental shelf and none of the deep bays and harbours that shelter ships today.

Australia was then indeed a 'desert' island, two-thirds covered in dunes which extended onto the exposed continental shelves of the western and southern margins. Rainfall in the desert area is believed to have been half or less than half of the present-day amount. An arc of semi-arid savannah lay to the east of the dunefields as far as the Great Escarpment, and to the north — where it spread over the great area of continental shelf that was exposed on the Sahul Shelf and Arafura Plain.

A DESERT ISLAND AT THE GLACIAL MAXIMUM 18,000 YEARS AGO

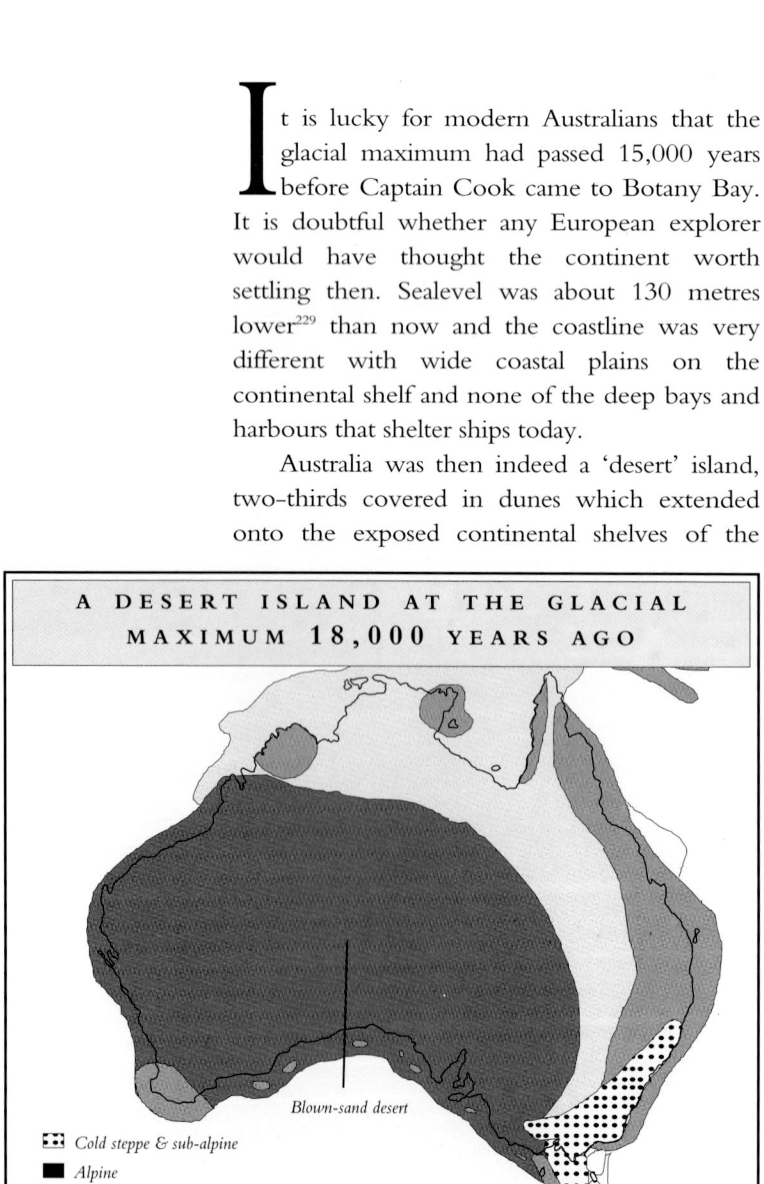

After Hope and Kirkpatrick 205

The semi-arid region in the southeastern part of the continent between the dunefields and the Great Divide was invaded by mallee, *Callitris*, *Acacia* and *Casuarina* woodlands, and relict stands of these vegetation types can be found in the eucalypt woodlands that now occupy the area.

Trees were excluded from 85 per cent of the continent.[205] Woodland (with pockets of forest) occupied the strip along the eastern margin between the Great Divide and the modern coastline, spilling onto the exposed continental shelf. Where the sands of Fraser Island were connected to the mainland at this time rainforest invaded the area and became established, to remain as the only remnant of complex rainforest on pure sand substrate. Woodland also extended down the western margin of the Bassian Plain into Tasmania, and was present in the southern corner of Western Australia.

The high country in the southeastern corner of the continent was cold steppe and alpine grassland, treeless right down to the coast. The lowered snowline had brought snow gums down into valleys at low altitudes. Inland, the river red gum forests had been driven out of the southern floodplains by frost and salination. Even river *Casuarina* had been driven away from upland rivers. The Bassian Plain, apart from its western fringe, had the same cold steppe grassland vegetation, extending down into Tasmania.

Refuge areas scattered over the continent must have maintained pockets of vegetation of types that had colonised the major vegetation zones before the coming of the extreme conditions of the glacial maximum.

Refuges must also have existed for the aquatic biota of the Centre[226], the most hard hit of all ecosystems during the glacial maximum. Obviously the organisms that live in salinas had evolved during times when fluctuating water levels, changes in salinity and desiccation were all part of their environment, and they had developed strategies to cope under extremely adverse conditions. Good dispersal mechanisms for eggs, cysts and seeds by deflation and transport by birds, and ability to tolerate desiccation were inherent characteristics.

Mound-springs are believed to have had a significant role in maintaining aquatic species in areas which were otherwise devoid of water. Their existence and flow were not completely influenced by local conditions as the water that emerged from them was travelling as a result of pressures built up far outside the region and long ago. Mound-springs have a highly specialised endemic fauna of gastropods, crustaceans and isopods which probably evolved in the springs. Other occupants of some mound-springs today can be recognised as 'refugees' which used to live in lakes where conditions were very different. They migrated to the springs when their habitats disappeared and have stayed there as waters with the conditions which they prefer have not returned.

Some water bodies presumably survived in desert areas even in the most arid times, in places where special conditions prevailed. The cosmopolitan aquatic herb *Ruppia maritima* has a patchy distribution in mound-springs, salt lakes and artesian bores today. It is part of a relict aquatic flora that was widespread in the Centre during better-watered times. Fossil seeds of *Ruppia* have been recovered in lake core sediments in Lake Frome though the plant does not occur there when the Lake fills today. Such disjunct distribution patterns occur in terrestrial as well as aquatic biotas due to fragmentation of

EXTENT OF POLAR ICE AT THE LAST GLACIAL MAXIMUM

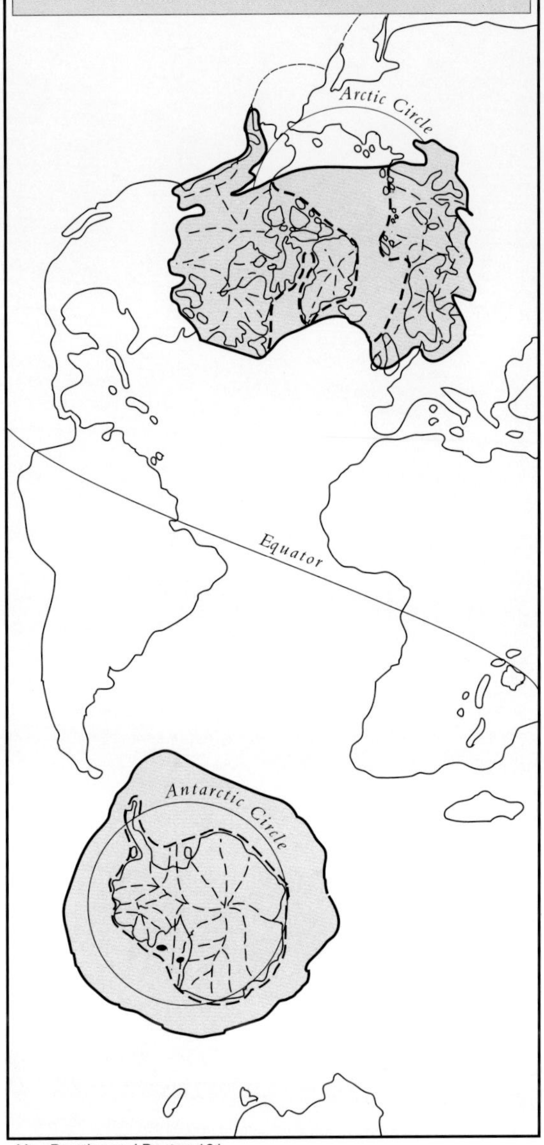

After Prentice and Denton 191

range and survival of isolated ecosystems in refugia when climatic and environmental conditions changed.

The driest-ever ancestral Australia created by the glacial maximum is the immediate progenitor of modern Australia. In geological time the few thousand years that separate us today from the desert land of yesterday is not even a blink of an eye. The evolution of environments up to the glacial maximum had been a progressive drying which involved salination of soils and groundwater over vast areas; creation of enormous areas of wind-blown sands; alteration of river behaviour and drainage patterns; weathering and leaching of ancient soils; modification of the environment by fire, more frequent with the coming of Aboriginals; and changes in the biota which fine-tuned it to the vagaries of an increasingly

Top right: Overgrazing, assisted by termites, has so degraded this land near Camooweel that it is little better than desert.

M.E.W.

REFUGES IN A DRY LAND

Top left: Bratchina Gorge in the Flinders Ranges.

M.E.W.

Left: Cycads in Stanley Chasm.

M.E.W.

Right: Kyabra Waterhole.

M.E.W.

inhospitable land.

What we inherited 200 years ago was a land convalescing from the most horrendous climax of aridity. The record of the last 14,000 years shows us how that recovery was achieved and what it has meant in terms of our modern landscapes and biota. What our management of it since then has done has been to set that recovery back by upsetting the frail balances and recreating some of the pressures that had characterised glacial times.

Fire is a major cause of disturbance and is

on average three and a half times more frequent under present-day regimes than it was when the Aboriginals were firing the land (and their efforts had brought a new dimension to fire frequency as compared with naturally occurring fires). Continual disturbance of forest and woodland by fire discourages long-lived taxa and allows elements of Pleistocene communities such as *Acacia* scrub and tussock grasses to expand at the expense of vegetation dominated by trees. Over-grazed and over-burnt areas are being turned back to the low-diversity open vegetation of glacial times. Now because of the destruction of whole ecosystems with clearing for farming, for cities, towns and roads, and destruction of forests or their degradation, the enclaves of diversity no longer exist in areas to re-supply them should conditions improve. Australia is, because of its history, the continuing natural processes which made it the driest vegetated continent and the activities of humans, set on a path to becoming ever drier and ever more fiery.

Casuarina *on a red dune, with spinifex, mulga and* Cassia *shrubs. The ground below the* Casuarina *is covered with a dense mat of fallen leaves, protecting the soil and returning nutrients to the tree in due course.*

M.E.W.

Native cypress pine in Stanley Chasm.

M.E.W.

CHAPTER SEVENTEEN

DESCENT FROM THE GLACIAL PEAK...

(14,000 years ago to the present day)

Glacial stages develop gradually and have many ups and downs before they peak, but it seems that the warm interglacial that follows sees rapid amelioration of climate.

Deglaciation was under way by 14,000 years ago. Globally sealevel rose rapidly, decreasing the continentality which directly affected climates. Large-scale reduction in the amount of sea ice indicates warming of the oceans, and polar ice volume decreased. A warming of the Southern Ocean by 8°C and the retreat of sea ice[229] to its present position is believed to have occurred within 1000 years[232D]. Increased rainfall resulted from the change in the water budget of the planet.

In the Australian sector the rapidity of climate change was naturally dependent on latitude. Reduction in continentality resulting from the flooding of continental shelves was dependent on shelf width and topography. For example, parts of the northern shelf coastlines have retreated by up to 700 kilometres while most of the northeastern Queensland coast has retreated by between 100 and 200 kilometres.

In the north the broad Sahul and Arafura shelves were flooded as the sea rose from minus 130 metres at 14,000 to minus 35 metres at 10,000 years ago ('minus' refers to the distance below present-day sealevel). It took a rise of 70 metres to create the Arafura Sea between Australia and New Guinea and at minus 53 metres flooding of the Gulf of Carpentaria started. At minus 65 metres the Bassian Plain was submerged except for the islands that span Bass Strait today.

The New Guinea Highlands, which had been extensively glaciated[232A] during the glacial maximum, are a good barometer of the climatic changes during the deglaciation which commenced there at 14,500 years ago. On Mt Wilhelm in the eastern highlands the snowline rose steadily from 14,000 to 9000 years ago, when treeline scrub reached 3600 metres. On Mt Jaya in Irian Jaya there was a minor ice re-advance between 13,300 and 11,800 years ago which has been interpreted as reflecting locally increased rainfall related to the flooding of the Arafura Sea.

Ice retreat in the Snowy Mountains, as evidenced by sediments with a pollen record in a lake at 1800 metres, shows a transition from bare ground (ice covered) prior to 15,000 years ago followed by alpine herbfield, tall herbfield and bog communities, in that order. By 14,000 years ago peat bogs were established. Tasmania's deglaciation was well under way by 13,000 and was complete by 10,000 years ago.

Changes of regional significance in river and groundwater systems occurred during the 15,000 to 10,000 year interval. It was then that the rivers of the Murray Basin, from the Darling with its southeast Queensland sources to the Goulburn which rises in the Victorian Alps, changed from broad, wide-meander, bed-load-dominated systems to the sinuous, narrow, suspended-load channels of today. Riparian forests returned to the rivers and watertables were lowered as more humid conditions returned.

Dunes in the semi-arid regions and those bordering rivers were stabilised. Those of the Strzelecki dunefield were stabilised by about 13,000 years ago and the saline watertable was depressed.

In the Willandra Lakes, only Mulurulu and those in the highest part of the sequence were filled during the initial humid phase 15,000 years ago which also raised lake levels at Lake Frome and at Lake George. When Lake Mulurulu dried subsequently there was no return to clay-dune building which had happened in each previous drying episode. The groundwater conditions had changed from those of arid times when the lake was a groundwater discharge type to conditions in which the regional groundwater levels had fallen and the lake was part of a groundwater recharge system.

Lowering of the watertable had many important effects (and in the modern-day context it allows us to understand the salinity problems Australia faces today). Deep-rooted vegetation could return to areas where groundwater was lower, and the return of trees would enhance or accelerate the lowering.

The 15,000 to 10,000 years interval was a time when hydrological stresses decreased, temperature variations were less limiting to the biota, and wind regimes were becoming less forceful. There is difficulty in correlating events and changes on a continental scale. Age determinations by different methods have different margins of error and many factors can complicate calculations. Added to this, local conditions can often be more important in creating changes than regional climatic variations. The interval was by no means one of continual improvement from drier to wetter.

It is becoming increasingly evident as more is learned about Antarctic ice in particular that the concept of Antarctica varying little in its ice volume and the changes in the northern ice cap being responsible for global changes is no longer unchallenged. Considerable fluctuations in the size of the Antarctic ice cap since the glacial maximum are now considered to have been a factor influencing sealevel and climate marginally right up to now.

Evidence of climatic fluctuations all over the world in the time since the glacial maximum shows that the situation is complex. Cyclic changes which operate on a very short time scale have recently been recognised.

The Younger Dryas was a cold spell of about 1000 years[230] between 11,375 and 10,470 years ago. It was recognised as a time of ice re-advance in Europe which affected the movements of Neolithic hunters. It was considered an 'odd' event in the general post-glacial warming. However it can now be recognised in Australia by a dust peak in sediments in Lake Carpentaria.

The record in the Lake Carpentaria sediment core for the 35,000 to 10,000 years interval shows that cycles of approximately 2250 years are composed of drier periods of about 600 years when blown dust was evident (and during which dunes were re-activated), and better-watered times for 1500 years when dust was absent. Seven such cycles are recorded between 25,000 and 11,000 years ago. Droughts occurred during the 600 year part of the cycle when increased strength of the trade winds and failure of the monsoons over the Gulf resulted from intensification of the Southern Oscillation over the Pacific. The Younger Dryas was one of these events, more extreme than others perhaps, or it may be that as it was one that left a good record it was more visible.

Control of the 2250 year cycle by pulses of solar activity has been suggested. That the cycle continues to the present is suggested by the cyclicity of the same order seen, for example, in glacier advance and retreat in the Northern Hemisphere; in fluctuations in the Great Salt Lake in Utah; meltwater pulses recognised in coral formation in Barbados; and lake level fluctuations in Ethiopa — indicating the global nature of the phenomena.

If the Antarctic ice cap has been showing cyclic fluctuations[229], undetected until now, the delicate balance of global systems is underlined. The fact that small variations in temperature can trigger off responses in the ice caps and start a meltdown and a rise in sealevel is one of the alarming ideas that arises from the discovery of any inherent instability of the Antarctic ice cap. We cannot be complacent about a man-made greenhouse effect that will change temperatures by 2°C and, as we see it,

Paperbark forests fringe the estuary at Bremer Bay in Western Australia, as they do in similar estuarine and river bank situations all round Australia's coastline.

M.E.W.

Yellow-flowered Cassia *shrubs are widespread in Australia, even thriving in arid areas.*

JIM FRAZIER

raise sealevel by a small amount. Melting of Antarctic ice shelves which act as a girdle round the ice cap, keeping it based on the Antarctic continent, would allow slumping and a sealevel rise of undreamed-of proportions in a very small space of time. If a glacial maximum can turn into a very much warmer interglacial in a thousand years, how long would it take for a global disaster to occur because a critical point had arrived and Antarctica rose and shed a little of its ice burden?

Rain gauge-type information on the amount of rainfall[228] at different stages in the last 10,000 years comes from several lakes[256] in the western Victorian Newer Volcanic Province. Here basalts have been produced in a volcanic event which was taking place between Late Pliocene and Recent times. The craters of vents in the basalt became lakes (known as maar lakes). The area is 170 kilometres west-southwest of Melbourne, 30 kilometres from the sea, close to the boundary of the semi-arid zone today.

The Lakes Keilambete, Gnotuk and Bullenmerri have recorded previous high and low water stands during the last 9000 years. Before 8300 years ago levels were at their lowest and salinity was highest; between 8000 and 7000 years ago levels were high in at least two of the lakes; between 6500 and 3800 levels were high, with the lake-full stage between 6400 and 5700 years ago in all the lakes; between 3000 and 2000 years ago lake levels were low; from 1800 to 1300 years ago levels in two of the lakes were high; between 750 and 300 years ago levels fluctuated and salinity increased. Levels have fallen significantly in the last 100 years and salinity has increased drastically.

During the dry stage indicated by lake levels between 3000 and 2000 years ago dune building was occurring at Cobar[245] in western New South Wales. (Dune building was at a minimum during the 8000 to 4500 years interval all over the continent, and re-activation then commenced, to peak during the 3000 to 2000 years interval.)

Between 11,300 and 8700 years ago, Kow Swamp[232A] in northern Victoria experienced a lake-full stage and Aboriginals inhabited its shores and buried their dead in its dunes. The lake-full timing here is an example of an out-of-phase event when compared with the maar lakes of southern Victoria and with the Younger Dryas arid event that was recorded in northern Australia — an event synchronised with Northern Hemisphere climatic changes.

The Nullarbor was much drier than it is today during the glacial stage[247] and up to about 8000 years ago according to information from cave sites in which pollen records are preserved. Rainfall of about 180 millimetres is suggested during the 10,000 to 8000 interval, and after that it increased to stabilise at about 4000 years ago at its present 250 millimetre per year level.

The lack of humidity in the Nullarbor has led to some unique information on animals that lived there thousands of years ago. Two cave deposits, Thylacine Hole dated at 4600

Tall mangroves growing in Temple Bay, Cape York, reach 15 to 20 metres in height. Such mangrove forests existed in the Fitzroy River area of Western Australia during a wetter climatic interval between 6000 and 8000 years ago, and used to be far more widely distributed during Early Tertiary times which were more uniformly warm and wet.

Jim Frazier

years ago and Murra-el-elevyn dated at 3300 years, contain mummified remains of animals. Soft tissue on skeletons of two thylacines, *Thylacinus cynocephalus* the Tasmanian tiger, a dingo, a red kangaroo, a possum, and a nailtail wallaby are evidence of desiccating conditions. Gut contents in the mummified remains indicate that the plants which the animals ate were the same as those growing in the area today.

The Carpentaria Basin, in which Lake Carpentaria was situated[236], was gradually invaded by the rising seas between 12,000 and 11,000 years ago. Full marine conditions were only established by about 8000 years ago with the opening of the Torres Strait. Groote Eylandt is a large island in the western bight of the Gulf of Carpentaria and one of the rare records of climate and vegetation change in northern Australia during Holocene times is preserved in dune and lake sediments on the island. The island is the easternmost extension of the dissected sandstone plateaux of Arnhem Land that became an island when the sea invaded the basin. Four Mile Billabong is a perennial lake in the dunefield in the northeast of the island.

The lake originated as a seasonal swamp prior to 10,000 years ago. A progressive rise in watertables established it as a permanent water body by 9000 years ago. When it was a swamp, paperbarks (*Melaleuca*) surrounded it and water lilies and water milfoil (*Myriophyllum*, HALORAGACEAE) grew in it.

Yellow daisies with succulent leaves and stems cover the dunes in the desert near Stuart Creek after rain. The daisy flowers all face the same way at any given time, turning to face the sun in the morning and following its passage across the sky.

N.F. Alley

The surrounding dunefield was open grassland. Between 9200 and 7500 years ago paperbarks still fringed the lake and palms grew in the swampy ground. Between 7500 and 5000 years ago *Eucalyptus tetrodonta* open forest had replaced the grasslands and *Acacia* scrub had appeared. (Today this vegetation type covers 70 per cent of the area.) Between 5000 and about 3800 years ago lake levels were high, and the surrounding area was

VEGETATION ZONATION AT THE 9000 YEARS AGO 'GREENHOUSE'

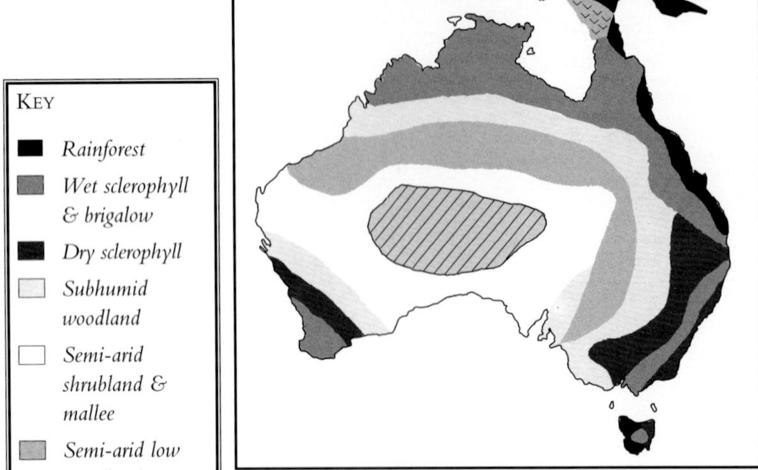

After Ross 232g

flooded. From then until 1000 years ago the lake level fluctuated, and at times it dried up. Fire frequency increased during this phase. In the last 1000 years fire frequency increased further, the open forest of *Eucalyptus tetrodonta* became more open with much more grass as understorey — a result of frequent 'cool' fires, which decrease diversity and progressively reduce the number of trees.

The desert centre of the continent provides little in the way of fossil records of vegetation to show the improvement in water balance during deglaciation. A core from Lake Frome indicates that from 10,000 years ago onwards communities of chenopods, hummock grasses, *Casuarina*, *Callitris*, *Acacia*, *Dodonea* (hop bushes), *Eucalyptus* and *Melaleuca* have been fluctuating in the area. The earlier phases show wetter conditions than at present with more trees. Later phases show a marked increase of ephemerals and an increase in fire frequency. The fire frequency is directly linked to the decline in trees, shrubs, grasses and sedges and their replacement by ephemerals.

In Western Australia, in the southwestern sector[247], the dry conditions of the glacial stage gave way to wetter times at about 10,000 years ago and by 8000 years ago dense karri forest (*Eucalyptus diversicolor*) had spread over large areas. The ratio of karri to *E. marginata* (jarrah) and *E. calophylla* (marri) varies according to rainfall, and on this basis pollen records in the area show that the region was wetter than it is today between 6000 and 5000 years ago, a drier phase followed until 2500 years ago, then it was wetter again until about 1300 years ago when a drying trend started. This drying peaked at about 500 years ago and since then the trend has been to higher rainfall.

In the northern part of Western Australia the uniformly wetter conditions shown by the extent of karri forest in the south in the 8000 to 6000 year interval is matched by the presence of tall mangrove forests in the Fitzroy River estuary. This implies a longer and heavier wet season then than at present. Melville Island in the extreme north has suffered a gradual decline in the amount of monsoon forest in the last 6000 years which may relate to the drier times since 6000 years ago when tall mangroves left the Fitzroy River area. However, Aboriginal fire on Melville Island, which was constantly inhabited, may be the culprit there.

Pollen floras are best known from southeastern Australia and they show that the open vegetation of glacial times was gradually replaced by woodland and forest, and communities were approaching those of today. As conditions changed the plants, and in some cases whole ecosystems, that had been confined to refuge areas spread back into areas that they preferred. In many cases the original range of many communities that had contracted during glacial times had been fragmented by dune invasion or alteration of physical conditions in some other way. Thus many disjunct distributions resulted.

Throughout the Pleistocene when climate was changing and fluctuating many communities had become isolated like little islands within their original range. In fact, many such islands may have been present in their specialised niches since Tertiary times. (The relict Gondwanan rainforests in their refuges on the eastern margin of the continent are a special case of this sort with a history that goes back to the start of Tertiary times, but the position of the core of these forests and their composition had remained much the same during glacial stages.)

Palm Valley and a number of similar oases in the Centre in which relict flora with *Livistona mariae* has survived are examples of islands of vegetation from better-watered times in a sea of aridity. Remnant communities of *Livistona alfredii*[254] occur in valleys in the Cape Range in Western Australia in a spinifex-sand dune region. In the Pilbara more isolated occurrences of the palms have been recorded, and recently a new species of *Livistona* has been discovered in the Bungle Bungles.

Some strange faunal examples exist where rainforest inhabitants[255] have been left behind in an area when the forest disappeared and have been forced to make drastic changes in their lifestyles to survive. In the Cape Range, on the mid Western Australian coast, limestone caves harbour an invertebrate fauna of spiders, pseudoscorpions, micro-whip scorpions, millipedes, cockroaches, slaters, beetles and crickets. These creatures have all become soft-bodied, pale, and eyeless, with enlarged non-optic sense organs. They are a relict tropical rainforest leaf litter community that has taken to the caves where humidity is constant as it was in the ancient forests. They cannot leave the deep caves to which they are now completed adapted. Their closest relatives occur in northeast Queensland. They must have been in the area since the Miocene, when last remnants of suitable complex forest were in the area.

There is no plant life in the caves because it is dark, but the creatures feed on dead plant material that is washed into the caves following heavy rain. Such inflows may occur every few months or may not happen for years at a time, depending on the characteristics of the cave. The only caves occupied in the big complex that riddles the limestone are those where humidity is over 90 per cent. (Caves on the Nullarbor also harbour invertebrates that live elsewhere in well-watered places.)

The coastal plain bordering the Cape Range is an interesting place for those who delight in weird and rare creatures. A subterranean aquatic fauna is present. It contains the entire vertebrate troglobite (obligate cave dweller) fauna of Australasia — two species of blind fish, a gudgeon and an eel — as well as some crustaceans.

Evidence from Australasia suggests that the interval between 9000 and 7000 years ago[232H] was one of the optimum times for rainfall and temperature — a greenhouse phase. It has therefore been used by some scientists to construct a model for conditions which are likely to prevail by 2030 AD if present theories on the effect of increasing carbon dioxide levels are correct. A doubling of atmospheric carbon dioxide levels would raise mean annual temperature in northern Australia by about 2°C and that of southern Australia and New Zealand by 4°C. Increased rainfall would occur between 15° and 37° latitudes and there would be a smaller decrease Polewards. Tropical cyclones might be expected to form further south than they do today, while frontal systems embedded in the mid-latitude westerlies might not extend as far north. A more positive Southern Oscillation would characterise the greenhouse climate. Reduced north to south temperature gradient implies weaker westerlies in middle latitudes and the latitude of the subtropical high pressure belt would be further Polewards. In Australia this would mean higher rainfall and lower temperatures on the east coast and lower rainfall and higher temperatures on the southern coast including all of Tasmania, and probably a water crisis in Perth.

Although the theoretical modelling of a greenhouse at 2030 AD[257] matches well with the conditions that are known to have applied during the climatic optimum of 8000 years ago, more recent research suggests that a better analogue might be the 5000 years ago optimum period. The 8000 years ago optimum was not recorded all over the Australian continent and sealevels were then about 20 metres lower than today. The 5000 years ago optimum showed highest temperatures uniformly over all sites and the sealevel was the same as it is now.

It is believed that the carbon dioxide level during the glacial maximum at 15,000 years ago was 200 parts per million, that 100 years ago the figure was 300, and that today it is 350 parts per million and ever-rising. While we might appreciate a warmer Tasmania if its drier state did not render its hydro-electric schemes less viable, and more rain in parts of the mainland might be nice — if our degraded and salinated soils can get any benefit from adding more water — we are not going to appreciate the rise in sealevel that will accompany this quite modest temperature rise. Add to it the awful prospect of collapse of ice shelves round Antarctica and much of our present civilisation is at risk.

Scientists drilling cores from the King George VI Ice Shelf have found evidence that it is rotten at depth — the slightly warmer water that rises under it from the abyss is having its effect. In addition, measurements show that in the last 40 years it has retreated by 40 kilometres!

The fine details of climate control are so

LAKE LEVELS, SEALEVEL, TEMPERATURE AND DUNE-BUILDING IN THE LAST 40,000 YEARS

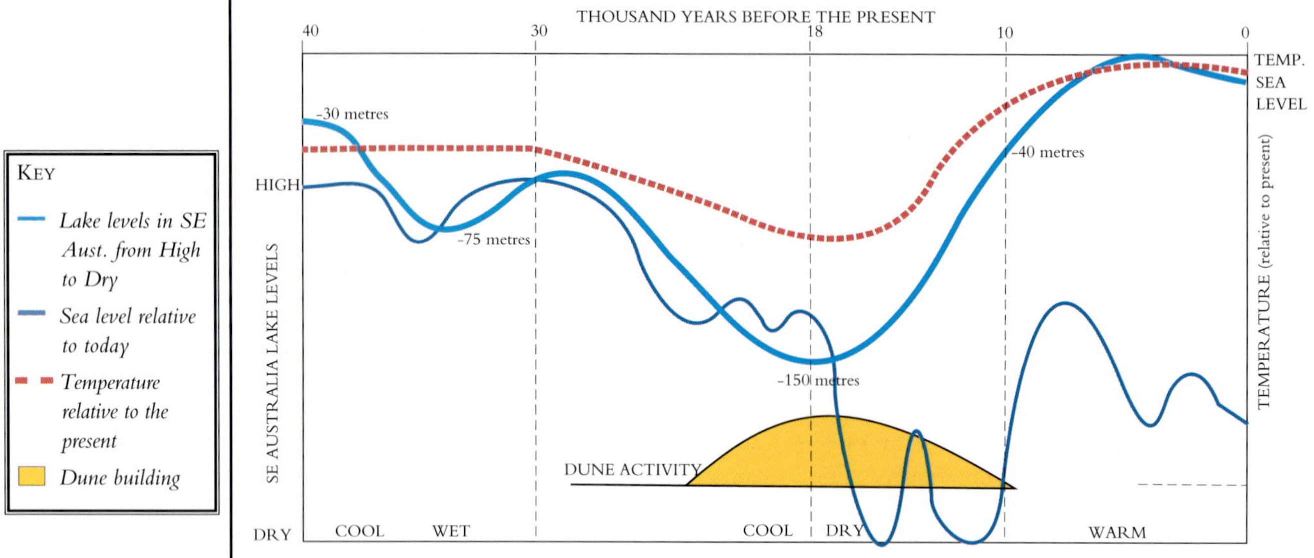

After Chappell 2321

A blind hunting spider (family Ctenidae) from the Cape range, Western Australia. This spider is closely related to African species which live in caves, in forests, and in drier regions. It is also related to species which live in forests in southwestern Australia. When the North West Cape was forested in prehistoric times it was a member of the Gondwanan forest community. It remained in the area after the forest had disappeared with the drying of the continent, moving into caves where humidity was constant and high and becoming modified for living in darkness.

PHOTO: MIKE GRAY.

little understood that it is impossible to say whether normal, possibly cyclic, events are involved or whether human activity might be disturbing the balance. Does anyone know if depletion of the ozone layer — a phenomenon that is definitely aggravated by Technological Man — and enlargement of the hole over Antarctica could hasten ice melt?

Sealevel changes over the last 40,000 years[2321] have been estimated for Australia. There is much variation in data but five time-slice levels have been established and they show agreement with temperature, lake depth and dune-building curves (which all relate to climate and thus to size of ice caps which principally control sealevel).

At 40,000 years ago sealevel was 30-40 metres below today's level and descending. A slight rise between 35,000 and 30,000 years ago saw levels at between 40 and 45 metres lower than today 30,000 years ago. Then sealevel plunged to reach its lowest level of 150 metres below the present value at the glacial maximum of 18,000 years ago. A rapid rise accompanied deglaciation. By 9000 years ago sealevel was at 20 metres below and rising steadily to equal the present day level at 6500 years ago. It continued to rise, peaking at between 5500 and 5000 years ago at between 1 and 2 metres above the present level, in what was a climatic optimum. Then it fell smoothly to its present level. The extra couple

THE HENBURY METEORITE CRATERS

About 5000 years ago a shower of meteorites collided with Australia at a site about 145 kilometres southwest of Alice Springs. Thirteen craters of different sizes resulted. The Main Crater is oval and about 200 metres long and 15 metres deep. It is in fact two craters which touch each other and the rim is about six metres high. The smaller craters are round, with diameters between 30 and 80 metres, depths between one and six metres, and rims about two metres high if they have not been eroded away. The Water Crater has its rim breached and a drainage line in the landscape channels water into it.

Fragments of iron and nickel have been collected in the area and it is possible to see a pattern of rays and loops of ejected rock around craters, like one sees on the moon. An interest was taken in the craters by American scientists planning moon landings because such recent and comparatively unweathered craters would approximate to moon craters which suffer no weathering.

The deeper craters provide microclimates in the dry and rocky terrain and whitewood (*Atalaya*, SAPINDACEAE) makes leafy groves in them, contrasting with the *Acacia* scrubland of the surrounding hills.

Aboriginals saw the impact, or so one assumes from their legends. The Henbury crater rocks are referred to as 'sun walk, fire devil, rocks' of their Dreamtime.

The landscape at the Henbury Crater site is gibber plain with scattered mulga and a backdrop of ancient hills.

M.E.W.

*The main impact crater provides a micro-environment where yellow-green whitewood trees (*Atalaya *of the family* SAPINDACEAE*) and greybush chenopods and other plants find conditions less harsh than those outside the crater walls.*

M.E.W.

of metres, which left traces in the form of dunes inland of modern estuaries and inshore of the present coastal fringe is a good indication of what greenhouse will do by 2030 AD if not amplified by Antarctic deglaciation.

Torres Strait, whose sill depth is only 9 to 18 metres, was only breached by between 8500 and 8000 years ago — the last major event in establishing the margins of mainland Australia. With Bass Strait flooding for the last time 13,000 years ago when sealevel was 65 metres lower than at present Tasmania had become the Island State.

When the continental shelf of the northeast Queensland coast was exposed during glacial times of low sea stands during the Pleistocene the coral reefs that had grown there sporadically since Miocene times died. As soon as rising seas reached a satisfactory depth the stumps of previous reefs that had grown in the region during interglacials were re-colonised. The Great Barrier Reef of today's Queensland coast began its present phase of development not more than 8000 years ago, restocked with a biota from coral reefs further afield in the Pacific Ocean. The adjacent Queensland Plateau once supported an enormous area of coral reefs when water depth over it was shallow, and dead coral there stands as a stark reminder of what happens when sealevel changes. The underwater topography off the north Queensland coast from the Capricorn Basin to the Coral Sea Basin is very complex. This is because it comprises rift valley systems that aborted during the passive margin rifting in the area associated with the opening of the Tasman and Coral Seas. Troughs and plateaux alternate and contort the ocean floor and subsidence of large areas has resulted.

Water depth is crucial for the living part of reefs and the rate of rise of sealevel must not outstrip the rate of growth of the reef. The Timor Sea is another graveyard of drowned reefs, and the effect of rapidly rising seas in a greenhouse world would be to put the Great Barrier Reef at risk.

It is only in the last 50 years or so that any attempt has been made to monitor the small changes that are occurring in modern sealevel on a global basis. Because of this it cannot be said with certainty what the overall trend is over a long time span and what are the 'normal' fluctuations that occur on a shorter scale. There has been, according to most observers, a rise in sealevel of 10 centimetres or more in the last 50 years worldwide.

Australian evidence is meagre but a 66 millimetre rise in 22 years between 1966 and 1988 has been recorded at Cairns by researchers at the Flinders University. The rise there is blamed for the death of paperbark forests between Cairns and Cardwell. Forests adjacent to every major estuary are dying because of saltwater intrusion and are being replaced by mangroves. Sealevel all round Australia has risen on average by 1.22 millimetres a year in the same period. Levels are still rising.

Research into recent climatic change takes many forms and involves many disciplines. One new approach is the study of tree rings which show the rate of growth and hence indicate climate. In this field some interesting results have been obtained in Australia.

A group of Huon pines was discovered growing at over 1000 metres altitude[268] on Mt Read in northwestern Tasmania.[270] The usual altitude at which these trees grow today is between sealevel and 300 metres. The Mt Read trees are up to 700 years old, while King Billy pines, pencil pines and celery-top pines in the same area are 1000 to 2500 years old. The trees are growing in a valley sheltered from the cold westerly winds of winter and from the hot northerly winds of summer. They only grow actively in summer. Because they are a relict group and are at the limit of their range they are very sensitive to temperature change, making them ideal for tree ring determination of past temperature.

The information obtained from studies of these unique trees indicates that temperature has risen more rapidly in the last 25 years than at any time since 900 AD — warning us of the accelerating greenhouse advance. The mean annual temperature has risen from just below 15°C to 16°C. The record also shows a cold snap in the early 1900s when pack ice in Antarctica is known to have expanded; the 'mediaeval warm event' of the 12th century AD; and the 'Little Ice Age' of the 17th century — which should convince us of the accuracy of the record.

In the estuary of the Stanley River in northwestern Tasmania[269] sub-fossil logs have been found in the river gravels and sands. Fifty-eight logs of Huon pine and celery-top pine have so far been discovered. Those exposed on the surface are dated at 7000 years old, those deeper in the deposit are between 9000 and 10,000 years old and 12,000 to 13,000 years old. When the investigation of all this material is complete a great deal should be known about climatic fluctuations since deglaciation in Tasmania.

With the stabilising of sealevel about 5000 years ago Australia had assumed its modern form. Its sandy beaches, sand bars, tombolos and drowned valley systems like Sydney Harbour were as we know them. Its major regions of the continent with their specialised biotas were established, though fire frequency was altering the composition of vegetation and selecting species for their fire-tolerance.

The great variability of climate that had characterised all of the Pleistocene and the first half of the Holocene continued. A dry period with re-activation of dunes and hydrological stress occurred during the 3000 to 2000 years ago interval. It and the 'Little Ice Age' of historical time in the Northern Hemisphere between 1650 and 1850 AD represent the cold/dry parts of the 2250 year cycles which are controlled by solar activity. (The cold spells last 600 years and are times when solar flare activity is absent or reduced.)

Throughout the Holocene, and probably through at least the last half of the Pleistocene, Australia has been a land of 'drought and flooding rains' because the immediate control of variation in climate on a year to year time scale has been by the Southern Oscillation. To understand many of the environmental problems that face Australia today it is necessary to understand the working of the Southern Oscillation and its two extremes — El Nino of the droughts at one end, and La Nina of the flooding rains at the other.

CHAPTER EIGHTEEN

Drought and Flooding Rain

ENSO calls the tune[258-263]

The year to year variations in climate experienced by the eastern half of the Australian continent and the northern monsoonal region are not random. They are orchestrated by the Southern Oscillation, a quasi-cyclic phenomenon operating on a two to three year time scale but varying in the intensity of its cycles.

The Southern Oscillation involves a global pattern of oceanic and atmospheric temperature and pressure fluctuations with much intensified events occurring approximately every ten years. Extreme anomalies in the pattern result in dislocation of rainfall patterns in the tropics, bringing drought to some regions and torrential rains to others. These anomalies typically last about a year. Related anomalies of atmospheric circulation extend high into the atmosphere and spread Polewards into the temperate zones, especially in the Southern Hemisphere. Some major changes in sea temperature and in ocean currents are also related to the Southern Oscillation. The best known of these is El Nino, which is characterised by a marked warming of a body of water in the eastern equatorial Pacific Ocean. El Nino events are linked to extreme anomalies of the Southern Oscillation that result in drought and the two are jointly referred to as ENSO (an acronym for El Nino-Southern Oscillation).

During the opposite extreme of the Southern Oscillation the east equatorial Pacific is cold and a 'La Nina' phenomenon is recognised in which heavy rainfall and flooding is observed over areas that were affected by drought in the El Nino event. ENSO results from interactions between the tropical oceans, particularly the Pacific, and the atmosphere. It amplifies the climatic variability in the areas that it affects relative to the variability elsewhere even after the effects of latitude and mean rainfall are taken into consideration. A characteristic of phenomena during intense cycles is the rapid switching from drought to flood within the quasi-biennial cycles.

ENSO is phase-locked into the annual cycle in Australia, in which seasonal aridity and frequent droughts are already a pattern without the super-position of ENSO. Each El Nino or La Nina event operates for about a calendar year. Thus if an extensive drought or wet cycle is established by the middle of a year it is unlikely to 'break' until the following year.

Some examples of the ENSO cycle at Bourke in western New South Wales serve to illustrate the quasi-biennial cycling and the extremes that are juxtaposed at times when the amplitude of the phenomenon is great. During the La Nina year of 1950 Bourke had 841 millimetres of rain (instead of its average 347) and next year it received only 217 millimetres. In 1982, which was an El Nino year, rainfall was 162 millimetres but in the following year 552 millimetres of rain fell. In 1976, 418 millimetres resulted in floods in February and only 98 millimetres of rain were recorded in the next six months — half the average for the season.

In order to maximise the relationship between the SOI and drought, the mean of the values for the SOI between April in one year and March in the next is used. As a result only severe and long-lasting droughts are seen as extra-high peaks. Epidemics of Murray valley encephalitis correspond to La Nina years when there was sufficient water for mosquitoes to breed in numbers sufficient to spread the disease.

THE SOUTHERN OSCILLATION BETWEEN 1940 AND 1992 FOR AUSTRALIA

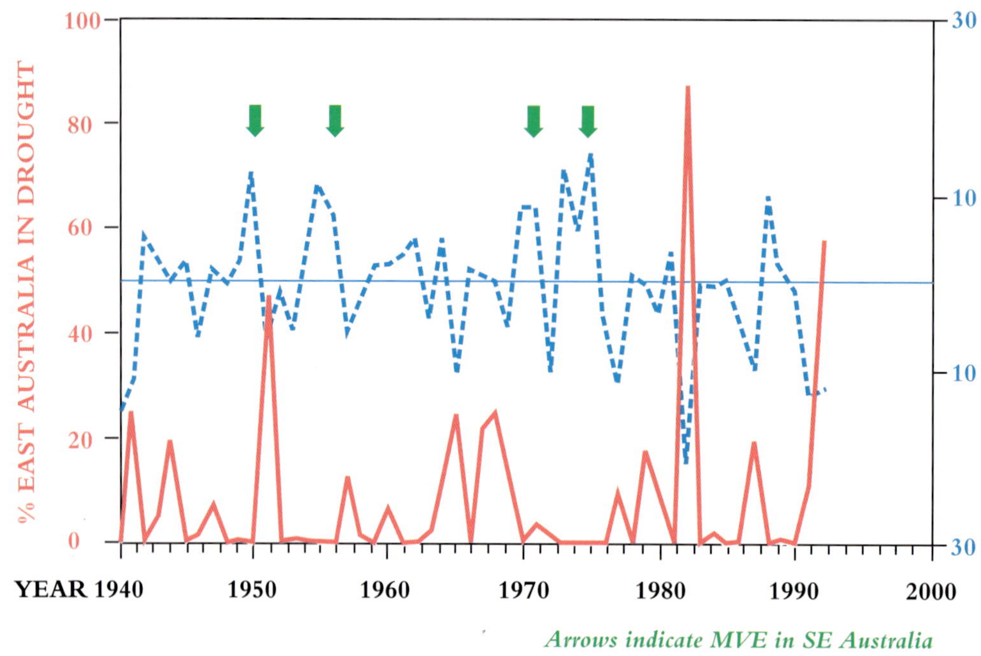

Graph kindly prepared by Neville Nicholls

Above: Rain turns a gibber desert into a watery landscape.

Jim Frazier

Above right: The chick of a letterwing kite, in its rat-fur-lined nest, is a product of a La Nina season of plenty when rat plagues enabled the kites to breed successfully.

Jim Frazier

ENSO's effects are not felt to any extent in the western half of the continent where the pattern of sea-surface temperatures in the Indian Ocean influences variations in climate. These patterns are only weakly related to ENSO.

In the eastern half of Australia the lack of mountain ranges leads to uniform responses to ENSO over very large areas. ENSO affected half the Australian continent severely during the 1982 to 1983 drought. The ranges of the Great Escarpment are the least-affected region.

In El Nino years frosts are more widespread and maximum temperatures are higher. Winds north of 25°S latitude are three to four times stronger than during La Nina years and these factors added to the decrease in rainfall increase the stresses for the biota.

The rapid change from drought to flood and back again to drier conditions increases the likelihood of fire. During wet years undergrowth flourishes. It desiccates in the drier years and is readily ignited. Degradation of the vegetation with each fire makes it more vulnerable to the next dry episode and the next fire. The worst fire years experienced in early settlement days[295] correspond to end of La Nina events when there had been a build-up of fuel in wetter times and the following drought saw conditions ripe for burning. A

Mud cracks and white bleached bones epitomise this land of boom and bust as orchestrated by ENSO. M.E.W.

Ephemerals like these button daisies depend on the La Nina seasons of good rains to produce a large enough seed bank to ensure survival through the many years between when they may not achieve satisfactory seeding. Jim Frazier.

Paddy melons rush to germinate, flower and fruit during the brief intervals when rain has fallen in the desert. When the drought returns and the mud dries into a mosaic of cracks, they, and the other opportunists which have taken advantage of the good times, die. The paddy melon is an introduced plant, brought by Afghan camel drivers to feed their animals, and many a native animal has been grateful for this food source since. M.E.W.

Above: When a claypan between desert dunes in the arid red centre of Australia becomes a temporary lake after flooding rains, pelicans and all sorts of other water birds appear as it seems from nowhere and there must be food for them in these ephemeral waters.

Jim Frazier

Left: A letterwing kite, an opportunist which takes advantage of boom years to breed abundantly.

Jim Frazier

Below left: Kangaroos and mulga — two species whose reproductive patterns are influenced by ENSO.

Jim Frazier

few examples remind us of the horror of fire in this ENSO orchestrated land: 'Black Thursday' on 6 February 1851 saw 76,000 square kilometres of Victoria burning with fire from 'the Wimmera to the sea, from Portland to the fringe of Gippsland, practically no area escaped.' The Plenty Ranges, Western Port and Dandenongs, and all the way to the South Australian border were burning. 'Red Thursday' on 1 February 1898 saw another major fire emergency when 260,000 hectares burnt, after Gold Rush days and a five-times increase in Victorian population which had seen greatly increased overall fire frequency while settlers cleared and burned the land. Even most of the mountain ash (*Eucalyptus regnans*) forests had been cleared by 1898 in the rush to get rid of the 'useless bush' and exploit the forests in the first enthusiasm of conquering a new land.

In more recent times fires following La Nina fuel build-up after the 1920 to 1923 major wet periods have become part of legend, so terrible were they. Tales of fire burning from 'Oodnadatta to the Barkly Tableland' and even to the Gulf of Carpentaria are hard to authenticate and probably represent smaller fires in many locations. In Victoria another catastrophic fire day on Friday, 13 January 1939 saw much of the State blazing. Nationwide there were awful fires in the 1970s. In the Alice Springs Pastoral District with the fuel build-up of wet years between 1973 and 1978 a third of the area, 240,000 square kilometres, burnt in 600 separate fires.

Loss of ground cover in droughts (and/or fires) makes soils vulnerable and the rapid switch to flood after drought leads to soil erosion. It is ironic that in an arid land the major erosion agent is water! The increased wind strength in El Nino droughts also contributes to soil erosion. Dust storms and 'rains of blood' where the red dust contaminates the welcome rain after drought are manifestations of ENSO extremes.

The 'natural' cycle of responses to ENSO is tending to amplify the forces which have been creating a general trend towards a drier more fiery land ever since fire and aridity became factors in the environment in Miocene times.

All these 'natural' cycles, problems and responses are the 'background noise' to which human intervention in the scheme of things has been added. Where people have increased fire frequency, overgrazed the land and reduced its plant cover, and otherwise subjected it to inappropriate use, a new dimension is added.

It is not possible to say how long ENSO has operated. Logically one would assume that during the time when free flow from the Pacific to the Indian Ocean was possible before Australia came too close to Southeast Asia and modified that circulation, it would not have applied. On that basis it would seem possible that ENSO has been operative to some degree since Pliocene times and has contributed to turning Australia into the driest vegetated continent throughout its drying phases.

Certainly many of the adaptations that have occurred in plants and animals are so tuned to the rapid and unpredictably spaced climate switches that one is led to assume that many millions of years may have been involved in their evolution. It can be argued that regular seasonal aridity alone without the total unpredictability which is added by ENSO would not have resulted in the same set of adaptations. No other continent has so many 'opportunists', able to take advantage of brief good times and to last out the hard times.

The amazing reproductive strategies of kangaroos[271] are a case in point. Red kangaroos inhabit the open arid and semi-arid plains that cover much of the continent. These plains include mulga lands, mallee, shrubland, grassland and desert. Female roos respond to good conditions by producing young in rapid succession and may have a young joey at heel, one in the pouch and another fertilised egg-cell in an early stage of division (a blastocyst) ready to take its place. The female mates and stores another blastocyst in quiescent state ready to develop and enter the pouch after the birth of each embryo. When the joey leaves the pouch it continues to suckle for four months using the teat which has nourished it from the start and which now produces milk suited to its size and needs. The new pouch-young attaches to another teat and obtains different quality milk suited to its early developmental stages. The ability to produce two kinds of formula at the one time is an amazingly advanced response. (Why marsupials tend to have been considered primitive when compared with placentals is not at all clear — 'different' they are, but certainly highly evolved.)

If drought follows the good times a high percentage of joeys die. Prolonged droughts result in cessation of breeding and the delaying of onset of sexual maturity in young animals. Rain immediately starts a hormone response and the rush to re-populate starts again.

The long-haired rat which lives in the arid Simpson Desert, Channel Country, Barkly Tableland and southwest Queensland is another opportunist. In La Nina years it is capable of breeding at a quite devastating rate and creating a plague. In 1860 when Burke and Wills arrived at Cooper Creek to recover after their desert journey their camp was invaded by rats at dusk and the swarming hordes ate everything in sight and forced the explorers to abandon camp and move. Since then plagues have followed years of exceptionally heavy rain and floods in this region where drought is a normal condition. The rats spread out into areas where they do not occur under normal conditions and when the boom is over most die and a few remain in refuges where there is always some plant food and water. On some occasions the introduced house mouse and the native spinifex hopping-mouse also reach high numbers in company with the long-haired rat.

During rat plagues the letterwing kite, and black kite, dingo, snakes, foxes and feral cats breed up in great numbers in response to the food provided by the rats.[260] The letterwings in particular show the opportunistic responses that characterise many desert birds. They produce multiple broods in La Nina years when the rats swarm and to do this in the treeless desert which offers no traditional nesting sites they can be found nesting on the ground, in old tin cans and anywhere at all in the rush to produce as many offspring as possible while the going is good. The profligacy which characterises breeding in boom times in arid Australia without consideration for the millions that are doomed to die is astounding.

Birds of arid parts of Australia show many adaptations that can be attributed to the unpredictable environment of ENSO's boom and bust regimes. Nomadism is one strategy and a high proportion of Australian birds are true nomads. Nesting in many inland species is not a seasonal phenomenon but is triggered by sudden falls of rain. The ability to produce multiple broods in wetter times is widespread and many do not breed at all unless conditions are favourable, and clutch sizes vary in accordance with the availability of food. Some birds, like zebra finches, are capable of breeding at a very young age which enables them to take advantage of good years, and some bird species have developed social structures where raising of nestlings is left to the male and juveniles so that the female can lay and brood another clutch of eggs.

Ecologists regard many of the adaptations of Australian animals to variable rainfall as unusual when seen in a global context. More nomadic birds are present in Australia than in other arid lands and the link between rain and breeding is highest. While in other arid lands breeding is linked to day length and therefore seasonal, in Australia there is aseasonal linking to precipitation.

Some characteristics of Australian vegetation appear to be, at least in part, attributable to ENSO's influence on climate. While arid areas in other parts of the world have a proportion of succulents in their vegetation, Australia has just about none in arid and semi-arid areas. Although succulents elsewhere require little moisture they do need regular rain. The unpredictability of climate in Australia requires plants to have the ability to withstand a year or more without water and be able to take advantage of the La Nina that follows. This does not apply to succulents.

About a quarter of Australia is colonised by mulga (*Acacia aneura*)[260] and bladder saltbush (*Atriplex halocarpa*) is also widespread. These plants are examples of many that require the boom times of La Nina to reproduce. Mulga only flowers when there are heavy rains in summer and it requires subsequent heavy winter rain to set seed. Germination of seed takes place in summer and heavy rain is needed again to establish the seedlings. La Nina episodes characteristically last from one summer, through a winter and into the next summer and it is only in these episodes that mulga re-establishes.

Uniquely Australian problems and responses can be seen to be partly at least controlled by ENSO's drought and flooding rains.

CHAPTER NINETEEN

Fire—Irregular and Inevitable

Wildfires caused by lightning and volcanic activity[288] have been an environmental factor throughout the 400 million years of geological time when there has been life on the land.[282] Certain vegetation types in regions where seasonal aridity was pronounced were more likely to burn than were closed forests in humid places, then as now.

The ability of a fire to travel over large areas depends on continuity of ground-cover to carry the flames, and it might be argued that the coming of grasses, and to a lesser extent forbs, and their increasing role in forming understorey in woodland and continuous cover in grasslands has resulted in the greater impact of fire over larger tracts of country. In the Australian context the expansion of grasslands, open woodland with grass understorey, and change to sclerophyll vegetation and fire-promoting eucalypts in particular, gave fire a licence to burn over vast areas.

Burnt Kingia australis *in the Stirling Ranges, W.A.*

M.E.W.

The changes in vegetation that occurred during the last 45 million years have been documented in the Drifting section of this book, with the Drying section showing the aridification and the production of the driest vegetated continent. Our knowledge of prehistoric fire comes from the pollen record where carbon particles are preserved with the spores and pollen in samples of sedimentary rocks and lake-bed sediments. Information is scarce and not all carbon found in samples is the result of fires. It is only in Pleistocene times that the record improves to give some idea of the impact of fire on ecosystems.

During the Early Tertiary when the land was largely forested[293] and rainfall was high the incidence of fire was presumably low.[294] However, from the time when Australia started to be affected by aridity, fire has been a powerful influence in the environment. During dry periods from the Middle Miocene (about 15 million years ago) it probably contributed to the replacement of woodland with grassland and influenced the sifting and sorting of components of the vegetation in other ecosystems. That the speciation and spread of *Eucalyptus* was induced by increasing fire frequency in the change from mesic to sclerophyll vegetation as the continent dried is now an accepted belief.

Throughout ice age times, from 2.4 million years ago up to the time of the arrival of humans in the continent, natural fire increased in its frequency and impact because the vegetation types it promoted were increasingly fire-prone and fire-dependent.

Far left: Fire has damaged this jarrah tree, weakening it and making it susceptible to dieback and other problems.

M.E.W.

Left: Bluff Knoll in the Stirling Ranges of Western Australia was devastated by fire in 1990 when even the steepest slopes burnt. Recovery from such a cataclysm may take decades and the original climax vegetation may never reappear.

M.E.W.

Below: Burnt grass trees, Kingia australis, *in the Fitzgerald National Park, Western Australia.*

M.E.W.

The trend seems to have been towards a drier and more fiery land through the interaction of ice age global climate regimes, the ENSO drought and flooding rain syndrome[289], and the inevitability of fire. In the absence of a global climatic change that would make middle latitudes better-watered, the trend towards a drier and more fiery continent might be expected to cause it to revert over time to a state resembling the last glacial stage, when the continent was largely a desert land.

The fluctuations from glacial times of extreme stress to interglacials when stresses were relaxed resulted in a very delicate balance being maintained in many ecosystems — and unpredictability was the common denominator in all of them. The biota evolved through many millions of years to cope with the unique mixture of relatively predictable seasonal aridity coupled with the unpredictable ENSO extremes. An extra dimension was added to natural wildfires by the climate, the landscapes and the vegetation changes that resulted, and the biota had learned to cope with this factor as well.

Selection of plants over time for their fire-tolerance had been one of the on-going processes while the continent dried and became more fire-prone. The strategies and adaptations that enabled plants to survive were responses to the many different influences in the environment, not just to fire, and many of them had been characteristics of the plants concerned before the conditions were dictated by drought and fire.

In general, plants cope with fire in two ways[285], either by resprouting from protected buds after their aerial parts have been killed by fire[292], or by recolonising from seeds. The seeds frequently require fire to open their hard protective coats before they will germinate. In the resprouting category the development of underground structures like the woody lignotubers of mallee eucalypts means that though the above-ground part of the plant dies the same extensive root stock survives through many fires. (The name 'mallee' derives from the Aboriginal word *mali* which meant water.[274] Aboriginals dug up long roots of mallee eucalypts and obtained water from them.) The ability to develop lignotubers is an inherent eucalypt character and it became a valuable attribute when fire and drought became determinants of vegetation. The epicormic buds (buds below the bark along the trunk of trees) which enable resprouting of branches on eucalypt trees damaged by fire are another resprouting device.

Grasses, particularly the bunch type, survive fire well because they can resprout from basal buds which are protected by soil or the clumped habit of the plant. Their problem comes with the overgrazing that follows in rangelands which are now used for grazing more sheep or cattle than such arid ecosytems were designed to accommodate.

The plants which depend on seeds to allow them to regenerate after fire benefit from the ash-bed which supplies nutrients, provided that the fires are not so frequent that the nitrogen available from leaf fall round plants is lost. Mulga, which colonises a quarter of the area of Australia, is killed by fire and depends on seed to recolonise. It has been found that burning of mulga woodland which destroys the build-up of fallen leaves that surround the bushes reduces available nitrogen by more than 50 per cent, so frequent fire is degrading soil in spite of the ash-bed effect.

Long cores of lake sediments like those at Lake George[189] and Lynchs Crater show clearly that with the change of vegetation[294] from forest to sclerophyll the fire frequency increased. At Lake George the greater than ever increase at 120,000 years ago has been attributed to the arrival of ancestral Aboriginals as the climatic factors and nature of the vegetation had not altered in a way that would explain the sudden increase. Aboriginal burning doubled the fire frequency. The last four centimetres of the core which show the changes in fire regimes attributable to European settlement show a three and a half times increase in charcoal fragments, tapering off in the last centimetre when fire has been largely excluded. The great abundance of charcoal in sea-bottom sediments off the Queensland coast at 140,000 years ago supports the view that people came to Australia much earlier than had been thought.

The extreme unpredictability of rainfall over most of the continent required an adaptation to the environment by humans similar to that required by the biota. Nomadic, opportunistic, resourceful and tough — terms that apply to many living things in at least 70 per cent of this continent — they evolved a culture that enabled them to manage their territories to meet their needs. They were not 'noble savages living in harmony with their environment' any more than early humans were in other parts of the world. They exploited and changed the environment to survive and for their convenience and their tool was fire.

The very nature of the land to which they had come, particularly during two glacial stages and the lead-up to the last when aridity was much more acute than now, was suited to the walkabout approach. It did not concern them that their fires affected vast areas beyond their own territories whenever they ran wild. Whether they learned to 'manage' fire to create the effect that they wanted before they arrived in this continent or whether they already had practised 'fire-stick farming' in the course of their wanderings in Asia during the time when humans were spreading out across the world will never be known. However, it is known that fire had been a tool for the human race from its earliest dawning.

The use of fire by the ancestors of *Homo sapiens*[291] goes back 1.5 million years, when *Homo erectus* used bone torches, cooked his meat and protected himself from predators by lighting fires at the entrance to caves at Swartkrans in South Africa. The ape-men *Australopithecus robustus* were still living in caves in the area while *Homo erectus* was using fire. (Previous to the Swartkrans discovery the oldest evidence of the use of fire was in caves near Beijing in China where the date of fireplaces was 500,000 years.)

Man's mastery of fire was the first technology and the basis for all other technologies. Once tamed, fire enabled him to bake clay to make utensils and later to heat sand to make glass, and ore to make metals. The very evolution of the human species from cave dweller to modern man was thus dependent on the mastery of fire. Australia was the one continent where fire technology did not result in the progression from stone age to bronze age and on to increasingly technological societies, nor from nomadic hunter-gatherers to sedentary agriculture.

Australia has been described as the 'burning continent'.[302] Early explorers saw the smoke of countless fires as they sailed along the coasts. Major Mitchell, seeking a route from Sydney[306] to the Gulf of Carpentaria, wrote in 1848 'Fire, grass, kangaroos, and human inhabitants seem all dependent on each other for existence in Australia; for any one of those being wanted, the others could no longer continue.'[304] Ernest Giles in an account of five exploration journeys in central South Australia and Western Australia wrote '...the natives were about, burning, burning, ever burning; one would think they were of the famed salamander race, and lived on fire instead of water.'

Although we do not know what fire practices the earliest Australians used, we do know a little of their pyromania at the time when explorers and settlers came to the continent. And because we know how wildfire can spread and keep on burning over incredibly vast areas when unchecked now when conditions are a great deal more benign than they were in the lead-up to the last glacial stage, it is easy to see what a powerful force fire has been for at least 100,000 years.

The Aboriginals carried fire with them and used it with purpose as well as casually during their travels through the arid zone. This zone was much more extensive during the lead-up to the last glacial maximum and throughout that intensely arid and windy stage. Fire was used to signal, to singe the fur off meat and for other basic 'cooking' practices, to keep warm, and in hunting and manipulating the environment to produce and promote food supply. Fire drove out game, produced a green flush which attracted animals within range for killing, promoted the germination of many food plants, opened up the country by replacing thick scrub with grassland and maintained a mosaic of communities at different stages of regeneration.[387] This latter product of patch burning resulted in greater diversity within a territory.[388]

With burning at all seasons, whenever possible, the intensity of fires would have declined over time and large wildfires would have been less prevalent. Remembering that fire, irregular and inevitable, was part of the system before the coming of people, the fire regimes imposed by the Aborigines on the one hand increased frequency while on the other the impact of fire over large tracts of arid land was modified. The Aboriginals knew that survival depended on food supply so presumably they learned to control fires. Their aim would have been not to burn out the whole landscape at once. (We know, even today with technology to contain 'control burning', how different the aim and the reality can often be.)

The introduction of ancestral Aboriginals and their firestick farming into the continent where fire was already an enormous problem for the biota changed the balance in many different ways. In the well-watered areas, wet sclerophyll forest[297] replaced closed forest — a product of wildfires on a grand scale. In the tropical north open savannah replaced open forest. In the semi-arid woodlands a grass and forb understorey[298] and spaced trees and shrubs opened up the vegetation and woody shrubs were reduced. In the arid lands a mosaic of patches with vegetation at different stages of regeneration was produced and the timing of wildfires was dependent on sufficient groundcover having returned to carry a fire. In the chenopod shrublands groundcover was probably insufficient then as now to sustain extensive fires. Rainforest patches would only have burnt in times of extreme drought and even then some refuges remained where fire was excluded by the local microclimate and environment — and these refuges are the cores of the remaining remnants of such forest today.

When semi-arid and parts of the intensely arid lands[282] became rangelands under European management, wildfires were brought under control as far as possible and the patch-burning of Aboriginals was no longer practised. Overgrazing by domesticated livestock and by feral animals (rabbits, goats, pigs, donkeys, horses, even camels in some places!) and increased grazing from abnormally high numbers of kangaroos have greatly reduced fire incidence and returned fire frequency and regimes to much what they may have been pre-Aboriginals. When fires do occur they are likely to be of high intensity and to spread over vast areas.

As an example of the sort of changes that have occurred in semi-arid lands under changing fire regimes and the cycles of drought and flooding rains that characterise the seasonally arid zone, the Pilliga Scrub has an interesting history.[289]

Situated in north-central New South Wales the 400,000 hectares was open grassy country with about eight trees per hectare in the 1830s. Frequent burning by Aboriginals and grazing by native marsupials had produced an open savannah landscape. Seed of the trees and shrubs germinated after fire but rat kangaroos ate the seedlings before they became established. With the introduction of sheep the rat-kangaroos disappeared. The major El Nino event of 1877 to 1878 further reduced the numbers of indigenous marsupials. When La Nina followed with a very wet year the few large trees of *Callitris glaucophylla*, the native cypress pine, seeded well and the farmers burnt to stop the grass

Far right: The patchy groundcover in spinifex country tends to prevent frequent wildfires affecting large areas as there is usually insufficient fuel to carry fire. M.E.W.

Right: Mallee woodland of the Murray Basin is fire-prone because of the nature of the trees themselves and the amount of flammable litter that lies beneath them. Spinifex as an understorey increases the fire hazard. M.E.W.

The button-grass flats of Tasmania are a product of fire. Aboriginal fire opened up much of the vegetation, and on river flats the eucalypt woodland was replaced by button-grass. Less frequent burning since European occupation has seen some of the open areas revert to bushland and forest.

JIM FRAZIER

seed in the abundant understorey from spoiling the sheeps' wool. Seedlings of *Callitris* germinated and did not have to compete with grasses or suffer from grazing rat-kangaroos. Large numbers of young trees survived and the good times continued with several successive La Ninas during the next decade resulting in more trees becoming established.

Introduction of rabbits then prevented further seedlings from becoming established until myxomatosis reduced the rabbit problem in 1950. The disease was successful in that year because it was a wet La Nina and the mosquitos that spread the disease were able to breed in suitable numbers. Fire had been kept out of the forest between 1917 and 1951 when the undergrowth and the accumulation of forest litter through all the decades of no fires made it ripe for burning. The fire that swept through it then caused the large numbers of seeds that had accumulated to germinate and now there were no rabbits to eat the seedlings and no rat-kangaroos. In every following La Nina year more seedlings became established, so that the forest of *Callitris* matured and was able to support sustainable timber harvesting. Thus open savannah became forest as a result of ENSO, fire and animal interaction, in this special case where cypress pine was the tree involved.

In semi-arid areas generally when fire is excluded and grazing continues, woody shrubs replace grass and savannah becomes dense scrub. Most of the woody shrubs are unpalatable, though mulga (*Acacia aneura*) is useful as a drought fodder.[307] Its availability has too often resulted in stock being kept on drought-stricken land for too long so that overgrazing and damage to bare soil resulted.

As long ago as the start of this century there was already concern that the combination of grazing and the high variability of rainfall was leading to degradation of vegetation in semi-arid areas. It could also be seen that fire following on build up of fuel during wet years compounded the effects of the following drought on soils and vegetation. In good years optimism (or greed) resulted in over stocking. The result was overgrazing and the drought that inevitably followed the wetter years resulted in permanent impairment of the carrying capacity of the land.

Now when much of the damage done is beyond repair, management strategies are being worked out against a background of understanding of the unique problems that Australia presents. The complexities of the problems are immense. The dilemma posed by fire is perhaps the most insoluble. Fire is inevitable, it is possibly necessary to maintain certain ecosystems and to maintain diversity (certainly if shrub control and grazing in some areas is to continue).

CHAPTER TWENTY

SALT—DESERT MADE AND DESERT MAKER

In nature salt and aridity go together. Chains of salt lakes along ancient drainages where no rivers have flowed since Late Tertiary climatic changes started to turn Australia into an arid land are features of the Western Australian plateau country. Large playa lakes like those which form a chain along the centre of the Amadeus Basin from Lake Hopkins in Western Australia, through Lake Neale and Lake Amadeus and many smaller lakes to the Finke River in the Northern Territory are the result of groundwater discharge in inward-draining basins.

The great Amadeus valley has no surface drainage. Streams from its 90,000 square kilometres of catchment in the Petermann[364] and Musgrave Ranges in the south and the George Gill range in the north 'flood out' and disappear into the sands without reaching the salt lakes. Its lakes are salt because under the present climatic conditions the rate of evaporation is too great for water to accumulate. Its aquifers contain the underground water on which Alice Springs relies, as do the Uluru tourist complex at Ayers Rock, smaller settlements, and farms in the region.

These groundwater aquifers receive very little, or negligible,[360] recharge under the present climatic conditions where rainfall[363] of about 250 millimetres is outweighed by evaporation of 3000 millimetres a year. The waters being drawn on are palaeowaters and a finite and non-renewable resource. Their chemical load varies depending on which rock stratum is being tapped, and potable water is in short supply. Even it contains high salt levels and nitrates may be in higher concentrations than is desirable. (These nitrates are of biogenic origin, produced by soil micro-organisms, particularly the cyanobacteria of the 'living duricrust' and within the soil profile.)

A similar situation, where the groundwaters on which farms[365] and settlements depend for survival in other regions are palaeowaters and finite because virtually no recharge of aquifers is occurring under present climatic conditions, exists in the Ti Tree Basin. This small sub-basin has its centre about 200 kilometres north of Alice Springs.

Lake Eyre, Lake Frome and other large salt lakes in the centre of other inward-draining basins have rivers which run into them periodically, supplying them with surface water, in addition to their being groundwater discharge zones. For long periods in the geological past they were freshwater lakes, including during lake-full stages in the present interglacial.

Experiments have shown that an amazing amount of water[357] evaporates from Lake Frome while it is dry and salt-crusted. The watertable is very close to the surface and the sun draws off the groundwater by diffusion and capillary action to the tune of 460,000,000 cubic metres a year. (The same phenomenon no doubt occurs in other 'dry' salt lakes where the watertable is close to the surface.)

The concentrated pools of brine below the salt crusts on playas are being investigated as a readily available source of a number of chemical substances used in industry and currently imported at considerable expense.

Common salt, gypsum, zeolites, magnesium and potassium can be recovered by simple technology.

The raw materials for fertilisers, food additives, glass-fibre products, building material and industrial catalysts and many other useful substances are there to be exploited and could form the basis of a lucrative export industry.

In all the arid and semi-arid basins where rainfall is less than 250 millimetres, dryland salting results from the high evaporation rate opposing the low recharge rate. The Standing Committee on Soil Conservation in 1982 estimated that 4.3 per cent[356] of the Australian landmass is salinised, of which 13 per cent (or 4.2 million hectares) is due to the 200 years of land use, and the rest is natural. Natural salinity is known as primary salinity and results from the geological history and the present climate. The underlying hydrological cycles are not understood and increasing natural salinity might be a manifestation of cyclic changes relating to past changes in climate over many thousands of years. Groundwater responses can lag far behind climatic fluctuations and it has been suggested by a leading expert on salt lakes, Dr Aro Arakel, that increasing dryland salinisation is an on-going natural trend which human activities have only accelerated.

The many small salt lakes, saltpans and scalds in the Murray Basin Mallee are examples of primary salinity due to its inward-draining groundwaters whose only external escape route is via the Murray River to the sea. These arid-land features date back to about 500,000 years ago when, after the disappearance of Lake Bungunnia, aridity changed the area into a groundwater discharge zone. The salinisation that affects cleared mallee lands is a secondary, man-induced effect. (And it should have been obvious that such marginal land was delicately balanced and anything which affected the water budget was dangerous.)

Secondary salinisation is of two types in Australia — dryland and irrigation-related. Topography affects the incidence of dryland salting. In gently undulating country with a shallow watertable and an arid or semi-arid climate, optimum conditions exist. (And how much of Australia is in this category!) In this type of region the water balance is fragile (recharge versus discharge) and is controlled directly by the rainfall, the amount of surface and sub-surface water, and the amount of evaporation and transpiration. That is why plant cover is so significant and removal of trees has often been the trigger for degradation.

Bare soil is much more efficient at evaporating sub-surface water and concentrating salts in the remaining groundwater and the soil. This is particularly the case following a wet season when rising groundwater picks up salts from subsoils and weathering rocks. These additions to the salt budget are then concentrated in lower parts of the landscape.

Secondary salinisation in which the watertable rises[366] and salting of agricultural land occurs, with added problems of salinity increase in rivers, is the result of the 200 years of settlement. Throughout southern Australia the filling up of regional aquifers has resulted from the extensive clearance of trees from both the highlands and the plains and the introduction of new agricultural practices. The development of large-scale irrigation and the modifying of river systems has had a dramatic and accelerating effect on the problem.

Rising groundwater brings salt that was previously stored in aquifers to the surface. (It has been estimated that 100 billion tonnes of salt are stored in the Murray Basin.)[362] The process is occurring in almost every geographical and climatic setting, including the highlands, the Riverine Plain, and The Mallee of the Murray-Darling Basin. All major irrigation areas are affected. The Murray Basin is particularly susceptible because of its geological structure. It is a shallow saucer whose sedimentary strata become completely saturated and it has sub-surface topography that results in all its groundwaters flowing inward. Its only external drainage is by the main Murray River channel from Morgan to the sea. Some areas are more at risk than others, depending on the topography and the depth of the watertable. Watertables have been rising at two centimetres a year in worst-affected areas, and if this rate were to continue the western Riverine Plain[361] would be a salt wasteland in 50 years.

The Murray-Darling Basin is of the utmost importance to Australia's economy. It represents a seventh of the continent in area and is the productive heartland. A quarter of the nation's cattle and dairy industry, half its sheep, and half its crop land are centred here. Three-quarters of the continent's irrigated land is in the Basin. The Murray-Darling is

Top left: A salt lake on the Oodnadatta track near Marree.

M.E.W.

Top right: Salt degradation of the wheatlands of Western Australia, as seen here near Three Springs, is a major problem for which there is no immediate cure. Raising the watertable by clearing the land for cropping has brought salt to the surface.

M.E.W.

*Above right: Lake Indoon in Western Australia used to dry up in summer. Now it is always full because the watertable has risen as a result of tree felling in the surrounding country. The beautiful red gums (*Eucalyptus camaldulensis*) round its margins are starting to die because of the salt.*

Right: Forests are dying where the salty watertable has risen, as it has here near Lake Toolibin in the Narrogin region of Western Australia.

KEVIN COATE

*At Tammin the salt-affected soils quickly become waterlogged after even a light shower of rain. Once-fertile land has become a mosaic of badly affected areas which are bare, and less severely affected areas where barley grass (*Hordeum maximum*) still grows.*

N. J. DAVIDSON

A vast stretch of badly salt-affected land at Tammin with patches that are less affected and still capable of supporting barley grass is being sown with Atriplex *in the first stages of rehabilitation.*

N. J. DAVIDSON

Below left: Sheep eating Atriplex amnicola.

N. J. DAVIDSON

Below right: A one-year old planting of Atriplex amnicola *on a mildly salt-affected site at Tammin where bushes were planted on a 5 metre grid and attained an average 2.5 metre diameter and 1.5 metre height.*

N. J. DAVIDSON

the fourth longest river in the world, 3780 kilometres long, and the twenty major rivers which form its system carry three-quarters of the country's river water. With groundwater rising and salt being added to the river system, the deterioration of water quality is compounded towards the river mouth. It is calculated that 1.3 million tonnes of salt cross the border into South Australia each year, which is 2.5 tonnes a minute. The fact that the rivers also act as sewers and drains for the cities and towns along their courses, and that agricultural fertilisers and pesticides and sediments from eroding lands are entering them along the way, is inviting disaster. In drought years (and is there ever a year when there is not a drought somewhere along a river system which is as long as this?), conditions favouring the development of toxic algal blooms are bound to occur. High nutrient levels are becoming standard in rivers in low-water state because the mud and silt in their beds is permanently nutrient-charged, and laden also with the resting spores of cyanobacteria which are capable of making the water actually poisonous (not contaminated with something that can be filtered out).

Management of the flow of rivers so that they can supply the requirements of irrigation has had serious consequences for the Murray floodplains where the red gum forests depend for survival on annual flooding. This no longer occurs and there is little regeneration in the forests as a consequence. (In the developing cotton lands of the Channel Country diversion of river water for irrigation of cotton is having, and will increasingly have, a disastrous effect on the wetlands which are of such enormous environmental importance as waterbird and wildlife sanctuaries.)

Civilisations have collapsed because the soils of their agricultural lands[356] were rendered sterile by salt, if one is to believe the historians who claim that the Sumerian civilisation died in 2500 BC because of this problem. Even if it is only partly true, it emphasises the seriousness of secondary salinity and of inappropriate and unsustainable land management practices in general which have such power to cause degradation of soils and waters — both surface and groundwaters.

Problems of salinisation of land are not recent phenomena that have suddenly appeared in Australia. It is only now when they are costing the country an estimated $100,000,000 a year in Murray-Darling Basin productivity alone that they are causing panic. In Western Australia in 1897, only about 65 years after the establishment of the Swan River Colony, it was noted that removal of native vegetation had resulted in salting of stream waters in the Northam-Toodyay district. Salinity in areas of dryland agriculture in the State is still increasing and yet more land in salt-prone regions is still being developed today, and no way of controlling salination or curing the salt-degraded land has yet been found. Trenching round growing plots, creating evaporation ponds for the salt, and other engineering devices are used to counteract the problem. By re-planting trees on catchments and borders of cultivated areas to lower the watertable the situation can be partially controlled and, in time when the trees have grown, improved. It has been calculated[356] that remedial management in the wheatbelt of Western Australia will take a few decades in higher rainfall areas to reverse the trend towards more salinisation, and that several thousand years may be required in semi-arid agricultural areas.

When the technology for extraction of valuable salts from the brine in salt lakes is mastered and small-scale equipment is made available, it would theoretically be possible for farmers whose lands are suffering secondary salination to process the salts which they remove from their land. Dr Arakel, principal research scientist with the Waterboard (N.S.W.), is involved in research and sees the loss of the salt that is removed by draining to lower the watertable, or which is dissolved in run-off irrigation water, as a wasted opportunity, not a solution. The salt remains in the system and ends up spreading its effects on the land and in the rivers. If it could be removed permanently soil and water quality would improve and a financial profit could be made by selling the material for commercial use.

In Western Australia, where wheatbelt farmers are faced with lower yields as soils become more salt, areas of salt-destroyed lands increase annually. Rising watertables are killing trees, turning freshwater lakes saline and expanding the areas of natural salt lakes and saltpans. Farmers have formed cooperatives, opting to do something about their problems and not to wait for government initiatives. By planting trees, leaving areas of natural bush on their lands if they had not already been cleared, and sharing their knowledge and expertise they are in many areas already reaping the benefits of their changed approach. Government Landcare programmes are involved in this rearguard action to try to stem the tide of soil and water degradation.

Experiments are under way to select salt-tolerant trees[318] to plant in salt-degraded areas to lower the watertable. The problem is that in the interim before these slow-acting measures begin to have effect a great deal more of the land will be rendered useless. The productivity of other land steadily decreases as salt-content rises, and many vast areas not described as salinised are carrying up to 3000 tonnes of salt per hectare and have become marginal.

The source of the salt in the soils and groundwater is the weathered rocks and, according to the experts, the huge amounts of salt that have been carried onto the land from the surrounding seas over the multi-millions of years when the ancient soils were being rained upon.

Australia's soils, as we have seen from a study of the continent's geological history, are mostly derived from ancient landscapes as a result of long and uninterrupted periods of weathering and leaching. They are nutrient-poor for these reasons, and are among the world's most impoverished in terms of organic matter content. They are affected by salts or underlain by saline groundwater. They have proved to be fragile and susceptible to degradation. It is quite appalling to know that now Australia's soils are considered to be amongst the most degraded in the developed world. Salinisation is only one of the types of degradation that have brought our most basic resource and primary medium for food production to this sorry state.

The problems of increased run-off from land that has had its native vegetation removed, of soil erosion, of silting of rivers, of wind erosion and duststorms when drought leaves land bare, are well-documented and very depressing. The silt that is flooding out of estuaries of northern rivers on the eastern margin of the continent has the power to affect and ultimately to kill the Great Barrier Reef.

Experiments have shown that water draining from a forested area carries 20-30 tonnes of sediment per square kilometre, and that the loss from cleared areas may be between 10 and 100 times that amount. The South Johnstone River in north Queensland carried 100,000 tonnes a year between 1942 and 1983 from lands cleared for sugar; today the figure is 300,000 tonnes a year. In the Hunter Valley each bottle of wine produced costs the land a kilogram of soil. From each hectare of vineyard 53 tonnes of soil are ending up in the Hunter River.

We need to remember that soils are not renewable over the human timescale.

PART FOUR

UNBALANCING THE BIOTA

200 YEARS OF CHANGE

INTRODUCTION

A satellite view of the 'bare bones' of Australia introduced the continent at the beginning of this book. The story that followed showed the evolution of the driest and most fiery vegetated land from a green and forested piece of Gondwana. How useful it would have been to have had satellite imagery to map the changing vegetation through the 160 million years encompassed, instead of the fragmentary information available from the fossil and geological record!

In today's technological age it is at last possible to use remote sensing to construct maps of vegetation cover over a whole continent. The LANDSAT satellites which fly continuously over Australia detect the density of vegetation, useful in mapping different vegetation types, and the extent of areas cleared for agriculture, degraded by drought, fire or overgrazing, and in monitoring the ever-changing picture.

When the vegetation types of different density have been mapped, botanical information can be added from other sources and a map of the whole continent showing generalised plant cover[324] results. AUSLIG (the Australian Surveying and Land Information Group) in Canberra has compiled a Vegetation Atlas from LANDSAT information. In addition a vegetation map which shows what the density of vegetation was like 200 years ago when European settlement commenced has been produced by compiling recorded information in order to show the changes which have occurred in the major vegetation divisions of the continent.

AUSTRALIA—RAINFALL

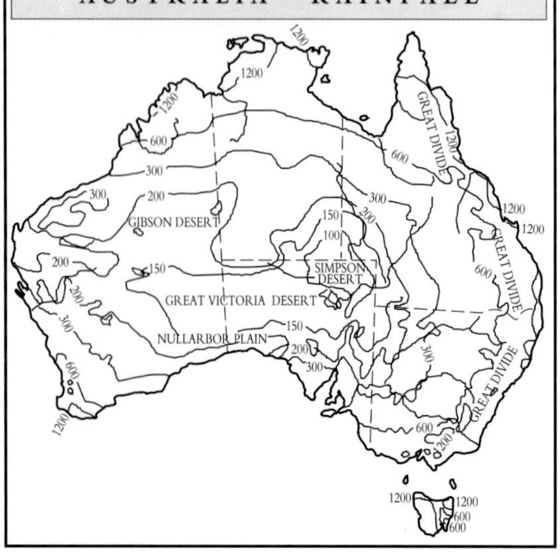

This pre-settlement map of vegetation is the best estimation we have on the status of the continent after 160 million years of evolution and change since rifting within Gondwana began for Australia, and it shows a land already altered by Aboriginal management by fire. The explorer Mitchell in 1848[306] wrote 'but for fire...the Australian woods had probably contained as thick a jungle as those of New Zealand and America, instead of the open forests in which white men find grass for their cattle.' In 1827, Hellyer wrote of northwestern Tasmania that the large open grasslands which extended southwards towards the rainforests were 'the resorts and hunting grounds of Aborigines who regularly burnt the grass.' When burning ceased there from 1935 onwards, wet scrub and *Nothofagus* forest came back into the area.[323] Similarly river valleys in eastern Victoria at the time of settlement were clothed in grass with a few big trees because of annual bushfires lit by Aboriginals, and when fire was controlled trees returned to the valleys.

European settlement has had a profound effect on Australia. The AUSLIG vegetation map for 200 years ago shows that all the major

AREA OF AUSTRALIA SUBJECT TO GRAZING
by introduced animals (controlled and feral) and distribution of rabbits

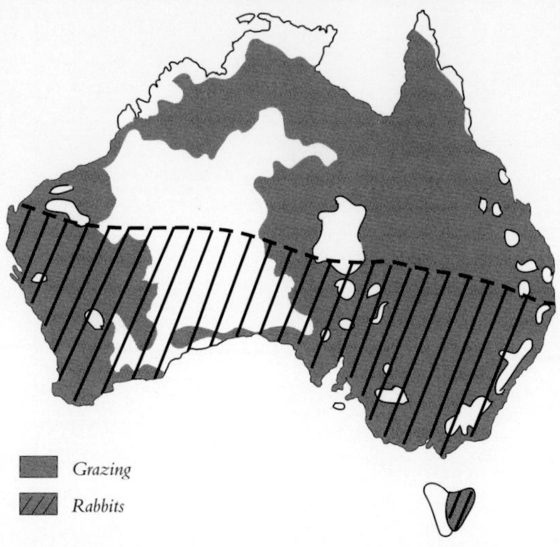

After Auslig 324

vegetation types which are subjected to grazing by introduced animals have suffered severe downgrading (as shown by decreased density of plant cover). In addition, of course, a very large proportion of those regions of the continent which have been cleared for crops and sown pastures are completely altered, and forestry operations have changed the composition of large areas of forest.

After the initial establishment of penal colonies, expansion into the relatively fertile, better-watered areas of the continent by early settlers was rapid. Spread into Victoria was fifty years behind New South Wales and when farmers brought sheep[295] there from Tasmania they did not know that they were coming to a region where fire was a greater hazard than anywhere else on Earth. Goldrush days brought hordes of people, rabbits, other feral creatures and a rush to buy property and land. By 1850 Victoria was a 'large unfenced pastoral run'.

Penetration into Central Australia was slower and by 1876[321] only 20,000 square kilometres were leased between the Northern Territory-South Australian border and 21°S, north of Barrow Creek. Two leaseholds in the Macdonnell Ranges were stocked. At the time of settlement in the Alice Springs area about 4500 Aboriginals were in the Alice Springs Pastoral District and they were practising patch-burning over the whole district.

In the southwestern part of the continent only 500 square kilometres[325] were cleared by

DISTRIBUTION OF INTRODUCED AND SOWN PASTURES AND CROPS

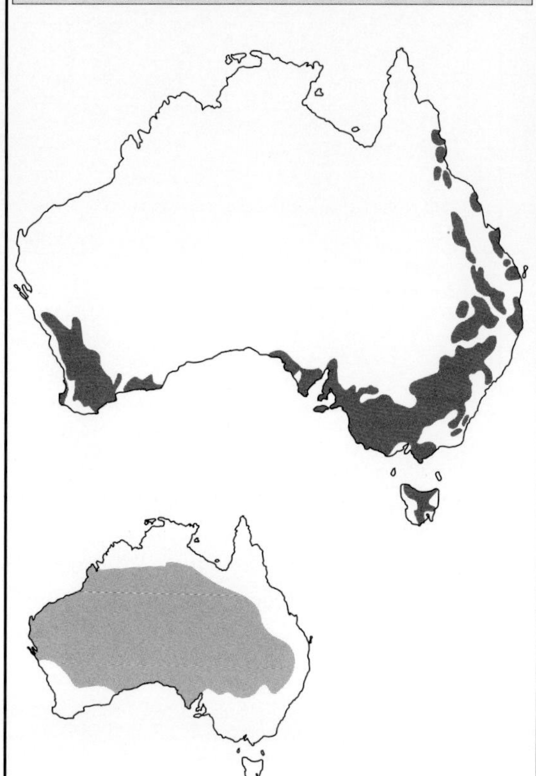

Semi-arid and Arid Australia 70% of the continent

After Auslig 324

1890, but the Western Australian wheatbelt now covers 140,000 square kilometres. Half of this was cleared between 1945 and 1982. The grants of land to farmers after the Second World War were conditional on the land being totally cleared within the first year, and the consequences of the mass destruction of vegetation are now painfully obvious with salinisation a major problem and 250 square kilometres going out of production each year due to rising salt.

The technology that allowed clearing by bulldozers, and the provision of water by percussion drilling of wells from the 1920s onwards, led to a greatly increased impact of grazing on the continent. Most areas suitable for farming had already been taken up by the turn of the century so there was little further encroachment into desert, and the most arid areas of the continent have been little changed in consequence.

The early settlers came from lands where mediterranean-style agriculture was successful, where soils were young and revitalised by glaciation, where climate was predictable and fire was not a factor to contend with. They were people of their time and they cannot be

FLORISTIC AND CLIMATIC ZONES AND RAINFALL PATTERNS

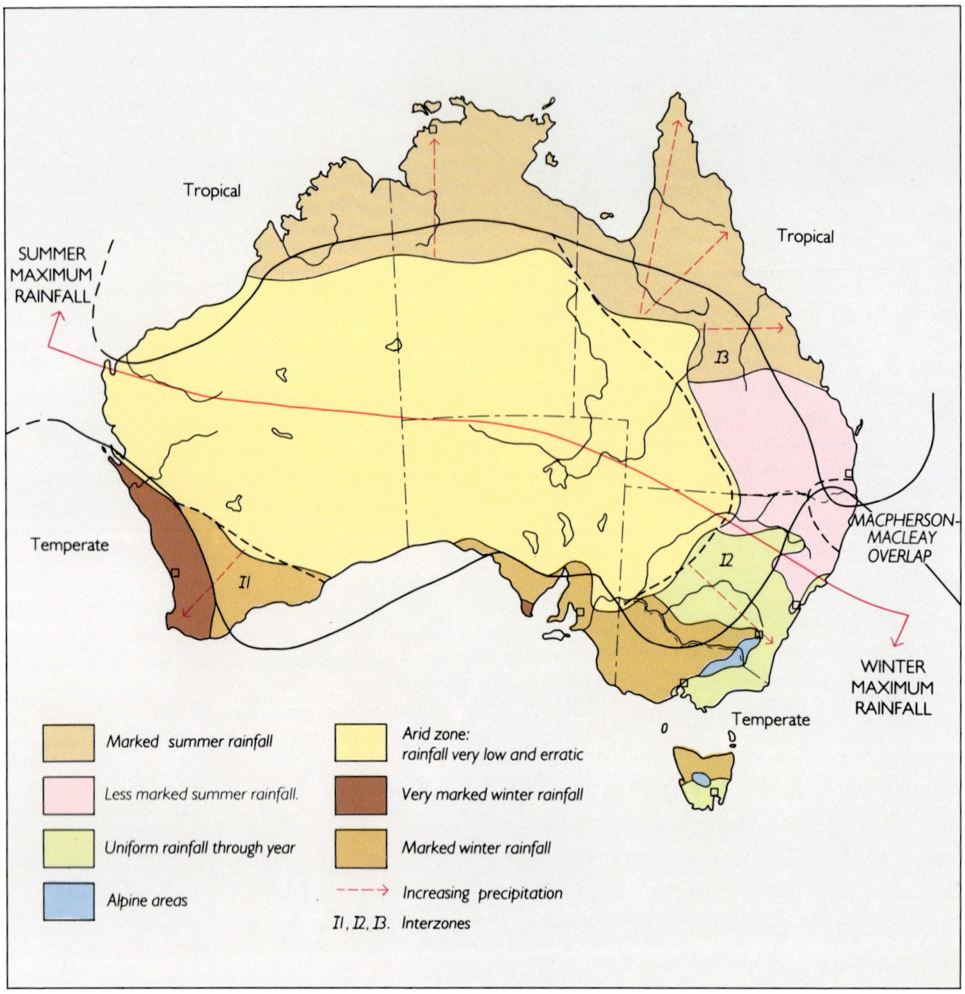

criticised for the effects that their inappropriate use of the land produced. What is surprising, or to the cynical not so surprising, is how long it has taken for the resultant degradation to be recognised and action taken to try to remedy it and achieve sustainable land use. While Aboriginal occupancy had altered vegetation types and provided 'sustainable' ecosystems that suited the purposes of hunter-gatherers, it was the superimposing of grazing pressures on a land where balances were already fragile that led to progressive degradation of all rangelands — and that represents 70 per cent of the continent.

It is a pity that no one listened to Charles Darwin[325] who visited Bathurst only 20 years after expansion beyond the Blue Mountains into inland New South Wales had occurred. In 1836 he noted that with the introduction of sheep 'the pasture was nearly destroyed....A few years since, this country abounded with wild animals; but now the emu is banished to a long distance and the kangaroo is become scarce....It may be long before these animals are altogether exterminated, but their doom is fixed.' He picked the wrong animals as heading for extinction, those being two that have often profited from changes brought by settlement (and in some areas reaching plague proportions as a result of the upsetting of natural balances), but the picture he saw was right.

It seems that only when financial ruin threatens, personal comfort is affected and an individual's standard of living is being lowered do we act to deal with problems that may have been visible for ages. If we do not realise that a nation is only as rich as its soil and water resources the future is bleak.

The physical environment is the controlling influence on the distribution of the different vegetation types that make up the Australian scene, despite the enormous impact that Europeans have had. Thus the problems encountered in different zones vary according to their history and specific characters. Because 70 per cent of the continent is subject to grazing by introduced animals, only the most arid parts of the land have not suffered alteration, frequently combined with degradation.

Apart from the millions of sheep and cattle introduced and farmed, the feral animal populations that rapidly expanded from the start of settlement onwards have been a significant contributing factor to degradation of the land.[325] It is recorded that the cattle introduced to Australia with the First Fleet escaped on arrival in Sydney and travelled to the Nepean River in 1788. From a start with four cows, a bull and a bull-calf their feral population grew from 61 in 1795 to more than 800 in 1804. The rabbit, whose domain now spreads all over the southern half of the continent, is perhaps the most accomplished desert-maker. Helped by feral cats and foxes, it also threatens the survival of native species of small mammals. In today's Australia the feral animal situation has reached disastrous proportions and 'feral' plants have contributed to undesirable change.

Maintaining the health of a land starts with maintaining healthy vegetation. It has

Above: An abandoned farmhouse in a man-made desert—testimony to 200 years of abuse of a continent.

JIM FRAZIER

Left: William Creek in the dead-red Centre where the desert has replaced once-productive saltbush country. A signpost for the future of all our marginal lands under present management policies?

M.E.W.

been vegetation that has created and held soils since the time of first life on the land 420 million years ago. In Australia where soils are ancient and mainly nutrient-deficient, native vegetation that had evolved with the land was completely adapted to the environment. In spite of the leached soils and other environmental constraints the vegetation and fauna maintained equilibrium up to the time of the coming of people. A natural balance resulted from the basic fact that animal life depends on plants and all food chains start with the ability of plants to photosynthesise. Balance between animal numbers and vegetation in each ecosystem was ensured. When people started using native pasture in arid land for populations of introduced animals beyond the 'natural' carrying capacity of the land, its 'doom was fixed', as Charles Darwin commented in a related context.

The response of different plant communities to the grazing and other land management practices imposed by settlement has not been uniform. Some systems were more delicately balanced than others and their capacity to adapt or recover has varied. A brief review of the major vegetation types, their status, and what is known of their requirements for sustainable use (or rehabilitation) follows.

CHAPTER TWENTY-ONE

A Continent Dominated by Two Genera

Eucalyptus and *Acacia*

It is an unusual characteristic for any landmass as large as Australia to have its vegetation dominated by two genera over 75 per cent of its area. The genus *Eucalyptus* dominates in better-watered regions with a rainfall of at least 300 millimetres, *Acacia* takes over dominance in the arid centre of the continent where rainfall is less than 300 millimetres. The two genera share dominance in interzones. It is undoubtedly the history of the land during the last two million years of ice age fluctuations that has resulted in this situation.

The eucalypts were promoted by the increasing seasonality with dry seasons and increased fire frequency in the Late Tertiary lead-up to the ice age, and replaced mesic vegetation progressively. During glacial stages, particularly the last, the dry woodlands of semi-arid regions moved towards the edges of the continent as the arid core expanded. (A map in the foregoing account of the last glacial maximum shows this stage.)

The genus *Acacia* had found its opportunity to attain dominance when the centre of the continent dried. The present pattern of distribution depends not only on the environmental conditions which apply now, but also on past distribution and evolution. The climatic and vegetation changes throughout the Tertiary have contributed to the present situation. *Acacia* is a mainly sub-tropical/tropical genus that has been around since the Late Cretaceous, when it spread over Gondwana.

In Australia in the Late Tertiary, distinct areas of speciation of *Acacia* occurred in response to climatic changes, and distinct northern, southwestern and southeastern provinces were produced. During glacial maxima, much of the vegetation of Central Australia was destroyed but refuge areas must have maintained core populations ready to re-invade when better conditions returned. A large gene pool was thus maintained by the variety of small isolated areas with favourable environmental conditions because selection pressures were different from one refugium to another. For instance, the range of variation within *Acacia aneura*, mulga, is striking. Many of the varieties are diploids or polyploids. Such varieties have multiples of the basic chromosome number for the species and their development or preservation in some populations is believed to have conferred an advantage when re-colonising bare areas. As a result a mosaic of variants was produced.

Hummock grasses — generally known as spinifex or porcupine grasses — form understorey or pure stands within the areas of arid and semi-arid zones where *Acacia* and *Eucalyptus* dominate the shrub layer. Chenopod shrubland (saltbush, bluebush) and tussock grasses occupy the largest, about equal,

NATURAL VEGETATION REGIONS OF AUSTRALIA

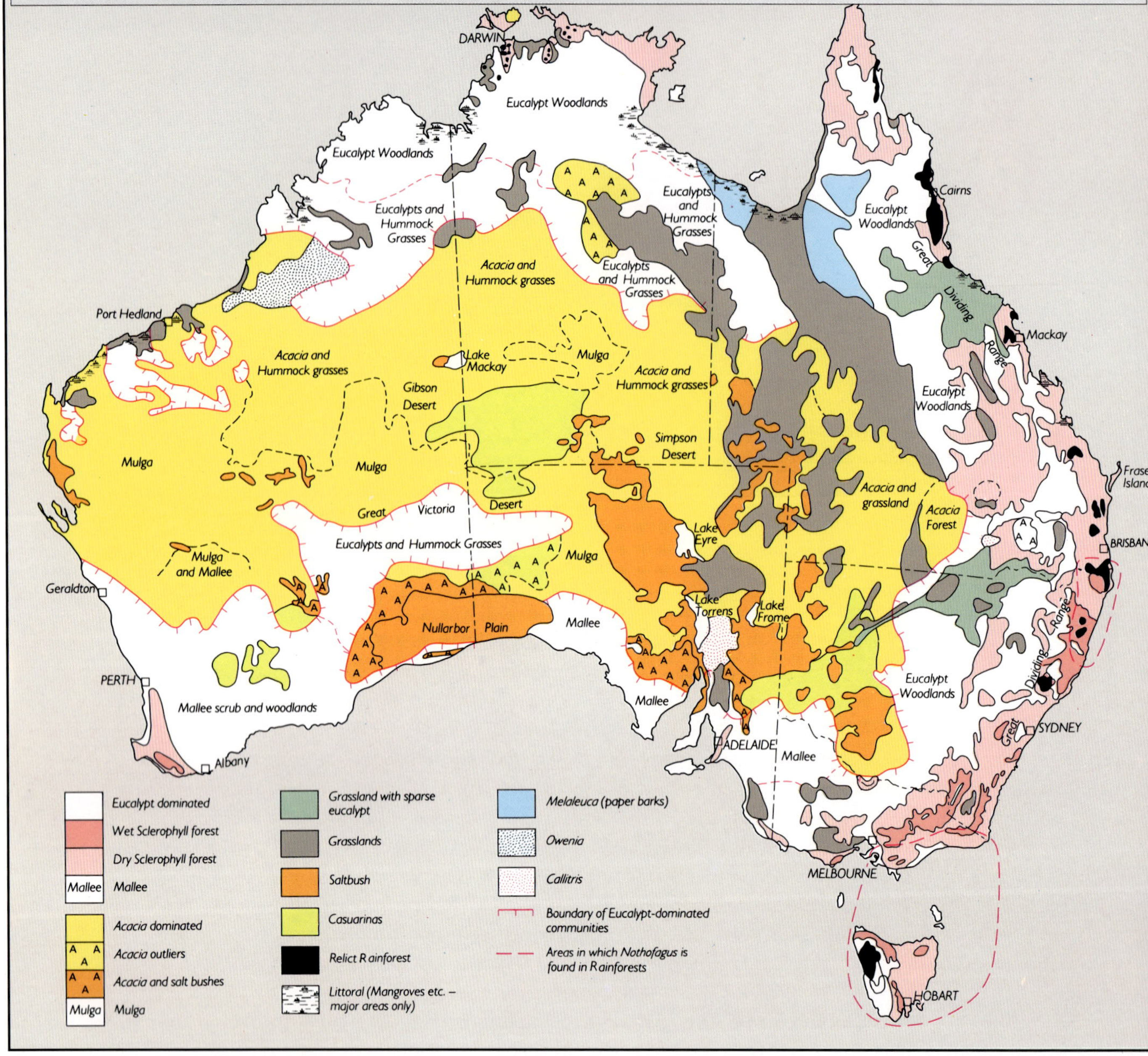

Top: Mulga, Acacia aneura, *covers 20 per cent of the continent.*

M.E.W.

Far left: Dry sclerophyll forest of tall eucalypts in the Border Ranges of New South Wales.

M.E.W.

Left: Eucalyptus papuana *on a dry stream channel in Central Australia.*

M.E.W.

areas in the 25 per cent of the continent not characterised by either dominant *Eucalyptus* or *Acacia*; *Casuarina* dominates in smaller areas in the southern half of the continent; *Melaleuca* paperbark forests of considerable extent are present in the Carpentaria Basin/Cape York area; and mixed or other floristic groups comprise smaller vegetation patches.

Only the fringes of the continent carry forest vegetation of a type with closed canopy, or open tall tree type, or with dense medium trees in open, layered formation with lower trees and shrubs. The drier woodlands, shrublands and scrub have grass and forb understorey. Rainforests, the last remnants of Gondwanan forest which exist in refuges along the east coast from Tasmania to Cape York, show variety of structure and content as a result of the sifting and sorting of the original mixed forest of the Early Tertiary. Rainforests of northeastern Queensland in the Heritage areas are closest to those early warm temperate forests, while the beech forests of Tasmania at the other end of the spectrum represent elements that have been sifted out and adapted to cool temperate conditions.

Wet sclerophyll forests, also called tall open forest, have suffered the most destruction from logging. The mountain ash forests of Victoria were largely felled last century and the jarrah and karri forests of southern Western Australia have suffered similar decimation. Open forest and woodland of the better-watered eastern sector of the continent have been largely replaced by pastures and cultivated lands.

VEGETATION REGIONS WHICH HAVE SUFFERED THE GREATEST CHANGE

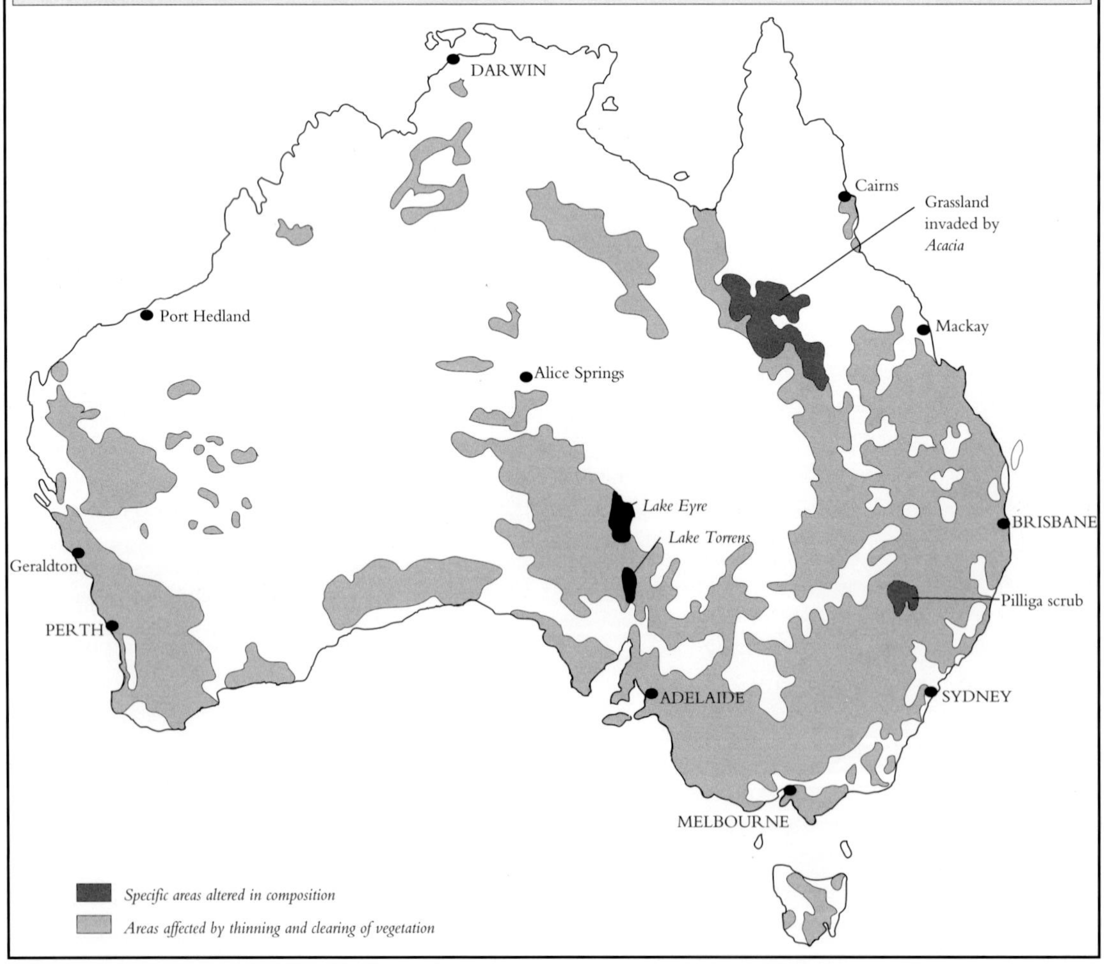

CHAPTER TWENTY-TWO

MULGA, MALLEE AND SPINIFEX— THE ARID-LAND TRINITY

1. MULGA

Mulga-dominated communities and hummock grasslands wooded[324] with mulga occupy about 20 per cent of the continent (more than 1.5 million square kilometres). Such communities occur in areas receiving a mean annual rainfall of between 200 and 500 millimetres, but mulga is conspicuously absent from semi-arid regions with a regular summer or winter drought.

Many different *Acacia* species are commonly known as mulga, but *Acacia aneura* is the widespread real mulga. It is most variable in growth form, size and in characteristics of phyllodes and pods and its varieties may one day be separated as species. The vegetation types it forms range from open forest up to 15 metres high in Queensland to sparse shrublands in desert areas. Its commonest form, however, is tall shrubland or low woodland.

Acacia aneura is wonderfully adapted to arid-land living. Its deep roots penetrate to depths where dampness remains when surface soil is dry, and it has root nodules in which bacteria fix nitrogen, helping to overcome the low nutrient status of soils. Its phyllodes (of various width and length in different varieties) are designed to minimise water loss. Thick cuticles, sunken stomata, abundant hairs and high oil content all reduce transpiration, and by shedding leaves in dry times the plants are enabled to outlast droughts and to recycle their own nutrients from the leaf-fall zone when rain comes. The phyllodes are borne in upright position so that a minimum area is exposed to direct sunlight and they act to channel any rainwater that falls on the plant. Rain runs down them, along the branches to the main trunk and straight down into the soil round the roots. This enables mulga bushes to get effective use of at least three times the rainfall that is registered over the general area.

DISTRIBUTION OF MULGA AND MALLEE

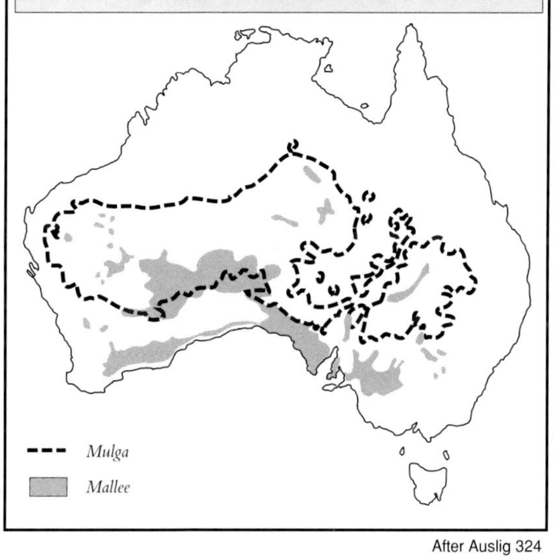

After Auslig 324

Top right: Mulga growing with saltbush on a dune near Hermit Hill in South Australia.

M.E.W.

Right: This dense stand of mulga near Kenmore Bore has virtually no understorey

M.E.W.

Even the grey-green colour of mulga phyllodes and branches works to the plant's advantage, reflecting more heat than dark green colour would have done.

Seed production, germination and establishment of seedlings[260] have been shown to be synchronised with La Nina events and to require these extra-length wet periods. La Nina events are often linked with fire because of the resultant build-up of flammable material, which makes mulga one of the plants born of fire, drought and flooding rain and uniquely Australian in its responses to environment. The whole zone occupied by the mulga lands is exposed to the greatest diurnal and seasonal temperature ranges experienced anywhere in Australia, with frosts common in winter. This is an added factor which contributes to the stresses on the biota. The sum of all the environmental factors explains why mulga lands have proved to be fragile ecosystems in which balance is easily upset, with adverse consequences.

Other 'mulgas' are *Acacia linophylla* (bowgada), *A. ramulosa* (horse mulga), *A. brachystachya* (umbrella mulga, turpentine), *A. stowardii* (bastard mulga) and *A. catenulata* (bendee). Their adaptations to arid,

A LIVING DURICRUST

Just as mineral duricrusts of iron- or silica-enriched materials cover parts of the landscape, preventing erosion and preserving ancient weathered profiles, so living 'duricrusts' of cryptogams form a protective crust on the weathered soils of the arid and sub-arid southern half of Australia. Cryptogams are plants that reproduce by spores and they include the mosses, lichens, ferns, algae and cyanobacteria that form the microscopic to often barely macroscopic flora of regions subject to winter rainfall. (The cyanobacteria are not strictly plants but their possession of chlorophyll which enables them to photosynthesise places them in an intermediate position between plants and bacteria.)

Because the duricrust flora is not obvious its importance is easily overlooked. The argument that hooved animals are responsible for soil degradation because they chop up the soil and expose it to erosion, while the woolly-footed kangaroos do not have this effect, is based mostly on the visible effects of sharp hooves disturbing the ground. The effect caused by breaking up and destroying the living duricrust is much less recognised, but the degradation that results is extremely important. Likewise, the effects of frequent fires on this barely visible stratum of the vegetation should be an important consideration when determining management practices in arid ecosystems.

Before grazing pressures from introduced animals altered[332] the balance, about 17 per cent of the continent was carpeted[337] with a cryptogam duricrust. Now trampling stock have destroyed it over vast tracts, and soil degradation, with related degradation of vegetation and decrease in carrying capacity, has followed.

Biotic crusts in which lichens dominate provide an interface between the atmosphere and the soil. They enclose air cavities, alter seed-beds and hydrology, and reduce wind and water erosion. They also play a most significant role in the fixing and conservation of nitrogen.

Lichens are long-lived and slow-growing life forms that are in fact composed of two sorts of organisms. The body of the lichen may be an encrustation of some size, or may be very small, or may be a foliose (leafy) structure. It is made of fungal threads matted to form its bulk and cyanobacteria and/or algal cells live within it. Fungi cannot photosynthesise, so the organism relies on its symbionts to feed it.

Soil lichens exist as patches of small scales each a few millimetres in diameter and coloured from pink to brown or black; as a continuous white crust looking like a paint spill; or as quite complex yellow-green plaques on the soil surface.

Lichens are tolerant of extremes of climate and can survive[335] long droughts in an air-dried state, but they are sensitive to heat when wet and this is what restricts their major distribution to winter rainfall regions. During hot summers they are dry and not vulnerable. Their absence from Australia's northern arid lands is explained by the summer rainfall patterns that apply there and their intolerance of heat when wet and active. They are excluded from real desert areas where droughts are extra-long because, though drought-tolerant, they cannot survive inordinately long periods without any water at all. They are not completely dormant in the dry state as some very low rate of respiration continues, depleting the energy reserves of the organism when it is not photosynthesising.

It appears that lichens from arid regions are very dependent on dew and that they can absorb it and photosynthesise early in the morning and be dry again by the time that temperatures start to rise later in the day. Thus they avoid the dangers of being wet and hot at the same time.

Studies in the distribution patterns of lichen genera[336] and species from the arid zone soil crust indicate that mean annual[333] rainfall, its seasonal distribution and soil type are the determinants. Lichens form a mosaic of different communities and associations over their range just like the vegetation does.

Although much has still to be learned about soil lichens, several observations emphasise the importance of establishing their role in vegetation. In arid South Australia the abundant[334] soil lichens on a reserve near Koonamore were studied. Lichens form a continuous carpet on the ground over large areas and grow on all the scattered pebbles, and 38 species in 25 genera have been identified. Endemic species comprised 19 per cent of the lichen flora. Of the other species, 48 per cent were shared with North American deserts, 37 per cent were also found in southern Africa, 28 per cent were shared with western Asian deserts and 21 per cent also occurred in the Sahara. It was found that an increase in lichen species in communities reflected the status of the vegetation. Climax vegetation had the most species and the best lichen coverage of soils, so lichens were contributing there to soil and floral stability. Areas bare of shrubs were without lichens, but regenerating areas where shrubs were beginning to become established had pioneer species of lichens appearing on their soils. The soils concerned were calcareous, so it can be stated that on such soils lichens only stabilise the soil after shrubs have returned, and they do not appear to be able to stabilise eroding surfaces.

Leafy ground lichens, small ferns and a green tinge from mosses form a protective ground cover on desert soil after rain. They will persist in a dry state when the soil dries.

Leafy lichens and a scum of moss cover this recently wet area of desert soil.

The effects of fire on the soil cryptogam cover were monitored[278] in experiments at Coolabah in New South Wales. The vegetation in the study area was eucalypt woodland on massive red earth soil and the woody shrub component of the vegetation was much increased (as in much of the degraded semi-arid woodlands as a result of the grazing and fire regimes under European management). Mature cryptogam communities in the area showed the relationship between lichens and soil mineral materials. The structure of each lichen patch has been described as like a jellyfish. The dark-coloured, usually greyish black, surface cap of continuous lichen (the bell) overlaid a zone of amorphous material in which sediment was cemented by organic matter (the skirt) and fungal hyphae formed a dense mass penetrating into the soil below (the tentacles). In between the lichen patches, micro-mounds of soil cemented with gel-like material were observed. Algal sheaths were associated with these, and it was obvious that they contributed to soil stabilising.

Fire reduced the cryptogam cover and increased the amount of mobile sediment and the area of bare surface. A single fire affected the stability of the top centimetre of sediment in the soil, by altering the cementing materials that the cyanobacteria and algae produce as gels in the crust. More frequent fires increased the amount of depositional material freed from the soil and there was less mixing of soil layers by the soil fauna. The infiltration rate of water falling on the soil changes when the cryptogam cover is removed and raindrop splash causes erosion, bare soil is readily washed away, and the stability of the soil is affected. Annual fires for seven years completely destroyed the biotic crust, but it recovered slowly and after four years without fire was more or less back to that of unburnt areas.

It can be seen that a combination of hooved-animal grazing, fire, drought and flooding rain is a recipe for disaster, and all these factors come together predictably, if at unpredictable intervals.

The role of the cyanobacteria that are part of the cryptogam flora[313] and living duricrust is still being investigated.[338] All of them produce a gelatinous sheath which is a soil stabiliser. All must be capable of desiccation without damage, and of reactivation with moisture. Several types are capable of nitrogen-fixing, a most important property in a land where soils are nutrient-deficient. They produce heterocysts, unusual cells which lack chlorophyll and are able to produce nitrates from atmospheric nitrogen.

Groundwater of much of arid Central Australia has high nitrate concentrations, in many cases exceeding the World Health Organisation's recommended limit for drinking water. These high levels may be the product of the activities of nitrogen-fixing cyanobacteria. No other source for the nitrates has been demonstrated.

DISTRIBUTION OF LIVING DURICRUSTS

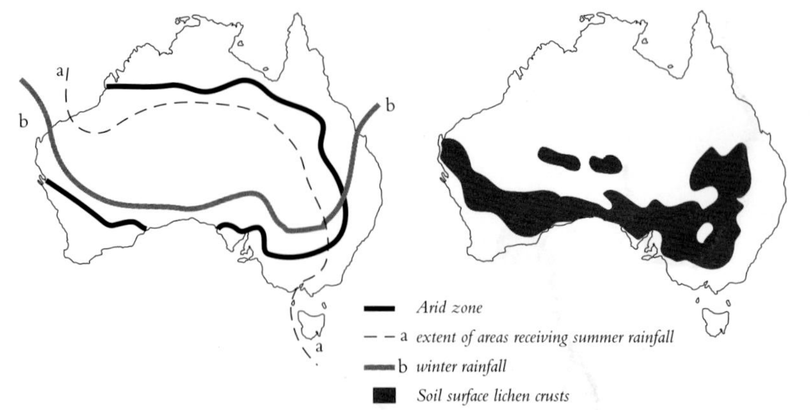

After Rogers 332

Soil cyanobacteria are easily recognised when one knows what the signs of their presence are. (We all know the black slimy film that makes paths slippery and causes black stains on buildings.) On soils they form a felt-like mat and when it is pulled apart and examined with a lens the filaments of cyanobacteria and associated soil fungi will be observed. Cyanobacteria occur commonly on the surface of pebbles as well. They prefer the underside of translucent quartz or chalcedony. The pebbles act as dew collectors, because moisture drains down their sides and collects below. The cyanobacteria use this moisture and the light transmitted through the pebble to achieve some quite substantial growth. In one paddock near Broken Hill, it was estimated that 8 per cent of the surface area of the ground was covered by cyanobacteria living below pebbles. (This behaviour is probably a primaeval one which was an adaptation to conditions facing earliest life on the land. It is also interesting in the light of recent discoveries of cryptogams living within the surface crust of Antarctic rocks, where shelter from the hostile environment is achieved in this way and light sufficient for life penetrates to the living cells because of the translucence of the crystals in the rock.)

Research into nitrogen-fixing organisms in soils of Central Australia resulted in the discovery that cyanobacteria and bacteria capable of nitrogen-fixing are widespread, existing free in soils and not necessarily associated with lichen or plant symbionts. Their adaptation to inhospitable conditions, desiccation and long dormant periods is remarkable. Some have special proteins which tolerate water stress.

From the point of view of botanists and earth scientists who are interested in the nature of first life on the land more than 400 million years ago, the living duricrusts and soil organisms of arid lands have a particular interest and significance. First life to colonise the eroding wastelands of a world with no land inhabitants was most likely a biotic crust of algae, lichens and cyanobacteria. Higher cryptogams, mosses, and the first vascular plants which evolved from green algae, soon joined them and started the conquest of the land. We should not be surprised, therefore, to find that ancient assemblages, virtually unchanged through all those multi-millions of years, are still colonising the bare earth, stabilising it and helping to feed the plants with which they co-exist.

Mosses are often components of the living duricrust, surviving desiccation or regenerating from spores when rains come. Very small fern plants that develop rapidly and take advantage of brief wet spells also occur in the crust in semi-arid regions.

The implications for the soil lichen flora[337] that is so vital in maintaining soil stability in arid southern Australia under predicted greenhouse regimes has to be considered. Under the generally accepted model, an enhanced greenhouse effect would mean a change of climate in southern Australia. It would be warmer and some summer rainfall would return to the region. This would bring the conditions which do not suit lichens — heat and wet together — and the cryptogam cover would be damaged. It would also probably mean that ephemeral grasses and forbs would grow in the summers, giving better grazing and inevitably more overgrazing. Without the living duricrust to help to stabilise soils, degradation of the pastures would result, neutralising the improved climatic effects in the context of grazing.

unpredictable environments are similar to those of mulga.

Mulga rarely grows on soils with a high clay content,[307] as it prefers the red sands, red hardpan soils, red earths and some red duplex soils, and calcareous soils wherever they occur. Exceptions to its absence from clay soils occur in some areas in Central Australia and in western Queensland where it colonises red and brown cracking clays. On some hillsides banding of spinifex and mulga occurs, reflecting the substrate. The spinifex grows where shale bands cause clay soil, the mulga where non-clay rock alternates. An inherent feature of Australia's arid land soils, and the soils of mulga lands in particular, is their infertility relative to soils in other arid areas of the world. A considerable proportion of the available nutrients in the soils is held in the top five centimetres and removal of this thin top layer removes more than 50 per cent of phosphorus and about the same proportion of the nitrogen.

In Western Australia about a third of the State, from the Tropic of Capricorn to 30°S, is mulga territory which occupies more than 700,000 square kilometres. *Eremophila* dominates as understorey and Western Australia has more than 100 species.

The mulga lands of Central Australia are separated from those of Queensland and western New South Wales by the floodplains of the Channel Country which support chenopods, grasses and forbs in open herbland. About 12 per cent of Queensland has mulga-dominated communities and only 19 species of *Eremophila* in understorey (though the genus still dominates), and *Cassia* and *Dodonea* (hop bush) are common understorey shrubs as well.

Apart from the mulga lands that are in the Great Victoria Desert[324] and the southern Great Sandy Desert (in the Officer and Canning Basins) this vegetation type is used for livestock grazing. Almost without exception grazing and management practices have resulted in degradation, and much of this took place from the 1880s to the 1930s. In particular the enormous enthusiasm and investment that resulted from the wet 1890s led to gross overstocking, as well as to expansion of wheatlands into saltbush country in South Australia and elsewhere. (As recently as the 1970s when La Nina years[321] resulted in good rains and availability of pasture in the Centre, cattle numbers in the Alice Springs Pastoral District were doubled. More than 600,000 cattle, and 60,000 feral horses and an extra 15 per cent of feral cattle were estimated to be grazing the ephemeral pasture with the inevitable result of land degradation when the good times ended and dry times returned.)

Because mulga foliage is a drought-relief fodder, mulga has suffered an extra dimension of degradation in the eastern mulga lands. Here farmers fell the trees in droughts allowing their animals to eat the leaves. In marginal areas this has meant that there has been little replacement of the trees that were felled because mulga needs good La Nina years to set seed and for its seedlings to become established. Young plants must survive the first critical years without being grazed. The infestation of these lands by rabbits which eat the seedlings is an added deterrant to re-establishment. In the 'soft mulga'[324] lands of southwestern Queensland, which are slightly better-watered, mulga may regenerate more readily, but this has often only meant that high stock numbers have been maintained and overgrazing has eventually resulted in thinning of the shrub, understorey and grass layers, leading to soil erosion and a great decline in the carrying capacity of the land.

Arid zone vegetation evolved under light or migratory grazing pressure and is poorly adapted to continuous grazing. Heavy grazing, particularly during a growing season flush is inimical to the survival of native grasses. Over time the seed-bank becomes depleted and edible forbs and ephemerals as well as grass disappear.

There is a saying in Queensland that cattle have the grace to die before the fodder is totally exhausted,[307] while sheep only die after the last bite has gone.

The discovery of the artesian water supply in the Great Artesian Basin which underlies the eastern mulga lands and the use of mulga itself as a drought fodder have been joint reasons for the serious degradation of these rangelands. A basic tenet for arid pasture management has been broken — the adjustment of stock numbers to match the grazing available.

Aboriginal fire, using patch-burning techniques[288] and usually low-intensity fire, had maintained much of the mulga land vegetation type as fairly open savannah with a

by intense fire or by lack of follow-up rain. For these reasons, intense fires at a frequency that does not allow seed-store replenishment in the soil cause long-term damage, and this applies particularly in the more arid and drought-prone areas.[283] In wetter regions fire may often promote massive seed germination and in some cases density of mulga can increase to 3000 per hectare, which is undesirable from the grazing point of view.

In the eastern semi-arid woodlands, problems with increased shrubs were evident only 20 years after settlement, and different management practices were tried to address them. Using goats that eat shrubs helped to create the feral goat disaster which is now the most serious feral problem over large areas of the continent apart from rabbits.

In Central Australia mulga communities have been variously affected by cattle and rabbits. Those with an understorey of perennials to the north and south of Alice Springs are not preferred by cattle or rabbits, while those with an annual understorey on limestone further south have been considerably affected. Mulga on red earths in the north and on sand in the south has been little affected by grazing. Because of the variations of soils, rainfall, vegetation composition, fire frequency under 'normal' conditions of inevitability and irregular response to ENSO, the problems of management of mulga lands are immense. So much is already badly degraded, some perhaps beyond repair. In the more marginal areas grazing has become uneconomic and 'natural' forces of economic ruin will cure some of the problems, though with policies of 'drought relief' in a land where drought is the only predictable factor much of the problem stems from assisting people to continue exploiting and damaging the environment in the face of all the indications to the contrary. The end result of degradation of vegetation and soils in semi-arid marginal lands is spread of desert, dust and mobile sands.

Research by the CSIRO into the control of the woody shrub[286] problem in eastern mulga lands[277] is gradually establishing practices which will allow sustainable use without degradation of soil or vegetation. A return to more frequent, low-intensity mosaic burning may be an answer, combined with control of grazing and browsing after fires. This may bring improvement in areas not so badly degraded that the native seed-bank has been lost. Reduction of grazing in the recovery period after drought is a major requirement. Rabbit and other feral animal control is vital as these animals breed in La Nina years and damage the vegetation in its most vulnerable recovery phases.

2: Mallee

Eucalypts with a multi-stemmed habit are known as mallee[309], as is the vegetation in which they dominate. The name is derived from an Aboriginal word *mali* which means water. The extensive woody roots of some mallees provide fresh drinkable water when cut — a useful attribute for people in a region usually without surface water, and for the plants themselves, giving them a store on which to draw when recovering from fire or surviving drought.

Mallees have underground lignotubers from which they generate new aerial parts[308]. Most eucalypts at some stage in their life possess a lignotuber[199]. Species in which lignotubers are persistent are most commonly found where the genus is close to its physiological limits and site qualities are lowest. Thus lignotubers are characteristic of communities regulated by drought and fire on infertile soils. Eucalypt species in general have the ability to modify their growth form to the local conditions.

Much of the natural vegetation of semi-arid southern Australia was dominated by mallee prior to the extensive clearing that accompanied settlement. Mallee has traditionally been regarded as occupying a transitional zone between the more humid sclerophyll woodlands to the south and the arid or eremaean region to the north. The mallee zone has predominantly winter rainfall.

Mallee root firewood was transported to Melbourne and other Victorian cities and towns in vast quantities as The Mallee was cleared. In South Australia Adelaide was supplied with mallee roots[326] and cut mallee wood for its domestic fires and for industry from the Yorke Peninsula until that ran out, and then from the Murray River flats and mallee lands. There is still a firewood industry based on mallee in South Australia although more than 80 per cent of the State's mallee has disappeared.

In South Australia mallee was brought into agricultural use with the development of the stump-jump plough in the 1880s. The mallee was flattened by rollers and burnt. Then the plough cultivated the area, jumping the stumps, and wheat was planted. Stubble was burnt and the fire killed off the resprouting lignotubers. It was found that the addition of superphosphate made the land productive. Between 1880 and 1886 half a million hectares of mallee were cleared on the Yorke Peninsula and the north Adelaide Plains; between 1907 and 1941 nearly three million hectares were cleared; after the Second World War mallee heaths of the southern Eyre, Yorke and Fleurieu Peninsulas, and Kangaroo Island were cleared, and the remaining mallee in the Murray Basin disappeared almost entirely.

The South Australian record is replicated elsewhere and since settlement three-quarters of the continent's mallee vegetation has disappeared and what remains is mostly degraded. The mallee in Western Australia has mainly suffered degradation by grazing. Clearing for agriculture has not been much practised as the soils on which mallee grows there proved unable to sustain cropping.

Core areas of mallee occur on calcareous soils[202] within the 250 to 380 millimetres rainfall zone, and much mallee occurs as tall open shrubland outside these core areas. The widespread mallee communities of the sandplains of the Great Victoria Desert north of the Simpson Desert and the mallee-heaths of southern coastal Western Australia are examples of mallee outside core areas.

Mallee shrublands and woodlands are the Australian equivalent[310] of the mediterranean ecosystems of other lands — the fynbos of South Africa, the matorral of Chile, and the garrigue of France and chaparral of North America. In common with these ecosystems mallee has extremely high flammability. A current hypothesis goes as far as to suggest that these communities have evolved traits which promote episodic fires which are necessary for the maintainence of populations of both plants and animals.

It has been found in these mediterranean ecosystems that their herbaceous floras depend almost entirely on fire for their perpetuation.[310] They exist mainly as dormant seeds until fire creates temporary gaps by removing the canopies of overstorey shrubs. These short-term gaps enable them to establish and reproduce, and to replenish their seed-bank before the shrubs return and the herb storey becomes dormant again.

In the more mesic mallee heathlands of western Victoria and southeastern South Australia in the first one or two years after fire there is a pulse in which *Stipa macalpinei* (speargrass), *Helichrysum obtusifolium* (paper daisy), *Stackhousia monogyna* (related to 'snow flowers' of the red dune deserts) and *Pimelia* spp. (rice flowers) are abundant before the canopy closes again. In more arid mallee, there is a flush of herbaceous species after fire, depending on rainfall, and native and feral herbivores are attracted. When major wildfires burnt the mallee lands of southwestern New South Wales between 1974 and 1977, it was noticed that there was a great increase in species diversity over the large areas burnt. Many plants that had been considered rare or even endangered appeared, and CSIRO developed a programme to try to establish what management practices would promote the grasses and forbs which improve grazing, maintain and improve diversity and stop degradation of the ecosystems.

There are over a hundred species of *Eucalyptus* that have the mallee habit.[324] Most of these are Western Australian though a number are widespread in both the eastern and western sectors. These include *Eucalyptus oleosa* (red mallee), *E. gracilis* (yorrell), *E. foecunda* (slender-leaf mallee) and *E. incrassata* (lerp mallee). Some species may occur in either tree or shrub form, and on rocky soils in some mountain parts of southern Australia the mallee habit is a response to inhospitable alpine terrain.

Most mallee grows to heights ranging from 3 to 9 metres, but on Kangaroo Island, off the South Australian coast, *Eucalyptus diversifolia* mallee grows to more than 10 metres.

The number of trunks arising from the lignotuber ranges from a few in 'bull' mallee, to many in 'whipstick' mallee. Understorey varies according to soils, rainfall and local conditions. Dense 'broombush' mallee may form thickets with *Melaleuca* or *Casuarina* in higher rainfall areas; or sclerophyll heath or sparse grasses, forbs, and chenopods form understorey on heavier soils. Spinifex hummock grasses form understorey on sand which is their preferred substrate.

Right: Spinifex forms a band on a hillside near Stanley Chasm, choosing that part of the slope where the soil is sandy.

M.E.W.

Far right: Spinifex, Callitris *native cypress pines and a euro in Bratchina Gorge in the Flinders Ranges, South Australia.*

M.E.W.

Below: Grey-green spinifex clumps and eucalypts festooned with mistletoe in the northern Flinders Range vicinity.

M.E.W.

Vegetative growth of mallee eucalypts occurs in late spring or summer although this is the hottest time of the year[308]. While they can flower at most times of the year in response to rain, the major flowering occurs in summer and may continue for two months in some species. Often the flowering of different species is stratified so that hybridisation between compatible species is reduced. Pollination is by insects, birds and bats which are attracted by the rich nectar and pollen, and bee keepers expect a good honey flow as often as every two to five years after rains. Large numbers of small seeds are produced, and the seed rain from ripe capsules does little more than feed harvester ants until fire causes a very heavy release of seeds. This fire-produced abundance satiates the predatory ants and leaves enough seeds which can wait for rain to germinate without losing viability for up to a year. Seedlings that germinate after rain and manage to survive the times of subsequent water stress can persist in a community in an 'advanced growth' seedling state, waiting until gaps appear in the plant cover and they have room to develop.

Left top: Reptiles are the most abundant animals in desert and semi desert areas.

JIM FRAZIER

Left: Termites are responsible for eating a great deal of grass and other pasture plants in arid regions.

M.E.W.

Below: Young Casuarina *trees with a columnar shape are associated with spinifex on the red sands of Central Australia.*

M.E.W.

The fuel properties of mallee determine the fiery nature of the vegetation[308]. Litter accumulates in large quantities under bushes and the strands or ribbons of bark that hang from stems and branches act as aerial fuel and promote the rapid spread of fire. Burning fragments are lifted vertically in the strong thermals that develop above a fire, and they are carried forward by the wind to cause spot fires ahead of the fire front. The oil-rich leaves also contribute to the fire as canopies burst into flame as soon as a critical temperature has been reached. The fact that in many regions spinifex hummock grasses form an understorey adds to the flammability. These grasses are resinous and fire-promoting and burn fiercely.

Because mallee-spinifex communities form a mosaic in semi-arid lands and not continuous cover, natural fire usually was restricted to limited areas, but after La Nina events when ephemeral plants had produced fuel that linked patches, massive fires through mallee lands resulted. Between November 1974 and March 1975, for instance, after a La Nina wet period, 19,000 square kilometres of mallee were burnt in New South Wales in fires burning for several weeks.

In the Murray Basin The Mallee is the region[202] where this vegetation type covered the dunes derived from the Parilla Sands and other sediments, freshwater and marine, after Lake Bungunnia had dried in Middle Pleistocene times. Linear dunes were first formed in the region probably about 500,000 years ago and were stabilised by vegetation. Reworked sands were blown into covering dunes in the last glacial stage and parna (blown dust from dry salt lakes and salt-pans) was spread across the land. Mallee mainly occupies the dunes, and the swales between carry the grasses and forbs on the clay-rich soil which contains parna and is not mallee's preferred substrate. Controlled burning in The Mallee can take advantage of the topography and distribution of the vegetation as cool fires burn the grass in the swales and have little effect on the shrub cover on the dunes.

Pockets of quite different vegetation are found within mallee[309]. Rosewood (*Heterodendrum oleifolium*) and belah (*Casuarina cristata*), sometimes with bluebush understorey, form a mosaic with mallee communities.

Of the 11,000,000 hectares of Murray Mallee that existed 200 years ago, only 15 per cent remains today. The Parilla Sands that underlie much of The Mallee are marine, laid down as the Miocene sea retreated from the Murravian Gulf, and today they form a generally saline unconfined aquifer (a water-mound) under the dune sands. In low parts of The Mallee, especially in the northern section, groundwater often reaches the surface and saline water seeps into the landscape. Salt lakes and salt-pans where salty dust is blown across the area down-wind lie in these discharge regions, and saltbush, not mallee, is the vegetation.

With the removal of the mallee and its deep root systems that stabilised the dunes and kept the watertable low in surrounding terrain, the watertable is rising and salination of the whole area is progressing inexorably. Irrigation of the cropped parts of the region since clearing has further raised the watertable, increasing the problem. The Mallee, like its semi-arid mulga counterpart, is a fragile ecosystem that had evolved with the region, adapted to the unique conditions that resulted from the geological history and the environmental stresses of seasonal drought, ENSO extremes and inevitable and irregular natural fire.

Although fire has played a part in the evolution of mallee[308], different fire regimes have a significant influence[309] on the botanical composition of mallee communities[311]. It has been found that the vegetation tolerates fire and recovers well if burning takes place in the spring and not more frequently than every five years. If more frequent, and particularly if burning takes place in autumn, the vegetation is degraded. The situation is complex because grazing pressures, whether there is rain after fires, and many other variables have to be considered.

In some mallee communities with hummock grass understorey and less than 300 millimetres average rainfall it is doubtful whether fire frequency can exceed one fire every 15 to 20 years, because it takes that long for spinifex to regenerate to a point where there is sufficient fuel to carry a fire.

It seems to be agreed by scientists experimenting with fire frequency[286], intensity, and different seasons for burning that as fire is inevitable and wildfires are hazardous and often enormously destructive, prescribed burning is the only option. If it can be found out how to balance fires at just the right time with improving grazing, increasing diversity and maintaining a healthy and dynamic vegetation that prevents soil erosion then what is left of the mallee vegetation type may survive.

3: 'SPINIFEX' HUMMOCK GRASSES

The two genera of hummock grasses, *Triodia* and *Plectrachne*[327], are Australian endemics. Endemism at a generic-continental level is unusual among the grasses, and the two genera are believed to be descendants of a Gondwanan ancestor. They exhibit characteristics of several grass families and are now classified with the Eragrostoideae, a group with Gondwanan origins. Evolution in isolation after the split-up of the supercontinent has resulted in two closely related endemic genera, adapted to the unique conditions that pertain in arid Australia.

There are about 35 species of *Triodia* and 11 of *Plectrachne*.[324] They can be divided into 'hard' and 'soft' types depending on their leaf anatomy. 'Soft' species have stomatal grooves mainly on the lower surface of leaves and produce leaf resin. 'Hard' species have stomatal grooves uniformly spread over both leaf surfaces and do not produce leaf resin. The leaves fold to form hard cylinders with sharp tips, and epidermal cells have grains of silica in them, adding to the rigid, sclerophyll nature of the plants.

While botanists and biogeographers try to make us abandon the name 'spinifex' because there is a genus *Spinifex* which belongs to another group, the Paniceae, and grows on coastal sand dunes in Asia as well as in Australia, 'spinifex' is to most of us synonymous with arid Australia and the magic of the Red Centre, and is here to stay.

The growth habit of spinifex, bristly cushions dotted on the sand or gibbers, and the soft grey-green and yellow-grey colours of the clumps, are an artist's delight. Aboriginal art depicts spinifex as fields of dots, a fitting symbol for an ancient, arid land where survival depends on integration of the animate and inanimate into systems uniquely Australian. The root system of each hummock is extensive and deep.

Spinifex occurs over the sandplains, dunefields and rocky ranges[324] of arid, and parts of semi-arid, Australia. It only exists as pure stands without shrub overstorey in restricted areas on some dunefields and on rocky ranges. As an understorey below very sparse to quite dense shrubs it covers about 25 per cent of the continent. We have seen how it forms understorey in The Mallee and in mulga lands. A large area east of Lake Eyre has a variation of spinifex/*Acacia* vegetation. Dune cane grass (*Zygochloa paradoxa*) and spinifex are sparsely distributed over low dunes with occasional *Acacia ligulata*.

Spinifex understorey in shrubland and open woodland occurs as far north as The Kimberley in Western Australia and Arnhem Land in the Northern Territory. Outliers from the main areas depicted on the distribution map show that the range of hummock grasses over Australia was probably much larger during glacial stages when dunefields were widely extended.

Three main spinifex alliances have been recognised in arid Australia[320], related to rainfall and soil-nutrient status. The *Triodia pungens* alliance occurs in the highest rainfall region with the better soils; the *Plectrachne schinzii* alliance in the mid-rainfall region with medium soils; and the *Triodia basedowii* alliance in the lowest rainfall regions on soils with a wide range of nutrient levels. *Triodia basedowii* and a few other less common species have a growth habit that results in spinifex rings. Plants grow from the centre outwards, and the centre dies. Kangaroos often camp within such rings, protected from the wind.

Topography can over-ride the general distribution pattern and, for instance, *Triodia pungens* can occur in low-rainfall areas when local conditions result in higher soil moisture. The palaeoriver systems that underlie dune sands and provide a high watertable are related to many of these occurrences, while areas adjacent to hills which receive run-on during rains also have higher watertables and provide local conditions at variance with the general area.

While shrubs are associated with spinifex over much of its range[319], trees are rare and the desert oak *Casuarina decaisneana* and the desert marble gum *Eucalyptus gonglyocarpa* are notable. Shrubs are represented by the genera *Acacia*, *Grevillea*, *Hakea*, *Thryptomene* (MYRTACEAE) and many others. Sub-shrubs are generally soft perennials with dense hairs on their leaves and stems, like *Crotalaria* (rattle pods, parrot beaks, FABACEAE), *Dicrastylis* (CLOANTHACEAE, lambs' tails) and *Solanum* (native tomatoes, SOLANACEAE). Grasses other than spinifex are usually annuals or short-lived perennials, while forbs comprise short-lived perennials or ephemerals. The ephemeral grasses and forbs that follow significant rains play an important role in the passage of fire through the systems, as the clumped spinifex with bare ground between does not encourage fire spread.

For arid ecosystems the spinifex grasslands carry a surprising number of species of plants. In the Simpson Desert 180 species are recorded, in the Central Australian sandplain 154 have been found. This reflects the high diversity of habitats and the marked change in species composition in the successions that follow disturbance. Many species only appear in the first year or two after fire or breaking of a drought and rely on long-lived seeds to retain a presence in the community. This is again a character of a flora which is adapted to the extremes of fire, drought and flooding rain. Opportunism is the keynote of success.

Spinifex at the Devils Marbles, Northern Territory.

M.E.W.

Post-fire succession in spinifex grasslands appears to be more regular than in other ecosystems. Short-lived grasses and forbs appear first, and most shrub and tree species resprout but form only a small part of areal cover in the first few years. Spinifex and woody plant cover increase with time and rainfall until maturity is reached. In northern regions maturity can be reached in less than 10 years, a reflection of monsoonal rainfall; in southern areas it may take 20 years for systems to mature. Spinifex is mainly a resprouter, but seed germination follows fire and La Nina rains.

Low rainfall is not an absolute requirement for *Triodia* species. An isolated population on the north coast of New South Wales grows on 'toxic' soils which overlie a body of igneous rock rich in magnesium and other elements (serpentine) in an area with a rainfall of between 1250 and 1500 millimetres. Sometimes it is lack of competition from other plants which cannot survive some extreme environmental factor in a locality that determines the local success of spinifex. Some spinifex species are salt-tolerant and some have a method of excreting salt through their leaves, implying a long period of adaptation to saline conditions.

Much of the spinifex country is desert and not used for grazing, much more is so arid that grazing pressures are light. Thus a good proportion of the hummock grasslands and shrublands are not affected by European settlement and even their fire regimes are natural. The impact of settlement on marginal and semi-arid areas is not negligible, however, and feral animals take a toll.

Population maintenance is dependent primarily on vegetative reproduction in both spinifex genera. Significant seed production[327] only occurs in La Nina years of extra rainfall. The seeds are large compared with those of other genera of eragrostoid grasses. As far as is known seeds do not have a long life.

Reptiles, termites and ants dominate the spinifex fauna[352]. The world's richest lizard fauna, with particular diversity in geckos, dragons and some specialist groups of skinks, makes use of the great variety of niches. The mosaic of slightly different ecosystems dictated by soil, water and other local conditions, and the successional stages present after fire account for the diversity. Just as the plants are opportunists, so are the reptiles.

The saltbush plains around Coober Pedy are surprisingly colourful, with red soil, pale mullock heaps and grey green chenopods.

M.E.W.

Below: A view across the saltbush plains from a dune near the Crowsnest Bore near Lake Eyre South.

M.E.W.

DISTRIBUTION OF SPINIFEX (HUMMOCK GRASS) AND SALTBUSH

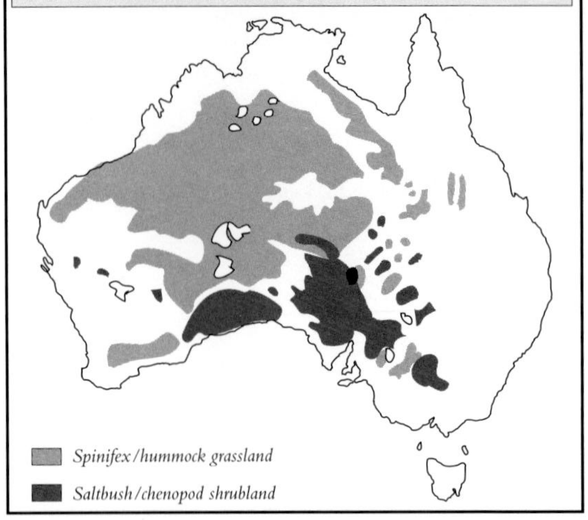

After Auslig 324

Termites are favoured by areas of infertile soils[312] which support large amounts of plant biomass. When the vegetation is too poor for persistent herbivores to use, a large proportion of the production goes straight into the detritivorous pathway, which is dominated by termites. Termites have symbiotic gut bacteria which enable them to use forage that is poor in nitrogen.

Ants and termites, both social organisms, are particularly prominent in the infertile, uncertain environment of arid Australia. Colonies represent the analogue of perennial plants because they have developed a persistent life-history strategy. The colonies act as storage organs which can buffer the pulses of production and they can forage more widely than individual insects — an advantage when resources are sparse and intermittent. While ants and termites represent the perennials, birds in arid regions are the ephemeral plant analogue. Nomadic and highly opportunistic behaviour characterises them.

Large mammals, like red kangaroos, are rare in the sandplains[319] or dune fields. Small mammals, however, take advantage of the micro-environments provided by spinifex clumps. Three known species of *Ningaui* (small marsupials) are restricted to spinifex habitats. Several species of native animals that were formerly common in the ecosystem are now rare or extinct. This appears to be due to several interacting variables, particularly the change from Aboriginal patch-burning to the occasional massive wildfire that burns out vast areas in a most destructive manner, and the impact of feral rabbits, cats and foxes.

The hare-wallaby *Lagorchestes hirsutus* has become rare[328] and endangered because of the change in fire regimes, and presumably other factors. It occupied dense spinifex and was protected there, and moved to areas adjacent to recently burnt patches where it could obtain fresh and digestible new plant growth. Bilbies (*Macrotis lagotis*) have suffered a similar decline in numbers probably for the same reasons, with competition from rabbits and predation by feral cats important factors.

Spinifex, perhaps the most uniquely Australian vegetation type of all, presents dilemmas in respect of fire management. The spinifex grasslands inevitably burn periodically and wildfires which occur after massive fuel build-up in good years are uncontrollable and devastating. It is known that small prescribed patch fires (lit when conditions are favourable for control), low-temperature burning, and predicted follow-up rain provide the most favourable recipe for species diversity and maintenance of stable and dynamic ecosystems.

Because water is a limiting factor for fauna, the palaeoriver systems with their core refuge areas must be maintained and protected. It has been suggested that mosaic burning in patches radiating from the core areas (which can be determined from satellite imagery) gives the best conditions for fauna. It would provide the different successional stages after fire that are exploited by the opportunists, both plant and animal, and in the case of animals these would be within their range as determined by water.

Implementation of such schemes in localised areas is said to have been successful. Whether such management is sustainable when it is so difficult to avoid damage by misreading the probability of drought or depleting the native seed-store by burning too often, and other interactions which are not yet understood, remains to be seen. Man is inherently a pyromaniac and must not abuse his power to 'manage' nature. Perhaps it would be better to use burning to create very limited fire-breaks to reduce the ferocity of wildfire, and to tackle the feral animal problem and stop grazing marginal lands. The land coped before Aboriginals and their patch-burning and would presumably still do so if introduced animals, including people, had not upset the balance.

CHAPTER TWENTY-THREE

Ecosystems at Risk from Grazing

The Low Open Shrublands and Tussock Grasslands

The Chenopod Shrublands

Saltbush, Bluebush and Samphires

This vegetation type has suffered greatly from grazing by introduced animals. While sheep have taken their toll and the carrying capacity of saltbush and bluebush pastures has declined drastically, it is rabbits that have proved to be the desertmakers in these shrublands.[321] They did a great deal of irreparable damage in the early days of settlement. Following La Nina years in the 1880s they had turned the saltbush plains of northeastern parts of South Australia into desert. Rabbits prefer the regions with calcareous soils, and chenopods have a similar preference.

Saline soils that cannot be used by other vegetation are the undisputed territory of saltbush and in the highly saline places surrounding salt lakes and areas of saltwater discharge, samphires find their niche. Some chenopod shrublands occur on red duplex soils in parts of the Northern Territory.

Chenopod shrublands form wide open plains steppe-type vegetation[324], and are understorey in arid *Acacia* and *Casuarina* woodland, over large tracts of land in southern Australia. They range from the western edge of the Nullarbor, across South Australia (where saltbush also occupies the central region as far as the Northern Territory border as the major vegetation type), into western New South Wales and the eastern part of the Riverine Plain. Plants from several genera of the family CHENOPODIACEAE dominate, *Atriplex* (salt bush) and *Maireana* (bluebush) being the most prominent. *Chenopodium* may be locally dominant in areas that are prone to flooding.

Most chenopods are low-growing shrubs, not exceeding 1.5 metres, but the 'old-man saltbush' *Atriplex nummularia* may reach three metres high. The bladder saltbush *Atriplex vesicaria* is economically the most important species, being the most palatable to stock. It is the grasses and forbs that grow between the chenopods which constitute much of the fodder when there has been rain, but when that has gone the chenopods are eaten. In places where conditions for survival of saltbush are marginal[321], grazing by sheep and cattle has eliminated this vegetation type, as in the area north of the Macdonnell Ranges. Elsewhere the effect of heavy grazing pressures, the slow

Top right: Mitchell grass, bleached white, covers the plains in parts of the Great Artesian Basin.

JIM FRAZIER

Right: A dingo hunts on the endless plain of white grass looking for its usual diet in these areas when nothing better offers — grasshoppers and other small creatures.

JIM FRAZIER

The tussock grasslands have suffered considerable thinning and quality loss as a result of heavy grazing pressures.

JIM FRAZIER

Rain in the tussock grasslands can no longer bring back the rich grass cover or the mulga where overgrazing has degraded the land.

Jim Frazier

re-establishment by seed only in La Nina years and, in parts, invasion by other plants like nitre bush have altered the structure of the vegetation. With the thinning of plant cover and the trampling of the ground, soil degradation has also occurred, as evidenced by the mounding that is often present round the bushes.

The saltbush and bluebush species are all drought-tolerant.[340] *Atriplex*, for example, has leaves designed to reflect as much heat[339] as possible by having surfaces covered in hairs and coated with crystals of the salt which they excrete. Stomata are sunken and can be shut when water stress is acute to minimise water loss. The inclination of leaves to the stems is also arranged so that minimal area is exposed to the sun, and there is seasonal change in leaf attitude so that maximum advantage is gained from the different conditions which prevail.

The seeds of saltbush, like those of nitre bush[340], have devised methods of coping with the unpredictable rainfall patterns of arid Australia. The seed cases and outer layers of the seeds themselves have high concentrations of salt which inhibit germination. It is necessary for them to soak for a considerable time before the salt is leached out and germination can proceed. Heavy rain is therefore required, and when that occurs (more often than not in a La Nina event) the embryo is capable of very rapid root production so that it can tap soil moisture for the seedling after surface water has dried. In addition an unusual feature of seed production is seen in this group of plants, conferring an extra advantage under unpredictable climatic conditions.

Plants produce two sorts of seeds, some small, hard and black; others larger, softer and brown. In *Atriplex holocarpa* water stress triggers the production of black seeds; in *A. hortensis* black, hard seeds are produced routinely late in the seed flush; in *A. heterosperma* black seeds are produced early in the flowering season; and in *A. subspicata* the small hard seeds are produced at fixed seasons related to day length. Having two sorts of seeds provides an option: to germinate soon or to delay germination and remain viable until really good rainfall comes along eventually, which is rather like hedging bets! Many of the soft seeds simply satiate the demands of ant and termite harvesters.

Chenopods are not fire-tolerant, and fire under natural conditions[286] is an unusual event in the chenopod steppe country as the sparse vegetation with bare ground between bushes does not carry fire. In woodland with an understorey of saltbush, fire risk is greater, and because the chenopods do not usually resprout and are not highly opportunistic seeders after fire, they tend to be replaced by grass or in some cases by woody shrubs. This may result in more frequent natural fires, because chenopods are not as fire-prone as their replacements are[285]. It is agreed by all the experts that there is no case for prescribed burning in chenopod-dominated vegetation.[292]

There has been no development of characteristics within the family to protect against herbivory. The plants are not armoured with thorns or equipped with poisons or unpleasant-tasting substances. They are long-lived and bladder saltbush may live for about 25 years, while black bluebush is believed to attain 150 years of age under favourable conditions.

Apart from some early experimental clearing of saltbush land to grow wheat[321], such as occurred in South Australia in the 1890s with disastrous results, little has been used for purposes other than grazing in its natural state. An exception is the eastern Riverine Plain in New South Wales where *Acacia pendula* woodland with *Atriplex nummularia* understorey has been largely replaced by short grasses as a result of clearing and grazing.

The Riverine Plain, like the floodplains of the Cooper and Diamantina[207] that drain into Lake Eyre, is a distinct geographical entity where conditions are different from surrounding regions. East of The Mallee the Plain is traversed by ancient and modern stream channels. The flat topography results in the streams which emerge from the foothills of the catchment breaking up into wandering distributory systems that rejoin and separate and eventually form the Murray River trunk stream near the Mallee-Riverine Plain boundary. The river channels are eucalypt-fringed, forming dark lines through the featureless terrain where saltbush, bluebush, open woodland and grassland form a mosaic. Similar vegetation occurs in the floodplains north of Lake Eyre, making a break between the mulga lands of the Centre and those of western Queensland.

The nature of saltbush and bluebush — slow growing, slow to replace by seed — makes chenopod shrublands fragile systems which need to be protected from overgrazing. Those that are marginal and close to the limits of their range will become desert if feral animals as well as domestic ones are not removed.

The Tussock Grasslands

The rolling downs country of central Queensland, on cracking clay soils derived from Great Artesian Basin sediments laid down in the Early Cretaceous epicontinental sea, is a sea of bleached grass. It is treeless, because the clay swells when wet, cracks when dry, and inhibits tree root development. A typical landscape is an endless plain, flat or very gently undulating, extending to the horizon. Geologists named the strata which underly this vast sea of tussocky Mitchell grass the Rolling Downs Group and it is possible to map the extent of the Cretaceous rocks by observing the boundaries of the grasslands. The Barkly Tableland of the Northern Territory is also Mitchell grass plains on cracking clay soils derived from sediments of the Cretaceous Eromanga Sea, outside the boundaries of the Great Artesian Basin with its Jurassic and Cretaceous aquifers.

Mitchell grass (*Astrebla* spp.) has proved to be remarkably tolerant of grazing. Over much of its range it has merely suffered degradation in that it is a more sparse cover of the ground now than it was before it was grazed by hordes of introduced sheep and cattle. In the northeastern downs in Queensland, however, an introduced thorny bush, *Acacia nilotica,* has invaded overgrazed areas and turned tussock grassland into open savannah.

The basaltic plains of Victoria, west of Melbourne, were extensive mixed tussock grasslands and savannah before they were cleared for agriculture or sown with exotic pasture. *Themeda*, *Danthonia*, *Stipa* and *Poa* were the dominant grasses. The loss of this major habitat to the eastern barred bandicoot and the contraction of its habitat elsewhere, together with predation by cats and foxes, now leaves this engaging animal on the endangered list with perhaps only 500 individuals surviving. This is just one example of the thousands which are the largely unseen consequences of environmental change that has followed European settlement.

Tussock grasslands were once widespread in semi-arid and better-watered parts, in coastal areas and in eastern and parts of southern Australia, including Tasmania. Much of this vegetation has suffered clearing or modification with sown pasture species, and the remaining natural grasslands have often had their species composition altered by the selective grazing of sheep or cattle.

The changing composition of pasture was already noted by farmers in the Riverina and elsewhere during the last century. Spear grass (*Stipa*) became abundant, replacing the more palatable *Danthonia* and *Themeda* and a consequence was the problem of the awl-like seeds of the spear grass burrowing into the skin and nasal passages of the sheep. The composition of grasslands on the Southern Tablelands has been shown by scientific observation to have changed under grazing by introduced animals. There the open nature of grasslands is not a condition that was induced by Aboriginal burning of woodland. The winter frosts that affect the area are the determinants of grassland, as is the case over much of the Highveld of southern Africa.

DISTRIBUTION OF TUSSOCK GRASSLANDS

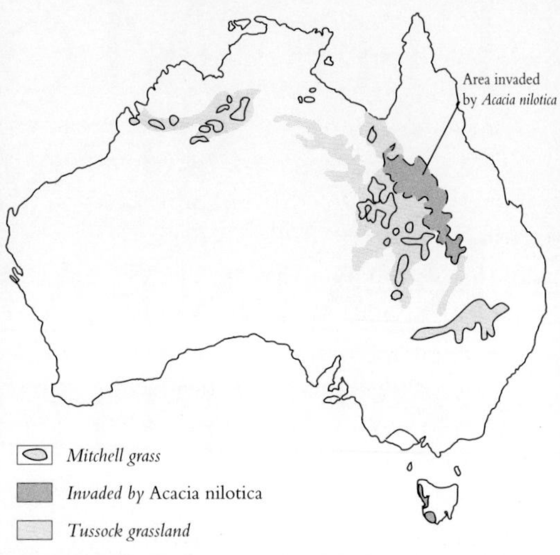

After Auslig 324

Today in the semi-arid zone near Blanchetown in the Murray Basin observations are being made on the biology of pasture grasses grazed by the now-rare hairy-nosed wombats and by grey and red kangaroos. Dr Rod Wells and a team from Flinders University have found that when no introduced herbivores are grazing the pastures, the native grasses do well and a balanced community exists. The changing mosaic of habitats and populations that existed in the past was a result of this balanced state. Heavy grazing by rabbits and sheep alters the nature of the vegetation. More of the exotic species become abundant and they have growth phases concentrated in spring and early summer and reliant on years of higher rainfall. Thus pastoralism and rabbits are pushing the lands towards a boom and bust economy with high productivity by exotics in good years and low productivity of the system in between. Times when palatable forms are available for grazing are out of phase with wombat weaning times under this altered regime. The combination of the altered species composition and altered sequence of productivity has a dramatic effect on the recruitment of young wombats.

Hairy-nosed wombats, and so many other native animals, have little chance of avoiding extinction in time because the added pressures imposed by people on a fine-tuned interrelationship between fickle environment and their specialised lifestyle are the final straw. Dr Wells believes there is a fine balance between pasture composition and the recruitment of other small grazing marsupials as well.

Aboriginals disrupted the balance by changing fire regimes, and modern Australians have accelerated the unbalancing of the biota. The semi-arid and arid zones are inexorably moving towards a boom and bust system and the end result is desertification. The native seed-bank declines and ultimately there is no regeneration. When vegetation is degraded extinction of animals accelerates.

The writing is on the wall for Australia's marginal lands — the chenopod shrublands, semi-arid grasslands and savannahs. To impose continuous grazing pressures on frail ecosystems whose balance was precariously held in the unpredictable environments of a land of fire, drought and flooding rains is a recipe for desert-making. Consider the northern Sahara — a man-made desert, and the eastern Mediterranean lands that once had the highest mammalian diversity in the world! Any complacence that one might feel because so far 100 years of grazing has 'only thinned' the arid-land pastures should dissipate!

CHAPTER TWENTY-FOUR

The Monsoonal Summer Rainfall Lands of the North

Gouldian finches are rare and endangered in their natural habitats in the savannah of the monsoonal north as a result of the burning practices there that see most areas burnt annually at their nesting time. The fact that they nest in low bushes increases their vulnerability.

Jim Frazier

The comparatively well-watered northern sector of Australia has 60 per cent of the continent's water resources and only 15 per cent of the population — a mis-match situation. With water the limiting factor for population growth and the Murray Basin region, for instance, with 90 per cent or more of its water-use capacity already committed, the main region capable of sustainable development with increased population is the north. That this situation has persisted reflects the original and still current attitude that mediterranean-type agriculture has been the way to go.

It is the firm belief of many of us who see Australia as the unique time-worn land with frail ecosystems that the continent is already quite grossly over-populated for sustainable development because so little of it can be used without causing degradation that will steadily turn it into desert. That even good and sensitive management will only slow down what is a 'natural' progression in that direction is the ultimate problem which technology may, or may not, be able to solve.

Even in the well-watered part of the monsoonal north the impact of introduced animals and of mis-management by fire has degraded, and continues to degrade, the landscapes. Nowhere does fire present more of a dilemma than in the savannah lands which characterise much of the region. They burn, inevitably, sooner or later under natural regimes. That they are being burnt annually is regarded as 'anticipating the inevitable'. Moreover, now that artificial fire regimes are imposed on virtually all of the region instead of in random patches which left some areas free of fire for long periods, the problem has new dimensions.

The most northerly Top End has areas of open forest with trees 10 to 30 metres high[313], and isolated patches of rainforest within these areas, mainly in a strip across the top of Arnhem Land and bordering Van Diemen Gulf. Woodland with trees 10 to 30 metres high forms a wide band adjacent to the coast from Melville Island to the Victoria River, and also occurs as a narrower band adjacent to the coast of the Gulf of Carpentaria. Low woodland surrounds these areas, merging into

Grass fires in the open savannah of the Northern Territory like this one in the Alligator River district, pass rapidly through the landscape and barely scorch the established trees.

BILL WHITE

Pandanus palms on the banks of the Alligator River in the Northern Territory form the margin of dense riverine forest.

BILL WHITE

spinifex lands with sparse eucalypt and *Acacia*.[345] The rainforest patches occur as small disjunct pockets in an otherwise vast expanse of eucalypt-dominated savannah. Most of the patches are less than a few hectares in extent, but occasional large areas up to thousands of hectares are still to be found along rivers and on coastal landforms. The viability of the small patches is doubtful as any degradation from whatever cause pushes such small isolates closer to oblivion. The larger tracts require protection and wise management if they are to survive.

Responses to fire of the different vegetation types[281] which exist under different climatic conditions[279] are often not the same. In semi-arid rangelands of the southern half of Australia too-frequent fire decreases the annual grass component of vegetation, and exclusion of fire for too long leads to increase in woody and inedible shrubs. In the north in the humid tropical savannah rangelands, perennial grasses are promoted by frequent fire. There is still much controversy about the effect of fire on the vegetation as a whole, and much experimentation is being carried out by the CSIRO. Because the Kakadu National Park[344] is involved in the experiments[290], many people are being made acutely aware of the fire dilemma in the north as a result of the increasing popularity of the Park with tourists. At least 85 per cent of the Park is burnt annually. Open forest, which represents 60 per cent of the Park, is burnt in two out of three years.

Natural fire in the north occurs most frequently towards the end of the dry season.[343] The *gunumeleng* — the first-storm time of the Gunwinngu-speaking Aboriginals — brings lightning, and the uncontrolled and uncontrollable fires which result are part of the argument for imposing a prescribed fire regime. Under Aboriginal management 'cleaning up' of the country starts in the late wet season at *banggerreng* — the 'knock 'em down storm time'. It continues until *gurrung* — the hot-sand time (April to November). The late dry season, which is the main time for the enormous fires which occur naturally, is a forbidden time for deliberate fire-lighting among the Aboriginals because of the intensity and dangerous nature of the fires which occur then. Their patch-burning in the lead-up to the driest and most fire-prone time serves to restrict the inevitable wildfires and to create fire-breaks, and it has been an Aboriginal practice to burn breaks round the patches of rainforest which provide them with fruits and other additions to their diet. In north-central Arnhem Land in the Blythe River region half the area is burnt annually including all habitats except the small pockets of vine forest.

Key

- *Open forest—trees 10-30 m high with some small patches of rainforest*
- *Woodland 10-30 m high*
- *Open woodland 10-30 m high*
- *Low woodland*
- *Hummock grassland*

DISTRIBUTION OF FOREST, WOODLAND AND HUMMOCK GRASSLAND

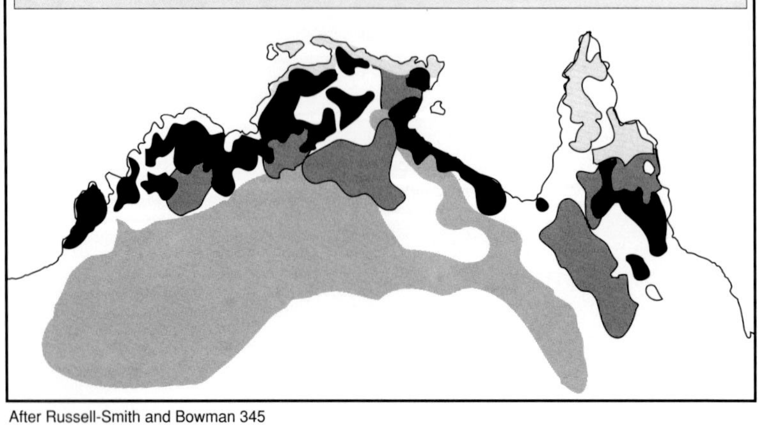

After Russell-Smith and Bowman 345

However, even in Arnhem Land under 'wise and traditional' fire management practices, the boundaries of the rainforest patches are being increasingly eroded by fire. It would seem logical to assume that the small remnants of rainforest that exist in the Top End today are what remains of larger areas that developed in even better-watered intervals of our present interglacial. The increased fire frequency that resulted from Aboriginal fire-stick farming has contributed to the near-disappearance of rainforest in the region.

A recent investigation of 1220 small rainforest isolates[345] in the Northern Territory found that a third were severely disturbed by fire; 20 per cent were badly affected by cattle and feral water buffaloes; 10 per cent were much damaged by feral pigs; 7 per cent were damaged by storms and 5 per cent by flood; and 22 per cent were weed-infested. Hardly a healthy situation — and it looks as though the monsoonal tropical rainforest isolates are on the way out.

In the eucalypt savannah only about 4 per cent of perennial grasses like *Themeda australis* and *Amphipogon caricinus* are killed by fire provided it is not an intensely hot fire due to a large build-up of fuel being burnt at the end of the dry season. However, annual *Sorghum* species[347] can be completely lost from grassland that is burned after the first rains of the Wet.

Intense fires affect the relative abundance of species and cause greater mortality even among perennial grasses. Re-establishment from seed usually compensates unless another fire intervenes before seed is set. If drought conditions follow fire, as is inevitably the case in the long run because of ENSO, degradation has to follow. *Callitris* (native cypress pine) forest patches are fire-sensitive and are destroyed by high intensity fires. The nature of fires in eucalypt woodlands in the north is different from that in the southern part of the continent as crown fires do not occur. The aromatic oil content of leaves is less under the summer rainfall, tropical conditions.

Experiments carried out on fire frequency were thought to show that there was no lasting effect on the vegetation[342] of the monsoonal savannah even with annual low-intensity fires, but it has since been shown that the experiments had flaws and the subtle, and some not so subtle, changes were not being detected. With increasing fire frequency many species otherwise relatively tolerant of fire

DISTRIBUTION OF ABORIGINAL LAND AND NATIONAL PARKS

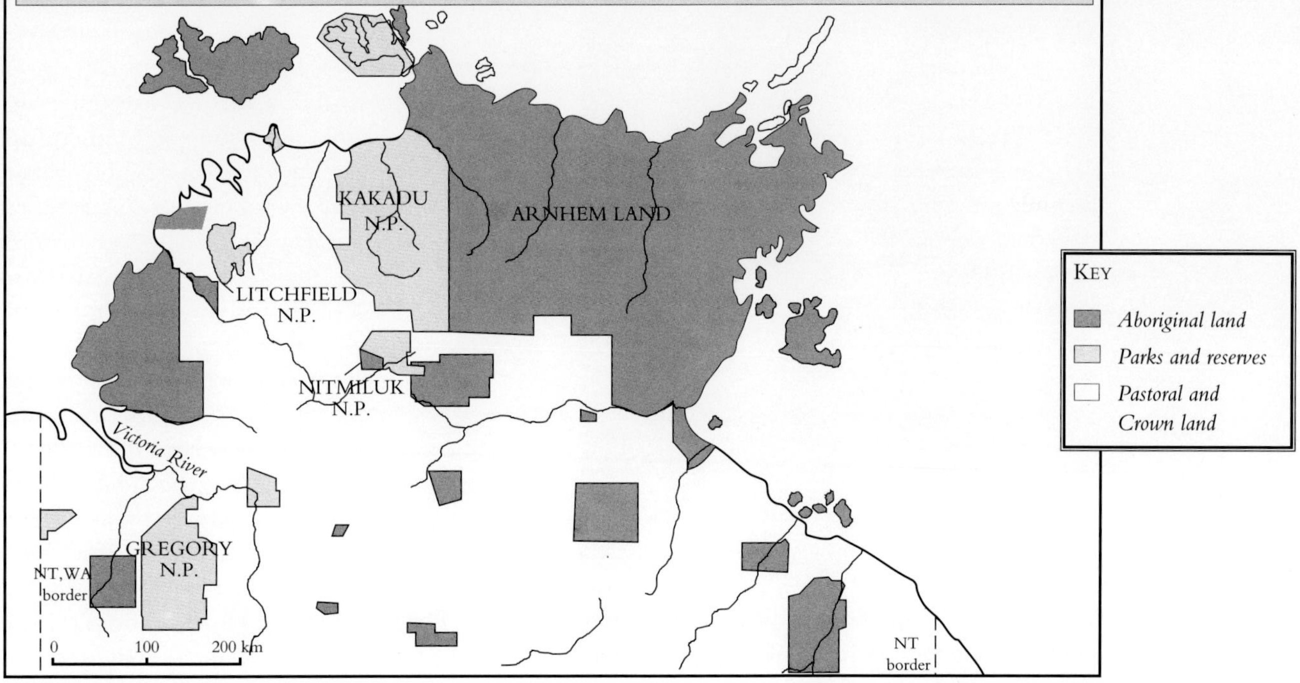

After Russell-Smith and Bowman 345

become increasingly susceptible.

An example of this can be seen on the Arnhem Land escarpment.[343] There *Allosyncarpia ternata* (MYRTACEAE), which is restricted to the region, is ecologically significant as the canopy dominant. It is tolerant of occasional fire but with increasingly frequent fires its saplings and mature trees with termite damage become susceptible. When trees round the margins of the forest patches are killed and the surviving understorey and regenerating plants are exposed, the boundaries of the vegetation type retreat and dry species like *Triodia* (spinifex) and *Acacia* take over. The seedlings of *Allosyncarpia* only manage to become established under the canopy, so opening up of the cover within the forest patches is unfavourable for the survival of this vegetation type. When margins have retreated as a result of frequent burning it takes 20 years without fire for the trees to come back.

That the frequency of fires is a critical ecological factor[342] is shown by the fact that in Kakadu National Park 37 per cent of the common forest species and 26 per cent of common woodland species showed significant response to fire in terms of cover value. Intense fires cause substantial loss of biomass and vegetation structure. If they were to become frequent, degradation would be severe. There is no agreement among scientists as to whether occasional intense fire plays a useful role in maintaining diversity. It may be that small intense fires (not the large all-enveloping ones that result from late dry season natural fires) may have to be considered under certain circumstances. For the rest, there seems to be no option but to use fire as a management tool in savannah where devastating dry season natural fire is inevitable.

The re-introduction of patch-burning[341] in much the same manner as used to be practised by Aboriginals would serve several purposes.[305] The dangerous and degrading wildfires would be prevented from burning over vast areas because the burnt patches would form fire-breaks. By burning early in the dry season and early in the wet season to establish the mosaic of burnt patches the least damaging and most controllable low-intensity fires would result. In order to maximise the diversity of the habitats in the region fire should be less frequent than the annual burning which affects so much of the landscape now.[290] It is also necessary to protect the water-seep and riverine areas on which the animal population depends.

Preservation of the rich floral and faunal heritage[341] of the tropical north is believed to depend in the now-prevailing circumstances on the wise use of controlled, low-intensity fire. This type of management meets the needs of the rangeland farmers and other human inhabitants by maintaining feed for stock and minimising the dangers of uncontrolled fire.

Long-tailed grass finches are particularly disadvantaged by the annual burning practices of farmers in the tropical north and are becoming rare and endangered in the wild.

JIM FRAZIER

At the same time it gives the best chance of survival for some of the animals and probably inflicts the slowest rate of change on the vegetation. That some species are becoming increasingly rare under human management of the complex ecosystems and that more will do so and many extinctions are inevitable results from the fact that different organisms are affected in various ways.

Small birds which nest in low shrubs at the very times when optimum conditions exist for 'safe and beneficial' burning are the first casualties. Finches and other seed-eating birds[347] are particularly threatened and it has been found that by protecting the water-seep areas in forests and woodlands where they congregate they will be less disadvantaged.

Deliberate selection of habitats created by fire by certain animals is a common occurrence in fire-prone landscapes in many parts of the world. Opportunists take advantage of particular stages of the regenerating vegetation after a fire, and many different animals have adapted their behaviour to benefit in different ways. For those that stay in their territory and have to face annual fires, obviously techniques for avoiding the flames and simply surviving are paramount. Small mammals are particularly vulnerable. Brushtail possums are becoming rare in areas where near-annual fire is the norm. After intense fires, native rats *Rattus tunneyi* may be exterminated and their niche[341] may be taken by *Pseudomys delicatulus*, leading to a local mouse plague.

In the category of creatures which have to develop strategies to survive fire, the reptiles show just how complex the management of the vegetation by fire to retain optimum conditions for the biota really is. Among lizards in Kakadu[346], some are fire-sensitive and any burning threatens their survival, some are favoured by early hot fires, some by low-intensity burns, and some by low-intensity fires that contain high-intensity hot spots. Thus in order to maintain the whole lizard fauna within the Park a range of fire regimes has to be maintained.

Among birds the raptors — the hawks, kites, falcons[344] and kestrels — are examples of opportunists for which fire means feasts of insects and small reptiles and mammals fleeing the blaze. Some of the birds fly into the burning fire front, even getting their feathers singed, to pick up prey. Their excellent eyesight and high flying allows them to see smoke from up to 20 kilometres away and they congregate at fires over a vast home range. They often remain after the fire has passed to collect any dead or injured animals. The hot-ash stage and the cold-black phase before any resprouting takes place are exploited by many bird species while their prey are visible and vulnerable. Crows and magpie-larks often forage immediately behind the fire front, while tree martins, black-faced woodswallows and little woodswallows forage in the canopy while fire goes through the understorey, and swifts fly above the canopy to catch insects which are caught in updrafts. Red-tailed black cockatoos and partridge pigeons forage for seeds that have fallen as a result of the fire.

Kangaroos and wallabies often come to newly-burnt areas[341] to eat the bits of green that have been revealed in the base of tussock grasses if the fire has not been too intense. Many different creatures, from invertebrates to large kangaroos, are attracted to the green pick that follows fire.

To avoid degradation of ecosystems when fire is used as a management practice the spacial and temporal mix of fires must be right. To do things right and take into account the known needs of important species requires the wisdom of Solomon. There is no way that changes will not occur over time. Species will become extinct and other problems will inevitably emerge. And fire cannot be seen in isolation because degradation of vegetation by overgrazing, of soils by trampling, of wetlands by feral buffalo wallowing, and other stresses imposed on the landscape will alter the 'predictable' effects of fire.

CHAPTER TWENTY-FIVE

OPEN FORESTS AND WOODLANDS

For the average urban Australian who does not see the continent as vast arid landscapes, red deserts and empty gibber plains, the 'typical Australia' is mainly gum trees in open forests and the eucalypt woodlands of the better-watered eastern sector. Together with the 'mediterranean-type farming is the way to go' syndrome, this myopic view has been responsible for most of the ills which afflict the land today.

It should be a prerequisite for politicians, policy makers and anyone who determines the future of Australia, to go on an extended trip to get a perspective on what the real Australia is before they are allowed to assume office. Their journey should preferably be in basic and reasonably uncomfortable transport which does not insulate them from the environment, not flitting in VIP aircraft or in airconditioned comfort of limousines! Travelling through the arid and drought-afflicted lands, the bad lands, the deserts and the amazingly wonderful empty or 'unproductive' parts of the land; seeing how marginal are the conditions and how unpredictability is the characteristic of most regions and that drought almost everywhere is inevitable sooner or later, would be a revealing experience for most. They might see that the carrying capacity of this land in terms of humans and their introduced stock is not determined by the size of this continent. Much of the marginal land is so degraded by inappropriate use that it should no longer be used for grazing. The area which is capable of something approaching sustainable use is already mostly developed and much of that fraction (which is perhaps a quarter of the total area of the continent) is also degraded and in need of special care.

Dry sclerophyll forest dominated by eucalypts is the most widespread forest type, in the reasonably well-watered parts of the continent.
M.E.W.

The 'typical' Australian eucalypt open forests and woodlands, seen with a perspective gained from knowing the whole continent, are holding back the desert at the margins of their range. How close this desert is in the absence of their protective cover is seen in the rising salt (the ultimate desert-maker), in soil erosion and dustbowl landscapes.

The open forests, mostly *Eucalyptus*-dominated, are mainly on the Great Divide and on the coastal fringe to its east. Today they comprise about 140,000 square kilometres of State Forests and 50,000 square kilometres of National Parks and reserves. This represents about half of the original forest that was here 200 years ago. Clearing for

DISTRIBUTION OF SCLEROPHYLL WOODLANDS AND FOREST

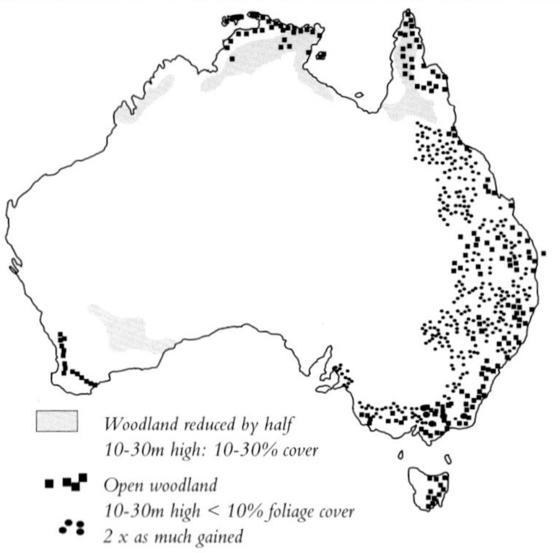

After Auslig 324

agriculture eliminated most of the open forest dominated by *Casuarina* and *Acacia* (particularly brigalow, *A. harpophylla*) as well as half of the open forest dominated by eucalypts. Regular logging of State Forests has altered them by modifying tree density and understorey, and fire practices have also changed their structure. Grazing of the grass understorey in forests has contributed to change.

In Western Australia the jarrah forests of the southwestern corner have been downgraded to woodland by the killing of trees by the dieback fungus *Phytophthora cinnamomi* which was spread originally by the logging activities within the forests. Not only was the fungus transported on the tracks and wheels of bulldozers and trucks and on the muddy boots of loggers, but the change in vegetation brought about by the burning that followed logging exacerbated the situation. *Banksia grandis*, a member of PROTEACEAE whose hard woody seeds need fire to germinate, was able to grow abundantly with the reduced competition and the fire-promotion of its seedlings. It proved to be susceptible to the disease and was instrumental in spreading it, and now many members of the PROTEACEAE are dying of dieback in the southwestern part of Western Australia.

The structure and composition of the eastern *Eucalyptus* open forests vary according to soil and rainfall and the understorey varies from low trees or shrubs to native pasture grasses (which are often maintained by regular burning) or spinifex. Grassy open forests are generally associated with more fertile soils and are increasingly common in the sub-coastal region of southeastern Queensland. Dry sclerophyll forest and woodland is characteristic of infertile soils south of the Tropic of Capricorn.

Narrow strips of fringing open forest and woodland of *Eucalyptus camaldulensis*, the river red gum, follow the rivers and flood channels into the dry interior of the continent. Forests of red gums spread out across the floodplains of major rivers. The deep roots of the red gums that follow the mostly dry watercourses of the interior tap sub-surface moisture which enables the trees to survive. In some places today their survival is being threatened by the introduced tree the Athel pine (*Tamarix aphylla*). Its extraordinarily rapid growth and very deep root system enables it to thrive at the expense of the gums. Competition for the limited available moisture puts the river red gums at risk. A great tragedy will result if the spread of the 'feral' plant is not controlled as the red gums are most beautiful trees and each represents a small ecosystem with its associated fauna — like an island in the sea of desert.

The open eucalypt forests of the better-watered regions merge into the layered woodlands, woodlands with low scrub, and woodland with grass understorey (from tussock to hummock). Much of the wheatbelt of New South Wales was layered woodland with *Eucalyptus microcarpa*, the grey box, over white cypress pine (*Callitris glaucophylla*). Eastern grassy woodlands have been much changed through clearing, thinning to promote native pasture, fire practices and degradation by overgrazing.

The species of *Eucalyptus* dominating in the open forests and woodlands differ in different areas. *E. populnea* is common in eastern woodlands. In the Darling Ranges of Western Australia the beautiful wandoo, *Eucalyptus wandoo*, with some jarrah associated, forms woodlands with a low understorey of sclerophyll plants and *Xanthorrhoea* grass trees. The wandoo woods extend further inland where they are mixed with York gum (*E. loxophleba*). Much of this woodland has been cleared for wheat. In Central Australia the white-stemmed gum, *E. papuana*, much beloved by painters for its silky, shiny white bark which contrasts so beautifully with the red rocks of the gorges, is widespread.

The grassy low woodlands of southeastern

Australia have largely been cleared and those of less altered parts merge into the mallee and mulga lands.

Dry sclerophyll forest is fire-prone. During the dry seasons[298], and particularly after a good wet season and build-up of fuel, the vegetation is highly flammable. Litter builds up in the forests, the hanging bark strips of many eucalypts act as fire brands to be carried by updrafts ahead of the fire front, causing spot fires, and the volatile oils in leaves are explosive in hot fires.

After fire, regeneration is by resprouting (as in eucalypts from lignotubers) or from seed. Some plants employ both strategies. Adaptations to survive fire include the production of thick insulating bark; the shedding of bark annually so that there is no build-up to burn and flash fires simply scorch the outside but cannot remain smoldering; the presence of epicormic buds which allow sprouting of branches and leaves along the trunk of trees that have been damaged by fire; and the production of hard seeds which require fire to germinate. Some hard seeds, notably those of some quite fire-sensitive legumes, can lie in the soil for up to 100 years waiting to germinate.

Fire frequency, intensity of burning and the season at which burning occurs influence the response to different fire regimes. The area originally occupied by dry sclerophyll has been greatly reduced by clearing for cropping and pasture and much of what is left is managed by low-intensity fire to reduce the hazard of the natural fires which would inevitably occur. Before humans intervened, the wildfires that must have swept across the land would have been patchy, leaving many areas unburnt and providing a mosaic even within the territory through which the fire swept because of different topographical and local conditions. Diversity would thus have been maintained on a broad scale. Aboriginal fire probably reduced the incidence of enormous wildfires and certainly changed the nature of vegetation generally by bringing more frequent but lower- intensity fire to much of the land.

Most studies on the effects of fire on the dry sclerophyll forests and woodlands have concentrated on simple interactions between fire intensity and species response. The regenerative responses of many dry sclerophyll communities may now be predicted, but the persistence of these responses is still unproven. It was found that in coastal woodland spring fires favour shrub regeneration, while autumn fire favours herbaceous species. Frequency of fires is important in determining the composition of the vegetation, and frequency to a great extent determines the intensity of fire, because big build-up of fuel increases intensity.

The situation is very complex because it has been shown that floral and faunal diversity is dependent on there being some disturbance of the system by fire. There is no uniformity of response to low-intensity fire, nor to intense fire, nor to fires at one season or another. To accommodate all the elements of the biota a range of fire regimes is required. General use of regular low-intensity fire disadvantages some species because high-intensity fire is eliminated and some seeds, for instance, only germinate after intense fire heating, even if they have to wait 20 or 30 years for such a fire.

A couple of examples of the complex interrelationships that exist between fire and the biota serve to show the sort of problems that have to be addressed when devising management strategies. In Western Australia the wallaby *Macropus eugenii* lives among bushes of *Gastrolobium bilobum*, a legume which has nitrogen-fixing root nodules. The nitrogen it makes available in the soil allows grasses, which are rare in the southwestern forests, to grow, and these are the food of the wallabies. The *Gastrolobium* is a fire-dependent species whose seeds require the heat of a high-intensity fire to germinate. Such fire is required every 20 or 30 years, the life span of the bushes under different conditions.

Also in Western Australia, the bettong lives on fungi which are dependent on high nitrogen in soils. This is produced by the leguminous plant *Bossiaea ornata* which shelters the bettong as well as enabling its food to grow. The bettong spreads the fungus, whose spores pass through its gut unharmed and are then dispersed by scarab beetles which eat the scats. The *Bossiaea* seeds also need intense periodic fire in order to germinate.

The web of life is so complex and the subtle changes that occur when humans try to manage natural systems can have effects in the long term, as well as in the short, which were not predicted. The very interference, imposing regular patterns on what were random events, dooms some living things to

Above: Open woodland with Eucalyptus wandoo *in kwongan-type vegetation on the coastal sandplain of Western Australia.*

M.E.W.

Right: Banksia coccinea, *a Western Australian endemic species much used by florists, is becoming rare and endangered in the wild because of over-picking by commercial flower sellers and its susceptibility to dieback.*

M.E.W.

Far right: Eucalypt woodland manages to survive within the sheltered microclimate of Stanley Chasm and other such refuges in the arid zone.

M.E.W.

extinction. When Nature cannot be allowed to take its course, and Nature and the aspirations of humans have been at loggerheads since the beginning, compromise has to be the answer. The recipe may be less frequent, low-temperature burning; occasional hotter fires of limited extent; and a mosaic of regeneration stages — or acceptance of a loss of diversity and more or less stable landscapes which provide what the managers want for their animals.

Brigalow—Going, going, almost gone!

The brigalow lands of central Queensland and northern New South Wales[324] have suffered more alteration than any other extensive vegetation type. Originally brigalow occupied about six million hectares. Now one per cent remains in National Parks and small, hastily declared reserves which resulted from the realisation that it was almost too late to find any remnants to preserve.

Brigalow is *Acacia harpophylla*. It formed dense uniform forests, or mixed woodlands with *Casuarina cristata* (belah), *Acacia cambagei* (gidgee), eucalypts and bottle trees (*Brachychiton rupestris*). Brigalow was hard to clear and regenerated prolifically from suckers, so although it was known that the soils on which it grew were fertile, little clearing was done until the 1950s. Prickly pear (*Opuntia*) had invaded the brigalow lands in the 1920s and 1930s but had been eliminated by the biological control method of introducing a moth whose caterpillars literally ate it to death.

The introduction of bulldozers into the southern sector of the brigalow lands in the 1950s enabled clearing on a large scale to succeed, and when the Commonwealth and Queensland governments introduced the 'Brigalow Scheme' to clear the Fitzroy River region and northern brigalow lands, this vegetation type started to disappear. More than 100,000 hectares were cleared annually until the mid 1970s. This early clearing was mainly intended to increase the carrying capacity of the land by promoting the native pastures.

The greatest impetus to the clearing of remaining brigalow resulted from the successful introduction of exotic pasture species. Buffel grass (*Cenchrus ciliaris*) in particular has been so successful that it is still the most widely sown pasture. Between 1961 and 1976 over two million hectares were sown to pasture and a dramatic increase in stocking rates for cattle resulted. Since then, progressive clearing has been mainly for cropping and the areas under fodder and cash crops, particularly oil seeds and cereals, has increased.

Former distribution of brigalow

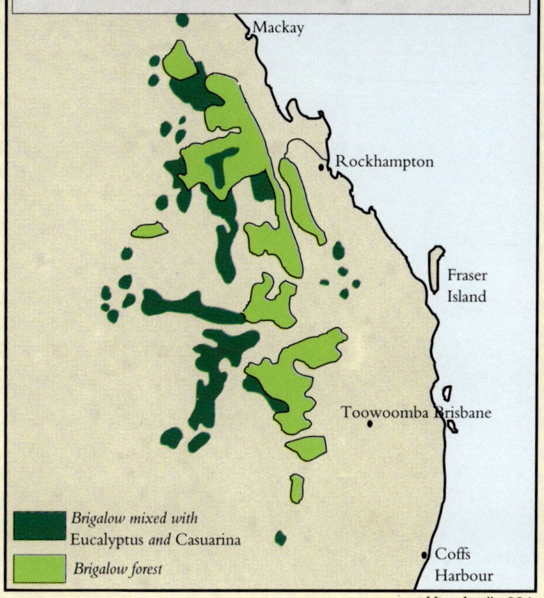

After Auslig 324

The brigalow lands' conversion to pasture and agriculture has been one of the few real success stories in which little or no land degradation results and sustainable high productivity is established. A combination of factors explains this success. The clay-rich soils are derived from old, deeply weathered shaly rocks, and vary from predominantly clay on the plains to duplex soils and red earths. In the Tara brigalow lands, for example, between the Moonie and Condamine River valleys, they represent part of a vast alluvial clay sheet of Quaternary origin. There is no problem with salinisation of the land after removal of the trees because, though the soils contain quite high levels of sodium chloride of ancient origin, the salt is distributed and held in the clays and does not accumulate in the watertable. Some of the soils have sulphates (gypsum), derived from oxidising of pyrite in the weathered marine sediments which underlie the region, and these are also constrained within the clays.

Brigalow forests have an understorey of species of essentially rainforest, not sclerophyll, nature. They represent a transitional vegetation type. Good rainfall and good soils, in areas which are not fire-excluding (as required by rainforest), determine the composition of the vegetation and the success of the use of brigalow lands for agriculture.

CHAPTER TWENTY-SIX

RAINFORESTS

THE GONDWANAN LEGACY: AN 'ARCHIPELAGO OF REFUGIA'

Rainforests are among the most complex ecosystems on Earth. Those in Australia are descendants of the ancient Gondwanan forest of the warm, humid Eocene 'golden age'. The newly freed island continent of 45 million years ago embarked on its independent journey northwards away from Antarctica with its wonderfully rich and diverse Gondwanan flora, its marsupials, birds, reptiles, amphibians and freshwater fish, and with a vast array of invertebrate creatures. Then, as now, there were plant assemblages for all environments and a mosaic of different communities from forests of different sorts with high diversity to specialised species or groups of species adapted to particular niches.

Because Australia lay in high latitudes and the gradient of temperature from the Equator to the Poles was comparatively slight, a great deal of the continent was probably well-watered and humid and forest of all sorts was probably widespread. The progressive drying of the continent while it drifted northwards and global climate changed towards an ice age saw the forests contracting as smaller and smaller areas remained suitable. Sclerophyll vegetation replaced ancient forest and became dominant over most of the land.

The composition of forests changed as elements that required perhumid and cool conditions, or needed good rainfall but could tolerate seasonal aridity, or species with other combinations of requirements, were sifted and sorted into their preferred environments. A warm temperate to subtropical climate prevailed during the time when the Gondwanan forests were evolving before Australia severed its connections with Antarctica and during the Early Eocene when it was first independent. For this reason it follows that the forests which occur today in areas that have similar warm-temperate to subtropical climates are closest in composition to their Gondwanan ancestors. They are the tall closed forests (complex mesophyll vine forests of Webb) and the closed forests of the World Heritage area of north Queensland.

It is a great tragedy that most of the coastal lowland tall closed forest growing on the rich alluvial and basaltic soils in the Innisfail region of Queensland has been destroyed and replaced by sugar cane. Such remnants as remain with giant sassafras (*Doryphora aromatica*) and other enormous trees which have escaped logging because they are in inaccessible places are reminders of the glory that has departed. This lowland forest type with trees more than 30 metres high and closed canopy corresponds to the rainforests of the Amazon in South America and of Zaire in tropical Africa, its Gondwanan cousins.

The other Gondwanan rainforests in Australia have trees 10 to 30 metres high and their content and structure vary. They form the 'archipelago' of island-like patches from Cape York down the eastern margin of the continent to Tasmania. Most of them occur in areas with a rainfall of at least 1200 millimetres a year, some having a dry season and others having rain evenly throughout. They occur on a wide range of soil types and at altitudes from sealevel to about 1200 metres. The cool temperate forests of Tasmania and southern

Victoria with *Nothofagus cunninghamii* dominating, the coastal vine forests, remnant *Nothofagus moorei* forest of the Border Ranges in the Lamington National Park, and the different forest types in Queensland all have their own rainforest magic.

Half of this unique closed forest vegetation type has been destroyed in the last 200 years, much of what remains has been logged (sometimes more than once) and is degraded. Botanically these are some of the most important forests in the world. They contain the greatest concentration of primitive flowering plants found anywhere, a fact which led to the belief that the cradle of flowering plant evolution was in Australia. The fossil record has shown that this is not the case and the ancestry of Australia's plants was almost entirely Gondwanan. It is the persistence of primaeval core forests in a direct line of descent from the Gondwanan ancestral forest here, and their isolation in the island continent that makes them special.

The closed forest ecosystem is an entity like a living organism with all its components interdependent and thus these marvellous forests are like living dinosaurs, aliens from a long-gone age, and deserving of complete protection. Their value to tourism alone in a world that is losing its rainforests at a rate that is threatening extinction of closed forest globally is inestimably greater than their commercial value as timber or woodchips for overseas markets. Their scientific value is beyond price.

Another forest type which is often classified as rainforest in Australia is the tall open forest where giant eucalypts dominate. These forests are perhaps better described as wet sclerophyll forests in order to distinguish them from 'true' rainforest which does not contain *Eucalyptus*. They occupy areas with at least 1000 millimetres of reliable rain in the southwestern corner of Western Australia, in Tasmania, Victoria and northern New South Wales. Outliers are found as far north as the Windsor Tableland in northern Queensland.

The wet sclerophyll forests are the optimum development of Australia's hardwood forest. The giant eucalypts are from 30 metres to 100 metres tall and few forests have escaped logging, sometimes many times repeated. Characteristically they tend to exist as even-aged stands because they are subject to devastating crown fires and regeneration after

DISTRIBUTION OF RAINFOREST AND TALL OPEN FOREST

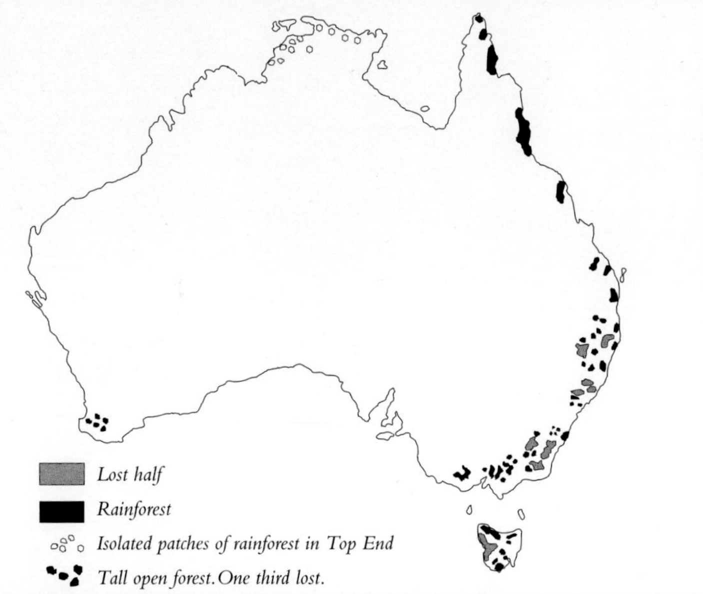

After Auslig 324

such events results in uniformity. (Early settlers in Victoria found out about the danger of horrific fires in the mountain ash (*Eucalyptus regnans*) forests which they were busily demolishing last century.)

The species of *Eucalyptus* dominating in the wet sclerophyll forests vary in different regions. In Western Australia the karri (*E. diversicolor*) and the jarrah (*E. marginata*) dominate; mountain ash forests are found in Tasmania and Victoria; forests of *E. delegatensis*, *E. obliqua* and *E. viminalis* occur in Tasmania, Victoria and New South Wales; mixed forests with *E. pilularis*, *E. saligna*, *E. grandis* and *E. propinqua* are found in New South Wales and Queensland. In total some 50,000 square kilometres of such forests remain, many heavily logged, few in their pristine state.

The understorey in the forests varies depending on local conditions like soil, regularity of rainfall and fire exclusion or regime. Most have a lower stratum of small trees and tall shrubs. In very moist situations, particularly in Tasmania and Victoria, tree-ferns are abundant and *Acacia melanoxylon* (the black wattle), *Pomaderris* (RHAMNACEAE), and *Olearia* and *Bedfordia* (daisy bushes, ASTERACEAE) are common. (Both the latter are interesting in that they have lignotubers, rare in the daisy family, to enable them to resprout after fire.) Palms occur in the northern forests and *Casuarina*, *Callicoma* (CUNONIACEAE) and *Syncarpia* (brush turpentine, MYRTACEAE) are

The wet tropical rainforests of northern Queensland are now largely protected by World Heritage Listing. They are the closest analogue of the original Gondwanan forest that was in Australia when it became an island continent.

Jim Frazier

widespread. Rainforest species occur as understorey in some forests which are transitional between wet sclerophyll and rainforest. In Tasmania some wet sclerophyll forests have emergent eucalypts and a closed canopy understorey of *Nothofagus*. These are forests in transition to beech forest.

On the Barrington Tops in New South Wales forests of *Nothofagus moorei* have emergent eucalypts. The eucalypts will die out if fire is excluded from these forests as they cannot regenerate under the closed canopy.

The wet sclerophyll eucalypt forests are products of major fires[297] and they present one of the most amazing and dramatic examples of fire-promoting, fire-requiring and yet fire-sensitive ecosystems. In this land where almost everything in the way of vegetation burns sooner or later under natural fire regimes, rainforest is the one exception. It occupies refugia where for reasons of topography and wetness fire is excluded. Fire defines its boundaries and under particular circumstances decreases the area occupied by damaging its margins. Areas like these margins and other places suitable for rainforest in terms of humidity and soils are sites for wet sclerophyll forest which is subject to infrequent but devastating fire.

The paradox of *Eucalyptus regnans* (the mountain ash) in particular being highly fire-sensitive, and the great flammability of the forest being the secret of its success is an amazingly Australian phenomenon. The forest needs a devastating fire every hundred, or maybe every two or three hundred, years in order to regenerate. There is no recruitment of seedlings except after an extraordinary fire because the light intensity on the forest floor is insufficient for seedlings to thrive. They are susceptible to fungus and insect attack, and the seed rain is barely more than satisfies the harvesting ants. This species does not produce seed which remains viable for any great length of time.

The ultimate, inevitable crown fire which kills also releases all the seed held in the small but obviously sufficiently protected and protective capsules, and the fire provides the conditions of soil and microbial activity that are required for germination. It has been calculated that 14,000,000 seeds per hectare can be shed after a crown fire. The litter of oil-rich leaves on the forest floor between fires inhibits the activities of nitrogen-fixing bacteria and when the litter is burnt and the composition of the soil micro-organism flora changes, nitrogen becomes available for rapid seedling growth. After a massive crown fire and total destruction of a forest at Noojee in Victoria, 2,500,000 seedlings per hectare of *E. regnans* germinated. They exhibited a superlative height-growth rate and an even stand of new trees was produced.

Crown fires almost seem to be promoted by the growth habit of the trees. Their ribbons of hanging bark make firebrands that can carry spot fires up to an incredible 30 kilometres ahead of the fire front; their open crowns favour the updrafts that promote crown fire; the very high content of volatile oils in the leaves makes them highly flammable; the abundant litter that accumulates on the forest floor feeds the initial fire.

Eucalypts in wet sclerophyll forest do not

The coastal rainforest of northern Queensland has mostly been cleared and replaced by sugar cane, as here at Tully where forest only survives on areas too steep to use.

JIM FRAZIER

Cool temperate rainforest with myrtle beech and abundant tree-ferns as seen here at Liffey Falls in Tasmania has none of the 'tropical' elements of the original Gondwanan forest from which it was sifted.

JIM FRAZIER

produce lignotubers from which to resprout after fire. Some species which have lignotubers when they are growing in lower, unreliable rainfall areas do not have more than rudimentary ones when growing under the uniformly wet conditions of the wet sclerophyll forest.

Wet sclerophyll forest....Phoenix forests, arising out of the ashes! Asking for fire when they are old — needing fire for rebirth. The end product, as it were, of the speciation of *Eucalyptus* that resulted from the increasing importance of fire as an environmental factor from the Late Tertiary onwards, and the rapid escalation of the process with the arrival of fire-stick farmers in an arid ice age-afflicted continent!

CHAPTER TWENTY-SEVEN

SCRUB-HEATHS, HEATHLANDS AND COASTAL VEGETATION

Heaths have a patchy distribution in coastal or near-coastal[324] situations all round Australia. They exist under various climatic regimes with widely different annual rainfall — from more than 1600 millimetres in Cape York to less than 300 millimetres in some places. They exist on a variety of soils but all the substrates have nitrogen, phosphorus, and often other element deficiencies.

Heathlands are so named because in the Northern Hemisphere they are characterised by the heathers, *Erica* (ERICACEAE), and other small branching shrubs with small and sclerophyllous 'ericoid' leaves. Plants of heathlands are typically subject to water stress, and in the heathlands of mediterranean climate regions they are adapted to cope with fire.

Western Australia's reputation as the Wildflower State rests on its sandplain heathlands. In many districts they are being managed by fire to promote maximum flowering for the tourist season. On the Swan Coastal Plain and as far north as Shark Bay, the broad zone of Quaternary sands of the Perth Basin bears these distinctive floras. Those of the Swan Coastal Plain near Perth have been almost entirely lost by the spread of the city from the coast to the Darling Scarp in a west to east direction and along the Plain to the north and south.

In the southwestern corner of the State and along the southern continental margin, where the Stirling Ranges and other mountains rise out of the sandplains, the shrub- and scrub-heathlands continue, and some remnants are left on the Yilgarn Block where wheatland has not swallowed up quite all the natural vegetation. There roadsides are often the only places left for wildflowers, and new and sensible policies demand that wide roadside strips are left for them to grow on and to act as wildlife corridors.

One of the most unusual and interesting vegetation types found anywhere is the kwongan open heath and scrub-heath[355] that occurs widely over Western Australia's mediterranean climate sandplains. It is more species-rich than rainforest and over large expanses its structure of low shrubs with emergent grass-trees gives a unique and primaeval appearance to the landscape. The Kalbarri National Park contains some of the best kwongan heath and a riot of *Verticordia* species and other wildflowers in season.

The mediterranean climate Southwest Province[354] of Western Australia has about 10,000 species of flowering plants and more than 70 per cent of them are endemics. Because so much of the well-watered, winter rainfall region has been cleared totally for wheat or affected in different ways by people, the number of extinctions of plant species is large and growing. Fifty-seven are officially extinct, 250 are endangered and about another 2000 are threatened.

The sandplain floras of the Southwest Province, in particular, are the product of evolution in isolation. This corner of the continent was like an island, separated from the rest by deserts. Because the extreme southwestern corner has had winter rains for about the last 5 million years and did not lose this pattern during the arid glacial stages, there has been a long time for the kwongan vegetation type to develop. The exposed continental shelves during low sealevel stands offered extra sandplain area for it at times when blown sand was making dunes over what is now most of its range. (As a result of this aeolian activity, particularly in the last glacial maximum between 18,000 and 14,000 years ago, 39 per cent of Western Australia is sandplain country.) The kwongan of the southwestern sector gives way to the *Acacia* scrub and spinifex of the Eremaean Province, and to the pindan of the Kimberleys northwards along the coastal strip. Pindan is savannah with scattered trees as upper storey[355], middle storey of *Acacia* and understorey of curly spinifex.

The kwongan is rich in PROTEACEAE, MYRTACEAE, FABACEAE (legumes) and EPACRIDACEAE (the Southern Hemisphere equivalent of ERICACEAE). The open heath form is found naturally in drier areas, and those with a better watertable below the sands carry eucalypt scrub and *Eucalyptus todtiana* trees and taller shrubs like *Grevillea leucopteris*. Frequent fire keeps shrub-heath in an open heath state, making it 'tidy' and much more photogenic for tourists.

There is a considerable risk that by imposing too frequent fires on it a number of the more fire-sensitive species are being lost and the seed-bank is declining. It is a fire-prone vegetation type and natural fire at unpredictable intervals is probably a requirement for maintenance of diversity. Many of the species recorded in it are only present in successional stages after fire. Its components are both sprouters and seeders. Fire eventually weakens the underground storage systems of the sprouters and they then depend on seed to re-establish. Seed may not be set if fire and drought coincide in bad seasons, and seedlings are fire-sensitive. So there is danger of a decline in diversity, of invasion by opportunistic weeds (like introduced South African grasses which are spreading on roadsides, and African daisies that turn the cultivated fields brilliant yellow) and of general degradation. Weakening of plants by fire heat and damage to the soil fungi which form a living protective shield in a zone of soil round root systems render the plants more liable to infections. The dreaded dieback, *Phytophthora cinnamomi,* attacks members of PROTEACEAE and its spread in the wildflower heathlands would be disastrous. It is already active in the Stirlings area.

The amazing species-richness of kwongan probably stems from the fact that its components are highly democratic. There are no dominants. Every space is filled with whatever one germinates, gets its roots down and establishes the right to exist. The niches are subtle and the complexity of interrelationships is enormous. The invisible mycorrhizal symbionts, the soil bacteria[317], the nitrogen-fixing organisms and the recycling of nutrients on the poor soils are the ultimate determinants of each plant's success.

Kwongan and sandplain plants show a wide range of adaptations and strategies which enable them to thrive on poor soils, under water stress and with recurrent fires.[317] Most have mycorrhizal fungi associated with their roots, some inside their tissues, others forming an interface between root hairs and the soil, to enhance nutrient uptake. Some have very deep roots to tap the water that lies in the water-mound type aquifers below the surface sands. Some have both deep tap roots and shallow surface ones which can utilise surface water when it is available. Many have lignotubers, tubers or bulbs from which they can resprout after fire and drought. Cycads have special nodular roots in a collar round the top of the root system containing cyanobacteria which fix nitrogen. Some plants, like *Nuytsia floribunda* the Western Australian Christmas tree, which is a member of the mistletoe family, are root parasites. Their roots tap into the roots of plants growing near them, everything from grasses to trees, and take water from them. Several species of *Olax* of family OPILIACEAE have a similar habit.

Three-quarters of the world's species of the fly-catching *Drosera* genus are found in Western Australia and many of them are in the kwongan and sandplain heaths. They obtain the nitrogen and phosphorus that is not

Roadsides in the kwongan heathland of Western Australia are like flower gardens in the spring.

M.E.W.

available from the soil from the insects that they digest. They have also devised ways of surviving drought by dying back to a resting bud which lies like a small silver spot on the dry soil until there is rain and it can come alive and develop into a new plant.

Where granite outcrops of the ancient Yilgarn Block rise out of the sandplains a 'resurrection plant' grows among the mosses and lichens. *Borya sphaerocephala* has crowded linear leaves in tufts which are desiccated and orange-yellow when dry and seemingly dead and which return to green and living with the first rain. It produces clusters of tiny white lily flowers.

Top left: Paperbarks fringe an estuary at Bremer Bay on the Western Australian southern coast.

M.E.W.

Top right: The Western Australian Christmas Tree Nuytsia floribunda *is a root parasite whose rootlets attach themselves to those of grasses and other plants growing around it.*

ELIZABETH GEORGE

Left: The ancient granite of the Yilgarn Block in Western Australia supports moss and lichen mini-gardens where the 'resurrection plant' Borya sphaerocephala *grows. It can remain desiccated and dormant when the shallow soil in the rock hollows is dry, and it revives rapidly after rain.*

M.E.W.

Drosera sewelliae, a sundew with brilliant flowers, obtains the phosphorous and nitrogen which it requires for healthy growth by digesting insects which it traps on the sticky drops exuded by its leaves. Western Australia has three-quarters of the world's described species of *Drosera*, as a consequence of the widespread sandplains with drastically impoverished soil and the richness of the specialised flora which evolved in isolation there.

A. LOWRIE

In South Australia east of Adelaide and in Tasmania there used to be extensive open heathland but it has been largely sown to exotic pastures and a great deal of fertiliser has been required to establish these. (One of Australia's major current problems, its water quality and the recent poisonous blooms of cyanobacteria in its major river systems, relates to the need for fertilisers to make soils capable of sustaining introduced pasture and crops. Not only the sandy soils of the heathlands are nutrient-deficient. Because of the geological history of the continent most soils have been leached. The nutrient-rich runoff from lands regularly fertilised causes enrichment of river water and microscopic organisms proliferate.)

Along the eastern margin of the continent heathland of scrub and shrub type characterises areas of sandstone rocks like the Hawkesbury Sandstone of the Sydney Basin; widespread coastal heath occurs wherever there was penetration of sea into low-lying regions during high sea stands during the Pleistocene, occupying the low-nutrient sands that resulted from these incursions. The extensive white sandplains near the coast between Port Macquarie and Forster are examples.

In river floodplains and estuaries the heathlands often merge into paperbark *Melaleuca* forests, sometimes of considerable extent. In northern Queensland extensive paperbark flood plains form a major vegetation type spreading inland from the Gulf of Carpentaria.

The largest mapped area of heathland-type vegetation in eastern Australia occurs in Cape York Peninsula. There from the dunes at Shelburne Bay heath extends into the centre of the Peninsula. The high rainfall and the conditions which apply in the Cape have produced unique vegetation with a unique fauna and this wonderful part of Australia is worthy of World Heritage listing and complete protection.

Recently there was a proposal to site a space base in the area and the environmental impact studies brought to light the special and largely unstudied nature of the biota. It also revealed that even a sparsely populated and relatively untouched piece of Australia such as this is suffering from the same problems as the rest.

The land has a very low carrying capacity and it is grazed by cattle which are doing damage in the sandplains and on the beach edges. Feral pigs abound, with very high concentrations in coastal areas, causing havoc in wetlands and damage over the whole of the Cape. Fire is used indiscriminately, including burning in the forests for no reason by Aboriginals. Even in protected areas, reserves and National Parks, Aboriginals are hunting with guns and killing wildlife that is protected from shooting by non-Aboriginal Australians. (This discrepancy between the rights of Australians, whether Aboriginal or not, is a matter which governments must have the fortitude to address. While traditional hunting practices, without modern weapons, in remote areas might be acceptable if there is no access to other food, it is totally unacceptable to allow rare and endangered creatures, like the dugong and sea turtles and many others, to be slaughtered at all.)

CHAPTER TWENTY-EIGHT

OUT OF CONTROL—THE FERAL PROBLEM

The feral animal problem in Australia today is out of control.[350] Rabbits, in particular, have proved to be better at desert-making than even humans are, and they and some of the other introduced animals which are running wild are steadily changing the face of the land. The feral grazers have a direct effect on vegetation and soil erosion and an indirect effect on wildlife; the feral predators have a direct impact on native animals.

RABBITS

The wave of rabbits that spread out over the continent in the 1880s,[351] 1890s and early this century[352] was an amazing phenomenon which may have no counterpart involving the spread of feral mammals in any other modern land. It is perhaps less surprising when seen in the context of Australia being a land that suits opportunists, and the rabbit being one of the best, as well as a generalist not restricted to any one food plant or niche. In those La Nina years that saw wheat growing on saltbush country in central South Australia, a rabbit swarm, like the army, marched on its stomach, devouring the countryside. (One wonders if perhaps in the geological past species which arose suddenly and possessed characteristics that made them successful opportunists in a changing environment suddenly became so numerous that they caused the same sort of havoc as rabbits have done and drove the original occupants to extinction.)

Rabbits had been introduced into Victoria in 1857. By 1880 about 8000 square kilometres of Victoria's mallee had been ravaged and abandoned. Rabbits were spreading northwards and outwards by as much as 100 kilometres a year. In 1887 it is recorded that 15 million were destroyed in New South Wales and the bluebush and saltbush country of northeastern South Australia had been turned into desert. In 1888, drought killed millions but in 1889 exceptional rainfall, like the 762 millimetres recorded in western New South Wales, set the scene for the big wave.

The rabbit plagues of 1890 defy description. Thirty-six million were counted on one property alone and the wave travelled up the rivers through the Centre, through the Simpson Desert to the floodplains of the Hay, Plenty and Finke rivers, and had reached the west coast by 1907. Millions had died in drought years, millions were born to replace them in wet years. Though first waves of rabbits crossed the Tropic of Capricorn, later the range was limited to south of this latitude except for the eastern sector of Queensland (except Cape York), for no clear reasons.

Rabbit fences were erected, every method was tried to kill the rabbits that were on farms, and drought was the only sure controller. The first breakthrough came with the successful spread of myxomatosis in 1950 (successful because the La Nina which created the rabbit plague also promoted the mosquito plague which transmitted the disease). About 90 per

Feral goats are extremely destructive. They browse as high as they can reach, ringbark by eating bark in hard times, and pull plants up by the roots, as well as causing erosion by breaking up the soil with their sharp hooves.

JIM FRAZIER

Goats stand on their hind legs to reach and eat leaves of shrubs.

JIM FRAZIER

cent of the population died, but control by this means was temporary because those with natural immunity were the survivors and they passed on their gene for survival of this pestilence to their offspring. That rabbit numbers have not reached the same mega-plague proportions since 1950 is probably because their ravaging of the landscapes has rendered the pasture less productive for the rabbits themselves.

Drought is still the only effective controller over large areas. Fumigating warrens, poisoning, shooting and trapping are all practised. If the warrens are ripped open during drought years when rabbit numbers are small, it has been found that predation by dingoes can keep the situation well under control for the following two years. The value of dingoes in keeping rabbits in check far outweighs their expense in terms of lambs and calves lost.

It is not only the amount each rabbit eats (and 100 eat as much as one cow; nine rabbits eat as much as one sheep) but how they graze that helps to make them the menace that they are. Chewing out the heart of tussocks, destroying the buds from which plants can regenerate, preventing the setting of seed, eating the roots and underground organs in droughts, climbing trees to eat the foliage

Above left: Feral pigs occupy bushy terrain where they can find shelter from the sun, and they prefer areas adjacent to farms with cultivated land and wetlands.

JIM FRAZIER

Left: A fox eating a road-kill kangaroo.

JIM FRAZIER

Feral cattle are found in many areas where they are within range of surface water or drinking troughs at bores.

JIM FRAZIER

when the ground is bare, and ring-barking trees by eating off their bark when all else has gone, enables them to use every last bit of food and do terrible damage. Because their plague years are followed by El Nino drought, and their warrens and scratchings for roots increase soil erosion, their impact is desert-making. Each plague, even with some recovery between, brings an additional degree of permanent degradation.

Vast barren areas that hardly produce any vegetative cover after rains now, like the Erldunda district of Central Australia, were well-vegetated not long ago. In that district residents remember rich saltbush and cottonbush (*Maireana aphylla*) plains with mulga and grasses, and the limestone ridges supporting healthy shrubs — before they were progressively degraded by rabbits and sheep. That the carrying capacity of native pastures and of sown pastures for cattle and sheep is impaired where rabbits are abundant goes without saying, and the more marginal the land the more pronounced is the effect of rabbits. (It was estimated in 1989 that financial loss to the pastoral industry as a result of rabbits was $90,000,000 and to South Australia alone about $17,000,000.) Damage to grain crops can also be substantial.

The impact of rabbits on native fauna is also devastating. By competing for rare green feed, altering the composition of pasture and destruction of habitat they are major contributors to the extinction of mammals in the arid Centre. Competition in the small refuge areas to which core populations of many animals retreat during drought is an important contributor to the loss of native species.

Two-thirds of the mammal fauna of the arid lands and deserts are endangered or extinct. The medium-sized marsupials have been worst affected with 67 per cent extinct and 26 per cent endangered. Predation by introduced foxes and cats also contributes to the tragic wildlife loss, and wedge-tailed eagles have been able to contribute to losses because of the increasingly easy access to prey that results as groundcover diminishes. In the Western Australian half of the Nullarbor a survey has shown that the decimation of native fauna has occurred there as well. Of thirty species of mammals (excluding bats) known to have been in the region, only five have been recorded in the last ten years. In western New South Wales, inland Queensland and northwestern Victoria 75 per cent of the native rodents are extinct. Rabbits have contributed to the loss and in many cases have been the major culprit.

Just as rabbits threaten the survival of, or cause the extinction of, native animals, so they inevitably cause equivalent effects on the land itself. The 'extinct' areas ravaged repeatedly are often beyond repair, the endangered areas need the complete removal of all grazers, feral and legitimate, and programmes of soil reconstruction and vegetation rejuvenation.

Goats

Feral goats are a major problem in Australia today.[350, 351] They have been creating a dustbowl desert in western New South Wales during the drought of 1990-1991 and have been in plague proportions in the Flinders Ranges and other rocky places in South Australia, The Mallee, and the eastern part of the Yilgarn Plateau in Western Australia.

Feral herds are mainly found in the southern half of the continent, extending as far as central southern Queensland. The damage that goats do to vegetation, particularly during drought, is extreme. They browse and eat the bark of shrubs and trees, ring-barking and killing in hard times; they graze whatever is there to be eaten and pull many plants up by the roots; they trample the soil with their sharp hooves, destroying the living duricrust and laying the soil open to erosion; they compete with sheep and cattle on native and sown pasture; they occupy the rocky ridges and gorges outback which are the refuges and life-sustaining places for native species; and they use the precious water that remains in waterholes in drought, trampling and degrading the surrounding terrain.

Because of their preference for rocky places and caves and overhangs they have ousted the yellow-footed rock wallabies from these safe havens in western New South Wales. The wallabies require the shelter to avoid the wedge-tailed eagles and dingoes that prey on their young, and loss of their habitat and competition with not only goats but also rabbits has made them an endangered species.

The great increase in goat numbers is a direct result of the destruction of the dingo in many areas. The feral goat population is a serious worry because not only are goats great desert-makers and endanger wildlife as well, but they are carriers of footrot disease which affects domesticated flocks. In addition they pose a serious threat should exotic diseases of sheep and cattle be introduced into Australia because they are susceptible to them and would be carriers and spreaders, making control almost impossible.

As potential carriers of exotic diseases they present a great danger to the nation's economy. They could carry and spread foot and mouth disease should that dreaded infection slip through the quarantine precautions that have so far kept it out.

PIGS

Feral pigs are the most serious pests of agricultural lands, as opposed to rangelands, in Australia.[350, 351] They are absent from areas which do not have surface water and fairly thick cover in which to shelter during the heat of the day. No one knows exactly how many feral pigs there might be in the continent because many of the places where they are most abundant are inaccessible. The number, however, is in the millions. In the Macquarie Marshes in New South Wales up to 80 per square kilometre have been counted; elsewhere in the swamps, river floodplains and agricultural lands which they occupy the number is lower — from about twelve in the wet season to two per square kilometre in the dry.

Their distribution covers the whole eastern third of the continent and across the tropical north with isolated populations in the Kimberley and coastal Western Australia. In agricultural areas they damage crops and pasture, damage fences, and are the greatest predator of lambs. In some areas up to 40 per cent of the lambs born in a season may be eaten by pigs, costing the sheep industry millions of dollars. Their wallowing and digging for roots in floodplains, swamps and stream edges causes serious damage, with soil erosion, silting of waterways and destruction of wetland habitat.

There seems to be no effective way of eliminating feral pigs. Their rate of multiplication in favourable seasons is great. In droughts often all the young die. The fact that pigs are such opportunistic eaters, consuming frogs, birds, easily-caught small animals, insects, carrion, bulbs and roots as well as their preferred diet of grass makes them great survivors. Shooting, trapping, poisoning, and rounding up from helicopters make no lasting impact on the problem.

HORSES

Herds of brumbies occur in many parts of Australia, with major concentrations round Alice Springs in the Centre; east of Birdsville in southwestern Queensland; in the northern part of the Northern Territory; along the edge of the Gulf of Carpentaria as far as the Mitchell River; on major river floodplains and in the alps of the southeastern part of the continent. In the Snowy Mountains they graze the high pastures in the summer and go down to the plains in winter.

Where the brumbies occur in grazing lands they compete with cattle. A major complaint is that they come to watering points in arid country and there compete for a precious resource in droughts and trample the area. In the arid regions where so many of them live they are regularly decimated by drought.

A great deal of emotion is stirred up by any culling operations and there seems to be no answer to the problem. In much of the arid land where they occur in low numbers the terrain they occupy is number-limiting because of poor food and restricted water, but under these conditions they are probably affecting the balance for native species.

DONKEYS

Feral donkeys are largely concentrated in the Kimberleys[351] and the Hamersley Ranges of Western Australia, the Northern Territory savannah, and in part of arid Central Australia. They occur in low density in pockets scattered elsewhere on the continent. They are grazers and browsers with a wide choice of food plants.

They tend to occupy remote areas and are well suited to arid lands. Because they compete for water and forage they must have an effect on native animals. They can increase herd sizes by up to 28 per cent per year in good times and spread to anywhere that has

Above: Feral horses are a problem in some areas, like the Mitchell grass plains of the Great Artesian Basin.

Jim Frazier

Right: Rabbits have contributed to turning this land into desert and their burrows cause soil erosion.

M.E.W.

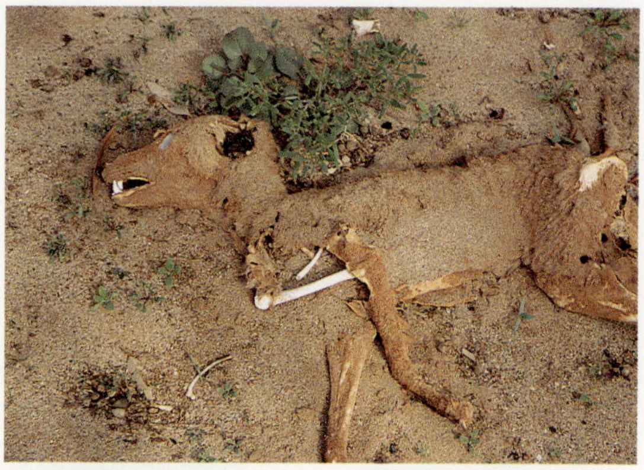

Top left: Cane toads are a spreading and deadly feral menace in Queensland.

JIM FRAZIER

Top right: Dingos are shot in their hundreds on some pastoral leases as on the property in the northern Flinders Ranges in South Australia where this young dog died and was mummified during a drought.

M.E.W.

Left: South African love grass and a species of Gladiolus *that is now rare and endangered in its country of origin thrive in the Western Australian coastal sandplains.*

M.E.W.

Below left: African daisies, now 'feral', colour the cultivated land in Western Australia.

M.E.W.

water, so overall there is potential for their occupation of increasingly large areas. Like other feral grazers they put heavy stress on pastures during dry times.

Thousands are shot from helicopters or on the ground and campaigns in the Kimberley region and the Victoria River district of the Northern Territory saw 76,000 killed between 1980 and 1982. This gives some idea of the size of the problem and the level of competition that donkeys pose for native fauna. They certainly pose a risk of degradation of vegetation because native pastures have been shown to be intolerant of more than infrequent grazing. In some of the very arid areas that they inhabit they are contributing to the advance of desert.

BUFFALO

The feral buffalo problem of the Top End[351] has recently been brought under control in a programme to eliminate brucellosis and tuberculosis from cattle. A very large proportion of the buffalo population has been shot and the remaining herds are mainly on farms or in the rocky parts of Arnhem Land, which is not their favourite wetland terrain.

There was concern about the environmental damage done by the very large numbers of buffalo that frequented the wetlands, swamps and floodplains. Their wallowing activities and the trampling of these frail ecosystems had caused much degradation. Where buffalo tracks had been cut between wetlands and estuaries, saltwater was invading and killing the paperbark trees and other freshwater plants.

Buffalo are grazers and browsers, camping in woodlands at night and grazing in the early morning and evening cool. They wallow during the heat of the day and thus their large numbers were concentrated in the limited surrounding areas with easy access to the water. Part of their grazing is on water plants and their rootstocks, so the overall impact of herds on vegetation of the wetlands and surrounding woodlands was considerable.

CAMELS

Who would ever have thought that Australia would be the only country in the world to have wild camels?[351] Maybe in time there will be a substantial export market to restock those parts of the Middle East that once were home to these amazing creatures — or is the day of the 'horse designed by a committee' over and its value as a circus, tourist and zoo oddity its only use?

Large tracts of the Simpson Desert, the area west of Alice Springs, and the Gibson and Great Sandy Desert, and even parts of the Great Victoria Desert adjacent to the western edge of the Nullarbor are camel country. Small concentrations also occur near some of the salt lakes of the arid zone. Most of the country occupied is spinifex-dune landscape with its arid shrublands and areas of saltbush and bluebush. Camels are selective browsers with a preference for shrubs and trees rather than grasses.

One might think that there is not much harm that could come from an animal adapted to similar dune deserts of other continents, and that the land occupied by camels in Australia is virtually useless as rangeland, so the feral camels can surely be left in peace (and perhaps farmed?). But this is an over-simplification and disregards the fact that Australia's deserts are not lifeless but support (or supported) a large population of different animals. Though it would be nice to think that extinctions in the desert are going to stop with the very large percentage of animals that have already disappeared there, perhaps it is time to accept that the demise of the rest is inevitable given the feral rabbits, donkeys, cats, foxes, fires, and droughts, and we should opt for camels, donkeys and anything that likes to live there while it is still habitable. This may be heresy, but we may find that even with plants we have to accept invasion by exotic species capable of colonising man-made desert because none of the original plants can do the job.

FERAL CARNIVORES: CATS AND FOXES

The feral carnivores that have a direct impact on native animals[351] may not be in the physical desert-making class but their role in degradation of the Australian biota helps to produce an empty wasteland. They, like the feral herbivores, are out of control. The extreme seriousness of the situation is becoming apparent. Where once there were flocks of budgies and hosts of other birds and small mammals round remote waterholes in uninhabited wilderness, now there are skeletons, and the last to die in many cases was a cat or a fox. Its last meal had been the last lone survivor of its predations.

The cat problem has crept up on us in the same way as a cat stealthily creeps towards its victim and surprises it with the end result. Suddenly there are many millions running wild all over the continent. Over vast areas of the semi-arid tropical north, the floodplains of the inland and areas adjacent to population centres and agricultural lands they are abundant. They tend to congregate in parts where rabbits are plentiful and their diet in such parts is up to 70 per cent young rabbits. They feed on small native mammals, up to the size of brush-tail possums, and to a lesser extent on birds, reptiles and invertebrates. In forested eastern Australia they prefer ring-tail possums, bush rats and small marsupials.

Domesticated cats also pose a threat to wildlife, because, as every truthful cat owner acknowledges, there is a wild thing inside the most docile cat.

Control of the feral cat problem or even the part-time-feral cat problem nearer to home is virtually impossible. It may be that the introduction of a horrible cat disease and the licensing and innoculation of all domesticated cats will be the necessary solution. There can be no room for sentiment when a rampant killer is at large.

Foxes, introduced for sport, now run wild over the southern half of Australia with much the same distribution as rabbits. They are virtually absent from areas where dingoes are in direct competition and in this respect the dingo fences that crisscross the country affect their distribution. Often these days travellers on the main highways see mostly foxes as road-kill. Foxes have contributed greatly to the extinction of small mammals and to the endangering of many others.

The list of feral creatures which are eating their way through the landscapes and decimating the native animal population does not end with the mammals. Cane Toads, introduced to 'control' cane beetles, on which they had no impact at all, are spreading in a manner not dissimilar to the rabbit. They are a deadly menace for native wildlife which has no resistance to their poison, and this is another feral animal out of control. The millions of its tadpoles, the remote ponds and swamps where it can breed unseen, make any sort of control virtually impossible.

In our rivers, introduced fish are ousting native species which, in the case of the Murray cod for instance, were already battling because of river pollution.

Among birds, the introduced starlings which form vast flocks over much of the land are displacing native birds by competing for food and for nesting sites; the Indian mynahs are taking over in suburbia; pigeons and sparrows in the cities.

Wild cattle on the beach at Temple Bay; a crocodile on a sandbank on the edge of the river, and the rainforest at the proposed space base site on Cape York.

JIM FRAZIER

CHAPTER TWENTY-NINE

Turning the Island Continent into a Desert Island

It is hard for most of us, the vast majority of the Australian population, in the comfort of our urban homes, in green gardens, with water on tap, to accept that our continent is in trouble. It may seem like an overstatement, and a highly emotive one at that, to suggest that our management of the continent is steadily turning more of it to unproductive desert and making the rest more fire-prone. We do not want to see that the environmental degradation we have caused has reached crisis proportions. We do not like to admit that we could be so accelerating the 'natural' processes which have been making a drier, more fiery and more salt-afflicted land that the time is not far off when the title of this chapter will no longer surprise us.

But a new awareness is obvious in our society and we are beginning to see that resources are not infinite, Nature is not unendingly forgiving and there is a price to pay for blundering on with the attitude that everything is here for our use and we have a god-given right to exploit and satisfy our greed.

Long ago, civilised people lost their connection with the web of life that supports them and maintains the planet as a life-supporting and life-enriching whole. Now, when gross global over-population is the basic problem from which most others stem, we are seeing the results that come from one species' dominance and the destruction of balance in the Earth's biota. In the Australian context we find we have a continent degraded by inappropriate use — its soils eroded, salt-affected or destroyed over large areas and its vegetation fast degrading in consequence; its precious water resources polluted; its native animals becoming endangered and extinct. We are taking notice at last because the degradation affects the economy, threatens our comfortable existence and in the case of water pollution (and atmospheric pollution in cities) affects our health.

James Lovelock's Gaia, the concept of the Earth acting like an enormous living organism with all its systems, organic and inorganic, in a state of dynamic balance, encapsulates for many of us the way we need to see the world and our place in it. A 'soft Gaian' approach, which an increasing number of scientists accept, removes Gaia from the realms of philosophy and theology with an implied consciousness that deliberately keeps conditions life-friendly — which was part of James Lovelock's original thinking. Accepting the basic tenets we see that the essential balances that control the system are the product of interaction of all living things and the environment through unimaginable lengths of time.

Only the immensity of geological time can have brought about the amazingly wonderful adaptations of living things to their individual environments and specialised niches and have woven them into a fabric that forms a living and dynamic whole. To fail to see and respect the importance of the interdependence of all the fragments from the sub-microscopic to the large macroscopic is a road to disaster. Gaia, the Living Earth, is only healthy when balance is maintained in and between all the systems it comprises.

It was my hope, when I set out to tell the story of Australia's relatively recent evolution from a green and forested piece of ancient Gondwana to the driest vegetated continent, that it would become obvious as the story unfolded why and how this continent is unique. This should enable each of us to understand why Australia has major difficulties today, and perhaps what might be the answers. Therefore such mention as has been made of its problems and the statistics of degraded land, salinisation and the rest are brief and in the context of their relation to the theme of a land and its biota evolving through time. Whole books have been written on soil erosion, rising salt, feral animals and the like. None to my knowledge has been written from the viewpoint of an earth scientist more involved in learning about the prehistoric world in order to understand the modern one. We need to see beyond the emotive and judgemental issues that cloud much of the fight for conservation and sustainable development.

With a perspective gained from understanding the prehistoric past, we see that the delicately balanced ecosystems in the three-quarters or more of Australia where unpredictability of climate and fire are paramount are a product of the geological past. The natural processes that had made this an arid land, the ENSO-related fluctuations of extreme climatic conditions and the natural fire that inevitably followed lightning strikes had determined a highly specialised biota. Being highly specialised is a disadvantage when environmental conditions are changed and patterns, even those into which unpredictability has been built, are altered. The introduction of the human species, their fire and their technology and their ability to make rapid changes to environments was bound to have serious effects in a land such as this.

The least acceptable part of the degradation and misuse of land has been that which results from careless and unintended side-effects of settlement. These range from garbage and toxic waste disposal to the enormous and seemingly unsolvable problem of feral animals.

The rabbits, goats, pigs, horses, donkeys, buffalo and camels constitute the feral herbivore problem, literally eating Australia to death with attendant damage to soils and impact on native fauna. All these feral animals are desert-makers. Their addition to the sheep and cattle that are doing the same thing in the marginal grazing lands (a large proportion of the rangelands) results in the physical advance of deserts into arid lands and of arid lands into semi-arid. The overall degradation of vegetation and soils in all the ecosystems involved is a cumulative process, a creeping malaise.

No longer does 'she'll be right when it rains', the perennial reply of pastoralists, apply. Anyone who has travelled along dingo fences in the dead centre of this land after rain can see how desert is made. The meagre response of vegetation on the protected side where dingoes roam and no sheep graze, and the lifeless side where grazing has destroyed the native seed-bank and there is nothing to regenerate the time-worn land, are stark reminders.

Over vast tracts of the inland deserts it is a case of 'where have all the flowers gone?', and where the desert still blooms we must protect ephemerals from grazing so that they may replenish their seed-banks and be there in other wet seasons for our children. For much of the land it is too late.

Our native wildlife is under siege, eaten or out-competed by invaders; dispossessed by the loss of its habitats; often disadvantaged by vegetation changes caused directly or indirectly by humans. Even the structure of invertebrate society is changed by aliens like the honey bee.

Much of the change we have to accept because we have no option and it is too late. Where we still have a chance to hold back the tide and even to win back a little of the ground we have lost we must fight with all our strength. A land without wombats; without wild free flocks of budgies wheeling over spinifex; without numbats busily scratching for termites in the forests of Western Australia to which they have been restored; without sugar gliders on *Banksia* flowers, would be for me a poor substitute for the wide brown land of my experience.

THE LAST PAGE

IF...

If you have travelled with me through the 160 million years of Australia's story of rifting, drifting and drying, and have arrived at this last page — thank you for being my companion on the journey.

If I have hurried you through parts of the scenery and happenings where you would have liked to take more time or to have explored further, or **if** I have stayed too long in parts of the route that you have found tedious, please forgive me. Voyages of exploration are very personal experiences and it was a mighty, and often awe-inspiring, journey. I was fascinated by the wonder of it all as it came alive for me and did not always remember that you were with me, step by step.

If the coming home to the present has been a let-down because we find ourselves surrounded by man-made problems that have sapped the strength of this time-worn land and threaten to impoverish it for ever, at least we have seen how it was, and how it may hopefully be again.

If now as much effort is put into restoring the health and vitality of our land as went into ravaging it in 200 years of inappropriate use that started in ignorance and has continued, to our shame, in spite of knowledge, it may not be too late.

'If only' are the saddest words of all.

Above: Nothofagus *forest remnant.* M.E.W

Opposite page: A river winds from the forest to the sea on Cape York—a region which should be protected as a National Park in its entirety. JIM FRAZIER

GLOSSARY

abyssal Of the deep ocean bottom.
aeolian Resulting from wind action.
albedo The ratio of the radiation reflected from a surface to the total radiation that is received by it.
alluvium Unconsolidated sedimentary materials transported by rivers — silts, sands, gravels, which may form fertile alluvial soil.
angiosperms The flowering plants, whose seeds are enclosed in a vessel.
Antarctic beech *Nothofagus.*
aquifer A porous layer of rock that is saturated and can supply water to springs or wells.
archipelago A chain of islands.
arkose A coarse-grained sandstone formed from the rapid disintegration of granite or gneiss.
artesian water Underground water which is stored under pressure.
azimuth The direction of the path in angular degrees from the South Pole.
barchan dune Crescent-shaped sand dune which lies transverse to the wind direction and has the horns of the crescent projecting down-wind.
basalt A fine-grained volcanic rock formed from a lava flow.
basin An area of continued subsidence of the Earth's crust where sediment accumulates over a long period.
bauxite An aluminium ore that results from the weathering of volcanic rock under tropical conditions and the action of bacteria on the weathered material.
biota The plant and animal life occupying a specific region and/or period of time.
blue-green algae Cyanobacteria.
braided streams Rivers which split into many meandering channels which separate and rejoin as they wander across flat terrain.
breakaway The edge of a plateau dissected by erosion.
calcarenite A sand cemented with lime. A coastal limestone.
calcrete Sedimentary rock cemented with calcium carbonate.
canopy The top layer of a closed forest.
chemical weathering Alteration, breakdown and decay of rocks by chemical alteration and leaching of components by percolating water.
chenopods Saltbush, samphires and bluebush. CHENOPODIACEAE.
circum-Polar current The ocean current that circles the South Pole.
clubmosses The common name for lycopods, a group of spore-bearing plants commonly known as fern-allies.
colloidal silica Silica in a gel state.
concretion A localised concentration of certain constituents of sedimentary rocks by aggregation around a nucleus — resulting in nodular structures.
conglomerate A cemented fragmental rock consisting of fragments of other rocks embedded in a finer matrix.
conifers Gymnosperms which bear cones.
continental extension Stretching of the continental crust during an early stage of rifting.
continent-ocean boundary The line on which continental crust meets basaltic sea-floor.
cosmic rays High energy particles of great penetrating power that bombard the Earth.
Cretaceous Period The geological Period between the Jurassic and the Tertiary, between 144 and 66.4 million years ago.
cryosphere Region of the Earth's surface that is permanently glaciated.
cryptogams Plants which reproduce by spores, not seeds. The algae, mosses, liverworts, lycopods, horsetails and ferns.
cuticle Waxy layer on the outer wall of epidermal cells.
cyanobacteria Also known as blue-green algae. Organisms which combine the characters of bacteria and plants, having simple cell structure yet possessing chlorophyll.
cycads A group of gymnosperms which retain some primitive features including motile sperm.
deep lead A palaeodrainage channel deposit.
deep weathering Weathering by leaching and chemical alteration of the regolith by groundwater.
deflation Blowing away by the wind.
depocentre An area into which drainage brings sediments.
detachment A shallow fault which results from stretching of continental crust during the early stages of rifting.
detritivorous Eating the products of breakdown of organic matter.
diatom Microscopic unicellular algal organism with a test made of silica.
diatomite A white, chalky rock made of diatom tests.
dinoflagellates Microscopic, unicellular algal organisms.
diploid An organism with a double chromosome number.
divide A drainage watershed.
dolerite Volcanic rock produced from magma that cooled sub-surface.
dolomite A carbonate mineral containing magnesium. Formed as an evaporite from seawater or by alteration of calcite rock by magnesium-charged solutions.
dropstones Rocks that were carried by glaciers or floating ice and deposited as erratics in unrelated sedimentary deposits.
duricrust An iron- or silica-cemented crust that now caps

landscape features, protecting them from erosion.

ecosystem A specific habitat and its living organisms which behave as an integrated unit.

endemics Organisms which occur only within a specific region.

ENSO The El Nino-Southern Oscillation climatic fluctuations which cause the drought and flooding rain syndrome in Australia.

Eocene An epoch of the Tertiary Period. From 58 to 36 million years ago.

ephemeral Short-lived or transitory.

epicormic buds Buds within the surface layers of tree trunks that enable resprouting after the crown of the tree has been damaged by fire.

epoch Subdivision of a Geological Period.

erosion A breaking down and wearing away of rocks by water, wind, ice, waves, etc.

escarpment A steep slope forming a linear feature, resulting from the erosion of gently inclined strata.

eustatic Related to worldwide changes in sealevel.

family In classification, the grouping of genera with similar characters.

fault A fracture along which rocks have been displaced in a horizontal, vertical or oblique sense.

feral Untamed or wild. (Organisms) which have been introduced and have become naturalised.

ferricrete Iron-cemented sediments.

flowering plants The angiosperms.

fluvial Spread by rivers.

Foraminifera An order of small unicellular marine organisms whose shells contribute to deepsea sediments.

forb Low-growing herb or small woody plant.

fynbos The heathland sclerophyll flora of mediterranean climate regions of South Africa.

genus, genera Unit(s) of classification into which species are grouped.

geomorphology A study of present-day landscapes and the explanation of their history.

gibbers Small scattered stones forming a surface covering on deserts.

ginkgophytes A class of gymnosperms of which *Ginkgo biloba*, the maidenhair tree, is the only survivor.

glendonites Carbonate mineral crystal structures which replaced other minerals under cold-water conditions.

graben A structure resulting from the subsidence of a strip of country between two normal faults.

gymnosperms A major division of the Plant Kingdom in which seeds are not borne in closed vessels. The conifers, cycads and ginkgophytes are living examples.

hardpan Clay-pan areas, like the swales between longitudinal sand-dunes in dune deserts.

herbivory Plant eating.

Holocene The geological epoch of the Quaternary Period, the last 10,000 years.

hot spot volcanoes Volcanoes that result from a weakness in the Earth's crust where molten magma is close to the surface.

inselberg A steep-sided eminence rising abruptly from a plain.

interfluvial Region between rivers.

Jurassic The geological Period between the Triassic and the Cretaceous in the Mesozoic era. From 208 to 144 million years ago.

karst Limestone country with etched surface, sub-surface rivers and caves.

lacustrine Of lake origin.

landscape inversion A feature produced by erosion where surfaces that were originally in the lowest part of a previous landscape (like ferricrete duricrusts formed at the watertable in palaeodrainages) now protect areas from erosion, leaving them proud in a landscape, capped by what was once lowest.

lateritisation Impregnation with iron.

leaching Process by which materials are removed from upper layers of soil or rock by percolating water and carried away in solution or suspension.

lichen An organism comprising a fungal body enclosing algal cells. A symbiotic arrangement.

lignite Brown coal with a high moisture content, intermediate between peat and bituminous coal.

lignotuber A woody underground structure which acts as a storage organ and provides buds which are a means of resprouting when the aerial parts of the plant are destroyed by fire or drought.

lithosphere The outer rigid shell of the solid Earth, overlying the less rigid asthenosphere. Comprises the crust (both continental and oceanic) and part of the mantle. About 50 to 100 kilometres thick.

loess A deposit resulting from the deposition of the fine-grained part of wind-blown sediments. In deserts the fines form the silts of claypans and the parna of swales; in glaciated areas, the loess is the finest grade rock flour.

lunette A crescent-shaped dune produced by wave or wind action on the downwind side of a lake or river.

maar lakes Lakes which develop in extinct volcanic craters.

magma Molten rock material originating in the Earth's mantle and crust and consisting of a liquid silicate base with or without a solid phase of suspended crystals. Magma cooled at the surface forms volcanic rock; cooled at depth it forms intrusive rocks.

magnetic excursion A deviation from the normal north-south polarity of the Earth's magnetic field, insufficient to cause a complete reversal.

magnetic reversal A periodic occurrence when the North Pole becomes the South Pole.

mallee habit Multi-stemmed growth of a eucalypt from a lignotuber.

Mediterranean climate Winter rainfall, dry hot summers.

mesa Flat-topped table-like hill, with at least three steep sides. Formed by dissection of a plateau, or, as in the case in Australian deserts by duricrust capping on elongated areas that are protected from erosion. (The narrow, elongated form here reflects the formation of the duricrust at an ancient watertable; the modern appearance being due to inversion of the landscape.)

Mesozoic era The Triassic, Jurassic and Cretaceous Periods.

mid-ocean ridges The linear volcanoes that produce basaltic sea-floor and widen the oceans that separate continents. About 80,000 kilometres of mid-ocean ridge mountains lie under the oceans.

Miocene An epoch of the Tertiary Period. From 23 to 5 million years ago.

'mixed flora' Ancient vegetation type in which there is a mixture of plants which have since been separated and sifted in modern floras to occupy tropical/subtropical and temperate

areas, and mesophyll and sclerophyll vegetation types.
nanoplankton Microscopic planktonic organisms (both plant and animal) which can only be seen under highest power microscopes. The minute shells of the organisms are an important component of deepsea sediments.
oceanic crust Basaltic seafloor, produced at mid-ocean ridges.
palaeo- Prefix denoting the geologically ancient.
Palaeocene The first epoch of the Tertiary Period. From 66.4 to 57.8 million years ago.
palaeomagnetism The record of the ancient polarity with which rocks were coded at the time of their formation.
Paleoflora mixta The mixed flora of the Late Cretaceous and Early Tertiary which had elements that have now been sifted into subtropical/tropical and temperate types, and sclerophyll and mesophyll vegetation.
pallid zone The white or very pale leached zone in a weathered profile.
parna Fine-grained aeolian sediment from the deflation of salt lakes.
Permian Period A geological Period of the Palaeozoic era, between the Carboniferous and the Triassic, from 286–245 million years ago.
photosynthesis The process by which plants with the green light-harvesting pigment chlorophyll synthesise organic compounds. Using sunlight as the energy source, they take carbon dioxide from the atmosphere and hydrogen from the water molecule, releasing its oxygen.
phyllode A leaf-like structure which is a modified stem.
pisolites Spherical, pea-like balls of ferricrete.
placer deposit Ore minerals carried by water and deposited and concentrated in an area.
Plate Tectonics The science of movement of plates of lithosphere by sea-floor spreading, resulting in continental drift.
plates Pieces of lithosphere on which the continents ride.
playas Salt lakes.
Pleistocene The first epoch of the Quaternary Period. From 1.6 million to 10,000 years ago.
Pliocene The last epoch of the Tertiary Period. From 5.3 to 1.6 million years ago.
podocarps Southern conifers of the family Podocarpaceae.
Polar reversals When the North and South Poles are switched.
polyploid Organism with a multiple of the usual chromosome number for the species.
potch Non-precious opal.
Primates The apes and monkeys, and ape-men and humans.
profile A vertical section through soil between the ground surface and parent rock showing its different layers (horizons).
proto- Prefix meaning the forerunner of.
Quaternary The geological Period comprising the Pleistocene and Holocene epochs. The last 1.6 million years.
radiocarbon dating Determining the age of rocks, using the rate of decay of radioactive isotopes.
refugia Refuges in which plants and animals are able to survive because conditions to which they are adapted are present in their micro-environments although surrounding areas are not suitable. If conditions change, recolonisation of wider areas can take place from the nuclei of refugia.
regolith The mantle of rock waste at the Earth's surface: weathered rock and soil overlying the bedrock.
saltbush Chenopod shrubs.
savannah Grassland with sparse overstorey of trees and shrubs.
scarp retreat The process by which the edge of a plateau or drainage divide migrates inwards due to erosion.
sclerophyll Plants with hard, leathery leaves, often with much reduced surface area, an adaptation to low-nutrient soils and water conservation.
sea-floor spreading The process by which new seafloor is generated on midocean ridges, resulting in the widening of oceans, and thus in continents moving further apart.
sedimentary rocks Rocks composed of sediments derived from the weathering of other rocks.
seif dune Linear dunes which stream down-wind in the direction of the prevailing wind.
silcrete Sedimentary rocks impregnated with silica.
southern beech *Nothofagus*.
southern conifers Araucariaceae and Podocarpaceae.
steppe Mid-latitude grasslands, treeless plains of middle Europe.
stomata The pores in the epidermis of leaves that allow gaseous exchange.
strewnfield The area over which ejected material is spread. Applied to the scattering of volcanic glass and tektites after asteroid and bolide impacts.
swale The hardpan strip of flat ground between dunes.
symbiosis An association for mutual benefit, as between a fungus and an alga in lichens, or a legume and nitrogen-fixing organisms.
tectonic Relating to processes which tend to build up various features of the Earth's crust — bending, warping and breaking; creating elevations and depressions.
Terminal Cretaceous Event The extinction event at the end of the Cretaceous 65 million years ago which saw the disappearance of the dinosaurs.
Terminal Eocene Event A sudden cooling about 35 million years ago which saw the start of a progression towards an ice age.
Terminal Miocene Event A dramatic cooling event at 6 million years ago which saw sealevels so much lowered that the Mediterranean sea was stranded and dried. This event may have initiated the evolution of the hominids from primate stock in Africa.
terrane A piece of continental crust attached to others whose origins are different.
Tertiary Period From 66.4 to 1.6 million years ago.
transpiration Loss of water vapour through the pores in leaves.
Triassic Period The geological Period between the Permian and the Jurassic Periods. From 245 to 208 million years ago.
troglobite An obligate cave-dweller. Animals which have adopted this life style lose their sight and become modified for life in the dark. (People who live in caves are known as troglodytes.)
water-mound aquifer A sand body which has a saturated zone acting as a water reservoir.
watertable The groundwater-saturated zone in the regolith.

REFERENCES

1: BMR Palaeogeographic Group: 1990: *Australia. Evolution of a Continent. Aust. Govt. Publ.* Canberra.

2: Coetzee, J.A. & Praglowski, I: 1984: Pollen evidence for the occurrence of *Casuarina* and *Myrica in* Tertiary of South Africa. *Grana* 23, 23-41.

3: Daniel, I.L., Lovis, J.D. & Reay, M.B.: 1990: A brief introductory report on the Mid-Cretaceous megaflora of the Clarence Valley, New Zealand. *Proc. 3 IOP Conf.* Melbourne

4: Dettmann, M.E.: 1989: Antarctica: Cradle of austral temperate rainforests? *in* J.A. Crame ed. *Origin and Evolution of the Antarctic Biota. Geol. Soc. Spec. Publ.* 47, 89-105.

5: Dettmann, M.E. & Thomson, M.R.A.: 1987: Cretaceous palynomorphs from the James Ross Island area, Antarctica. A pilot study. *Br. Antarct. Surv. Bull.* 77, 13-59.

6: Dettmann, M.E. & Jarzen, D.M.: 1988: Angiosperm pollen from uppermost Cretaceous strata of southeastern Australia and the Antarctic Peninsula. *Mem. Ass. Australas. Palaeontols.* 5, 217-237.

7: Dettmann, M.E. & Jarzen, D.M.: 1990: The Antarctic/Australian rift valley: Late Cretaceous cradle of northeastern Australasian relicts? *Rev. Palaeobot. Palynol.* 65, 131-144.

8: Dettmann, M.E. & Jarzen, D.M.: 1991: Pollen evidence for Late Cretaceous differentiation of Proteaceae in southern polar forests. *Canada J. Bot.* 69, 901-906.

9: Doyle, J.A.: 1977: Patterns of evolution in early angiosperms. *in* A. Hallam ed. *Patterns of Evolution.* Elsevier, Amsterdam.

10: Etheridge, M.A., Symonds, P.A. & Powell, T.G.: 1988: Application of the detachment model for continental extension to hydrocarbon exploration in extensional basins. *APEA J.* 167-187.

11: Etheridge, M.A., Symonds, P.A. & Lister, G.S.: 1990: Application of the detachment model to reconstruction of conjugate passive margins. *in* Tankard & Balkwill eds. *Extensional tectonics and stratigraphy of the North Atlantic margins. Mem. Am. Ass. Petr. Geol.*

12: Florin, R.: 1940: The Tertiary fossil conifers of South Chile and their phytogeographical significance, with a review of the fossil conifers of southern lands. *K. Sven. Vetensk. Handl.* 19, 1-107.

13: Frakes, L.A. & Francis, J.E.: 1988: Australian Cretaceous Ice. *Abs. 9th Aust. Geol. Conv.* Brisbane, p. 144.

14: Frakes, L.A. & Krassay, A.A.: 1992: Discovery of probable ice-rafting in the Late Mesozoic of the Northern Territory and Queensland. *Aust. J. Earth Sci.* 39, 115-119.

15: Francis, J.E.: 1986: Growth rings in Cretaceous and Tertiary woods from Antarctica and their palaeoclimatic implications. *Palaeontology* 29, 665-684.

16: Jefferson, T.H.: 1983: Palaeoclimatic significance of some Mesozoic Antarctic fossil floras. *in* R.L. Oliver *et al*, eds. *Antarctic Earth Science. Acad. Sci.*, Canberra.

17: Kemp, E.M.: 1978: Tertiary climatic evolution and vegetation history in the Southeast Indian Ocean region. *Palaeogeog. Palaeoclim. Palaeoecol. 24, 169-208.*

18: *Kemp, E.M.: 1981: Tertiary palaeogeography and the evolution of Australian climate. in* A. Keast ed. *Ecological Biogeography of Australia*, Junk. The Hague.

19: Mildenhall, D.C.: 1980: New Zealand Late Cretaceous and Cenozoic biogeography: a contribution. *Palaeogeog. Palaeoclim. Palaeoecol.* 31,197 233.

20: O'Neill, G.: 1992: We may have arrived 140,000 years ago. *The Age.* Melbourne.

21: Plumstead, E.P.: 1962: *Fossil floras of Antarctica. Sci. Rep. 9 Geology Trans-Antarctic Expedition 1955-1958.* London.

22: Powell, C.McA., Roots, S.R. & Veevers, J.J.: 1988: Pre-breakup continental extension in east Gondwanaland and the early opening of the eastern Indian Ocean. *Teconophysics* 155, 261-283.

23: Powell, C.McA., Johnson, B.D. & Veevers, J.J.: 1981: The Early Cretaceous breakup of Gondwanaland, the separation of Australia and India, and their interaction with South East Asia. *in* A. Keast ed. *Ecological Biogeography of Australia.* Junk, The Hague.

24: Quilty, P.G.: in press: The background: 144 million years of Australian palaeoclimate and palaeogeography. *in* R.S. Hill ed. *Australian Vegetation History: Cretaceous to Recent.* Camb. Univ. Pr.

25: Raine, J.I.: 1984: Outline of a palynological zonation of Cretaceous to Paleogene terrestrial sediments in west coast region, South Island, New Zealand. *Rep. N.Z. Geol. Surv.* 109, 1-82.

26: Rich, T.H. *et al.*: 1988: A high latitude terrestrial palaeoenvironment from the Early Cretaceous of southeastern Australia: Biotic and geochemical evidence for low temperatures and biotic diversity. *in* J.A. Crame ed. *Origins and Evolution of the Antarctic Biota.* London.

27: Rich, T.H. & Rich, P.V.: 1989: Polar dinosaurs and biotas of the Early Cretaceous of southeastern Australia. *Nat. Geogr. Res.* 5(1), 15-53.

28: Romero, E.J.: 1978: Palaeoecology and palaeofitogeography of cenophytic taphofloras from Argentina and vicinities. *Ameghiniana* 15,1-2; 209-227.

29: Romero, E.J.: 1986: Paleogene phytogeography and climatology of South America. *Ann. Missouri Bot. Gardn.* 73, 449-461.

30: Romero, E.J. & Archangelsky, S.: 1986: Early Cretaceous angiosperm leaves from southern South America. *Science* 234, 1580-1582.

31: Sheard, M.J.: 1990: Glendonites from the southern Eromanga Basin in South Australia: Palaeoclimatic indicators for Cretaceous ice. *Q. geol. notes, Geol. Surv. S.A.* 114, 17-23.

32: Stevens, G.R.: 1989: The nature and timing of biotic links between New Zealand and Antarctica in Mesozoic and early Cenozoic times. *in Origins and Evolution of the Antarctic Biota. Geol. Surv. Spec. Publ.* 47, 141-166.

33: Wilford, G.E. & Brown, P.J.: 1990: Maps of Late Mesozoic-Cainozoic Gondwana breakup: some palaeogeographic implications. BMR Canberra.

34A: Veevers, J.J. ed.: 1984: *Phanerozoic Earth History of Australia.* Oxf. Univ. Pr.

34: Veevers, J.J.: 1990: Tectono-climatic supercycle in the billion-year plate tectonic eon: Permian Pangean icehouse alternates with Cretaceous dispersed-continent greenhouse. *Sedimentary Geol.* 1-16.

35: Veevers, J.J.: 1991: Phanerozoic Australia in the changing configuration of proto-Pangaea through Gondwanaland and Pangaea to the present dispersed continents. *Aust. Syst. Bot.* 4, 1-11.

36: Veevers, J.J.: 1988: Seafloor magnetic lineation off the Otway/West Tasmania Basins: Ridge jumps and the subsidence history of the southeast Australian margins. *Aust. J. Earth Sci.* 35, 451-462.

37: Veevers, J.J.: 1987: Earth history of the Southeast Indian Ocean and the conjugate margins of Australia and Antarctica. *J. Proc. Roy. Soc. NSW* 120, 57-70.

38: Veevers, J.J.: 1990: Antarctica-Australia fit resolved by mapping of oceanic fracture zones. *Aust. J. Earth Sci.* 37, 123-126.

39: Veevers, J.J.: 1991: Mid-Cretaceous tectonic climax, Late Cretaceous recovery, and Cainozoic relaxation in the Australian region. *Geol. Soc. Aust. spec. publ.* 18, 1-14.

40: Veevers, J.J. & Ettreim, S.L.: 1988: Reconstruction of Antarctica and Australia at breakup (95+-5Ma) and before rifting (160Ma). *Aust. J. Earth Sci.* 35, 355-362.

41: Veevers, J.J. & McElhinny, M.W.: 1976: The separation of Australia from other continents. *Earth Sci. Rev.* 12, 139-159.

42: Veevers, J.J., Powell, C.McA. & Roots, S.R.: 1991: Review of seafloor spreading around Australia. 1. Synthesis of the patterns of spreading. *Aust. J. Earth Sci.* 38, 373-389.

43: Veevers, J.J., Stagg, H.M.J., Willcox, J.B. & Davies, H.L.: 1990: Pattern of slow seafloor spreading (less than 4 mm/year) from breakup (96Ma) to A20 (44.5Ma) off the southern margin of Australia. *BMR J. Aust. Geol. Geophys.* 11, 499-507.

44: White, M.E.: 1986: *The Greening of Gondwana.* Reed Books, Sydney.

45: White, M.E.: 1989: *The Nature of Hidden Worlds.* Reed Books, Sydney.

46: White, M.E.: 1992: *Time in Our Hands.* Reed Books, Sydney.

47: Benbow, M.C., Alley, N.F., Callen, R.A. & Greenwood, D.R.: 1992: Geological history and palaeoclimate. *in Geology of South Australia.* SADME.

48: Christophel, D.C. & Greenwood, D.R.: 1988: A comparison of Australian tropical rainforest and Tertiary fossil leaf beds. *Proc. Ecol. Soc. Aust.* 15, 139-148.

49: Christophel, D.C.: 1989: Evolution of the Australian flora through the Tertiary. *Plt. Syst. Evol.* 162, 63-78.

50: Christophel, D.C.: 1984: Early Tertiary Proteaceae: the first floral evidence for the Musgraveinae. *Aust. J. Bot.* 32, 177-186.

51: Christophel, D.C., Harris, W.K. & Syber, A.K.: 1987: The Eocene flora of the Anglesea locality. *Alcheringa* 11, 303-323.

52: Christophel, D.C. & Greenwood, D.R.: 1989: Changes in climate and vegetation during the Tertiary. *Rev. Palaeobot. Palynol.* 58, 95-109.

53: O'Dowd, D.J. *et al.*: 1991: Mite-plant associations from the Eocene of southern Australia. *Science* 252, 99-101.

54: Christophel, D.C. & Greenwood, D.R.: 1987: A megafossil flora from the Eocene of Golden Grove, South Australia. *Trans. Roy. Soc. S.A.* 111,3; 155-162.

55: Krieg, G.W. *et al* : 1991: Explanatory notes Curdimurka, S.A. *1:250,000 Geol. Series.* SADME

56: Christophel, D.C., Scriven, L.J. & Greenwood, D.R.: 1992: An Eocene megafossil flora from Nelly Creek, South Australia. *Trans. Roy. Soc. S.A.* 116, 2; 65-76.

57: Hill, R.S. & Christophel, D.C.: 1988: Tertiary leaves of the tribe Banksieae (Proteaceae) from southeastern Australia. *Bot. J. Linn. Soc.* 97, 205-227.

58: Rowell, A.I. & Christophel, D.C.: 1990: The dispersed cuticle profile of the Anglesea clay lenses. *Proc. 3 IOP Conf.* Melbourne.

59: Barrett, D.J. & Christophel, D.C.: 1990: The spacial and temporal components of two Australian Tertiary plant megafossil deposits. *Proc. 3 IOP Conf.* Melbourne.

60: Ambrose, D.J., Callen, R.A., Flint, R.B. & Lange, R.T.: 1979: *Eucalyptus* fruits in stratigraphic context in Australia. *Nature* 280, 387-389.

61: Christophel, D.C. & Lys, S.D.: 1986: Mummified leaves of two new species of Myrtaceae from the Eocene of Victoria, Australia. *Aust. J. Bot.* 34, 649-662.

62: Wilde, S.A. & Backhouse, J.: 1977: Fossiliferous Tertiary deposits on the Darling Plateau, Western Australia. *Ann. Rep. W.A. Geol. Surv.* 49-52.

63: Milne, L.A.: The occurrence of *Petrophile*-like pollen in the Eocene of Western Australia. (Thesis, unpublished.)

64: Milne, L.A.: 1988: Palynology of the Late Eocene lignitic sequence from the western margin of the Eucla Basin, Western Australia. *Mem. Ass. Australas. Palaeontols.* 5, 285-310.

65: Hill, R.S.: 1984: Tertiary *Nothofagus* macrofossils from Cethana, Tasmania. *Alcheringa* 10, 81-86.

66: Hill, R.S.: 1986: Lauraceous leaves from the Eocene of Nerriga, New South Wales. *Alcheringa* 10, 327-351.

67: Bigwood, A.J. & Hill, R.S.: 1985: Tertiary araucarian macrofossils from Tasmania. *Aust. J. Bot.* 33, 645-656.

68: Hill, R.S. & Macphail, M.K.: 1983: Reconstruction of the Oligocene vegetation at Pioneer, northeast Tasmania. *Alcheringa* 7, 281-299.

69: Hill, R.S.: 1983: *Nothofagus* macrofossils from the Tertiary of Tasmania. *Alcheringa* 7, 169-183.

70: Hill, R.S.: 1982: The Eocene megafossil flora of Nerriga, New South Wales, Australia. *Palaeontographica B* 181, 44-77.

71: Harris, W.K. & Twidale, C.R.: 1978: Revised age for Ayers Rock and the Olgas. *Trans. Roy. Soc. S.A.* 115,2; 109.

72: Twidale, C.R.: 1978: On the origin of Ayers Rock, Central Australia. *Z. Geomorph. N.F.* 31, 177-206.

73: Ettinghausen, C. von : 1883: Contributions to the Tertiary flora of Australia. *Akad. Wissensch.* Berlin.

74: Specht, R.L., Dettmann, M.E. & Jarzen, D.M.: 1992: Community associations and structure in the Late Cretaceous vegetation of southeast Australasia and Antarctica. *Palaeogeog. Palaeoclim. Palaeoecol.* 95.

75: Dettmann, M.E. *et al* : 1992: Australian Cretaceous terrestrial faunas and floras: biostratigraphic and biogeographic implications. *Cretaceous Research* 13.

76: Woolfe, K.J. & Francis, J.E.: 1991: An Early to Middle Jurassic glaciation. Evidence from Allan Hills, Transantarctic Mountains. *6 Int. Symp. Antarct. Earth Sci.* Japan.

77: McGowran, B.: 1990: Fifty million years ago. *American Scientist* 78,1; 30-39.

78: McGowran, B.: 1989: Evolution and environment in the early Palaeocene. *Mem. Geol. Surv. Ind.* 20, 21-53.

79: Christophel, D.C.: 1981: Tertiary megafossil floras of Australia as indicators of floristic associations and palaeoclimate. *in* A. Keast ed. *Ecological Biogeography of Australia* Junk, The Hague.

80: Mildenhall, D.C. & Pocknall, D.T.: 1983: Palaeobotanical evidence for changes in Miocene and Pliocene climate in New Zealand. *SASQUA Int. Symp.* Swaziland.

81: Ollier, C.D.: Global change and long term geomorphology. Unpubl.

82: Truswell, E.M. & Marchant, N.G.: 1986: Early Tertiary pollen of probable Droseracean affinity from Central Australia. *Spec. pap. Palaeontol.* 35.

83: Truswell, E.M., Sluiter, I.R & Harris, W.K.: 1985: Palynology of the Oligocene-Miocene sequences in the Oakvale 1 corehole, western Murray Basin, South Australia. *BMR J. Aust. Geol. Geophys.* 9, 267-295.

84: Truswell, E.M.: 1990: Australian Rainforests: the 100 million year record. *in* L.J. Webb & J. Kikkawa eds. *Australian Tropical Rainforests: science, value and meaning.* CSIRO Melbourne.

85: Ollier, C.D.: 1991: Laterite profiles, ferricrete and landscape evolution. *Z. Geomorph. N.F.* 35,2; 165-173.

86: Ollier, C.D.: 1988: Deep weathering, groundwater and climate. *Geogr. Ann.* 70A,4; 285-290.

87: Ollier, C.D.: 1991: Aspects of silcrete formation in Australia. *Z. Geomorph. N.F.* 35,2; 151-163.

88: Ollier, C.D.: 1988: The regolith in Australia. *Earth Sci. Rev.* 25 355-361.

89: Schmidt, P.W. & Ollier, C.D.: 1988: Palaeomagnetic dating of the Late Cretaceous to Early Tertiary weathering in New England, N.S.W., Australia. *Earth Sci. Rev.* 25, 363-371.

90: Ollier, C.D, Gaunt, G.F.M. & Jurkowski, I.: 1988: The Kimberley Plateau, Western Australia. A Precambrian erosion surface.

91: Ollier, C.D., Chan, R.A., Craig, M.A. & Gibson, D.L.: 1988: Aspects of landscape history and regolith in the Kalgoorlie region, Western Australia. *BMR J. Aust. Geol. Geophys.* 10, 309-321.

92: Senior, B.R.: 1979: Mineralogy and chemistry of weathered and parent sedimentary rocks in southwest Queensland. *BMR J. Aust. Geol. Geophys.* 4, 111-124.

93: Idnurm, M. & Senior, B.R.: 1978: Palaeomagnetic ages of Late Cretaceous and Tertiary weathered profiles in the Eromanga Basin, Queensland. *Palaeogeog. Palaeoclim. Palaeoecol.* 24, 263-277.

94: Twidale, C.R. & Milnes, A.R.: 1983: Aspects of the distribution and disintegration of siliceous duricrusts in arid Australia. *Geol. Mijnbouw.*

95: Eyre,E.J.: 1845: *Journals of expeditions and discovery into Central Australia and overland from King Georges Sound.* Boon, London.

96: Hall, G.C. & Kneeshaw, M.: 1990: Yandicoogina-Marillana pisolitic iron ore deposits. *in* F.E. Hughes ed. *Geology of the Mineral Deposits of Australia and Papua New Guinea.* AIMM Melbourne.

97: Bourman. R.P.: 1992: Modes of ferricrete genesis: evidence from southeastern Australia. *Z. Geomorph.* in press.

98: Twidale, C.R. & Campbell, E.M.: 1988: Ancient Australia. *Geo J.* 16,4; 339-354.

99: Harmsworth, R.A. *et al*: 1990: BIF-derived from ores of the Hamersley Province. *in* F.E. Hughes ed. *Geology of the Mineral Deposits of Australia and Papua New Guinea.* AIMM Melbourne.

100: Ollier, C.D.: 1977: Early landform evolution. *in* D.N. Jeans ed. *Australia: A Geography.* London.

101: Sluiter, I.R.K.: 1991: Early Tertiary vegetation and climate, Lake Eyre region, northeastern South Australia. *in* M.A.J. Williams, P. DeDekker, & A.P. Kershaw eds. *The Cainozoic in Australia: a re-appraisal of the evidence. Geol. Soc. Aust. Spec. Publ.* 18.

102: Jones, J.G. & Veevers, J.J.: 1982: A Cainozoic history of Australia's southeastern highlands. *J. Geol. Soc. Aust.* 29, 1-12.

103: Pain, C.F.: 1985: Morphotectonics of the continental margins of Australia. *Z. Geomorph. N.F.* 54, 23-35.

104: Coenraads, R.R. & Ollier, C.D.: 1991: Tectonics and landforms in the New England region. *in* W.S. Wilmott ed. *Regional and Economic Geology of the New England District. Geol. Soc. Aust. Qld. Div.* Brisbane.

105: Ollier, C.D.: 1982: The Great Escarpment of eastern Australia: tectonic and geomorphic significance. *J. Geol. Soc. Aust.* 29, 13-23.

106: Ollier, C.D.: 1982: Geomorphology and tectonics of the Dorrigo Plateau NSW. *J. Geol. Soc. Aust.* 29, 431-435.

107: Ollier, C.D. & Galloway, R.W.: 1990: The laterite profile, ferricrete and unconformity. *Catena* 17, 97-109.

108: Ollier, C.D.: 1986: The origin of alpine landforms in Australasia. *in* B. Barlow ed. *Origin and Evolution of Australasian Alpine Biota. Aust. Syst. Bot. Soc.*

109: Ollier, C.D. & Sutherland, J.A.: 1983: The Great Escarpment of eastern Australia. *Geogr. Bull.* 1983, 299-305.

110: Ollier, C.D.: 1984: Geomorphology and volcanism in eastern Australia. *Publ. Geol. Soc. Aust. NSW Div.* 1, 1-13.

111: Pickett, J.W. & Bishop, P.: 1992: Aspects of landscape evolution in the Lapstone Monocline area, New South Wales. *Aust. J. Earth Sci.* 39, 21-28.

112: Greenwood, D.R., Callen, R.A. & Alley, N.F.: 1990: The correlation and depositional environment of Tertiary strata based on macrofloras in the southern Lake Eyre Basin, South Australia. *SADME Rept. Book* 90/15.

113: Callen, R.A. *et al*: 1986: The Lake Eyre Basin - Cainozoic sediments, fossil vertebrates and plants, landforms, silcretes and climatic implications. *Australas. Sedimentols. Gp. Field Guide 4, Geol. Soc. Aust.*

115: Ollier, C.D.: 1985: Morphotectonics of passive continental margins. *Z. Geomorph. N.F.* 54, 1-9.

118: Ollier, C.D. & Taylor, D.: 1988: Major geomorphic features of the Kosciusko-Bega region. *BMR J. Aust. Geol. Geophys.* 10, 357-362.

119: Twidale, C.R. & Bourne, J.A.: 1978: Bornhardts developed in sedimentary rocks, Central Australia. *S.A. Geography* 6,1; 35-51

120: Harris, W.K. & Twidale, C.R.: 1991: Revised age for Ayers Rock and the Olgas. *Trans. Roy. Soc. S.A.* 115,2; 109.

121: Sweet, I.P. and Crick, I.H.: 1992: *Uluru and Kata Tjuta, a geological history*. AGSO, Canberra.

122: Barnes, L.C. & Townsend, I.J.: 1990: Opal deposits in Australia. *in* F.E. Hughes, ed. *Geology and Mineral Deposits of Australia and Papua New Guinea* AIMM Melbourne.

123: Ollier, C.D. & Pain, C.F.: 1992 in press: Tectonic and landscape evolution in Southeast Asia.

124: Hughes, R.J. & Senior, B.R.: 1973: Progress report on the geology of Bathurst Island, Melville Island, Coburg Peninsula and Fog Bay 1:250,000 sheet area, Northern Territory. *Bur. Miner. Res. Record* 1973/52.

125: Ollier, C.D.: 1991: A hypothesis about antecedent and reversed drainage. *Geogr. Fis. Dynam. Quat.* 14.

126: Barrett, D.: 1987: The Eocene flora of Golden Grove, South Australia. Thesis, Botany Dept. University of Adelaide, unpublished.

127: Alley, N.F. & Broadbridge, L.M.: 1992: Middle Eocene palynofloras from the One Tree Hill area, St Vincent Basin, South Australia. Alcheringa, in press.

128: van de Graaff, W.J.E., Crowe, R.W.A., Bunting, J.A. & Jackson, M.J.: 1977: Relict Cainozoic drainages in arid Western Australia. *Z. Geomorph. N.F.* 21,4; 379-400.

129: Kershaw, A.P.: 1985: A extended Late Quaternary vegetation record for north-eastern Queensland and its implications for the seasonal tropics of Australia. *Proc. Ecol. Soc. Aust.* 13, 179-189.

130: Alley, N.F.: 1990: Middle Eocene age of the megafossil flora at Golden Grove, S.A.: Preliminary report and comparison with the Maslin Bay flora. *Trans. Roy. Soc. S.A.* 111,4; 211-212.

131: Alley, N.F.: 1989: Preliminary palynological dating of macrofloras from Eyre Formation, Nelly Creek, Lake Eyre Basin. *Rept. Bk. 89/46, SADME.*

133: Benbow, M.C., Lindsay, J.M., Harris, W.K. & Cooper, B.J.: 1982: Latest Eocene marine incursion, northeast margin of the Eucla Basin. *S.A. Geol. Surv. Quart. geol. notes* 81, 2-9.

134: Woolfe, J. & Francis, J.E.: 1991: An Early to Middle Jurassic glaciation — evidence from Allan Hills, Transantarctic Mountains. *6th Symp. Antarct. Earth Sci. Nat. Inst. Polar Res.* Japan, September 1991.

135: Kamp. P.J.J.: 1992: Tectonic architecture of New Zealand. *in* J. Soons ed. *New Zealand, A Geography*. Christchurch.

136: Macphail, M.K. & Truswell, E.M.: 1989: Palynostratigraphy of the central west Murray Basin. *B.M.R. J. Aust. Geol. Geophys.* 11, 301-331.

137: Alley, N.F. & Benbow, M.C.: 1989: Late Eocene palynofloras from the Pidinga Formation SADME Ooldea Range 6, eastern Eucla Basin. *Quart. geol. notes Geol. Surv. S.A.* 111, 2-12.

138: Kemp, E.M.: 1976: Early Tertiary pollen from Napperby, Central Australia. *B.M.R. J. Aust. Geol. Geophys.* 1, 109-114.

139: Kershaw, A.P.: 1984: A prehistoric revolution. *in* P.S. Martin & R.G. Klein eds. *Quaternary Extinctions*. Univ. of Arizona Press. 691-707.

140: McNamara, K.J. & Scott, J.K.: 1983: A new species of *Banksia* (Proteaceae) from the Eocene Merlinleigh Sandstone of the Kennedy Range, Western Australia. *Alcheringa* 7, 185-193.

141: Frakes, L.A., McGowran, B. & Bowler, J.M.: 1987: Evolution of Australian environments *in* G.R. Dyne & D.W. Walton eds. *Fauna of Australia* Aust. Govt. Pr. Canberra.

142: Quilty, P.G.: 1980: Sedimentation cycles in the Cretaceous and Cenozoic of Western Australia. *Tectonophysics* 63, 349-366.

143: Kendrick, G.W., Wyroll, K.H. & Szabo, B.J.: 1991: Plio-Pleistocene coastal events and history along the western margin of Australia. *Quat. Sci. Rev.* 10, 419-439.

144: Lloyd, A.R.: 1968: Possible Miocene marine transgression in northern Australia. *Bur. Miner. Res. Bull.* 80, 85-100.

145: Martin, H.A.: 1989: Vegetation and climate of the Late Cainozoic in the Murray Basin and their bearing on the salinity problem. *B.M.R. J. Aust. Geol. Geophys.* 11, 291-299.

147: Stephenson, A.E. & Brown, C.M.: 1989: The Ancient Murray River System. *B.M.R. J. Aust. Geol. Geophys.* 11, 387-395.

148: Rozefelds, A.C. & deBaar, M.: 1991: Silicified Kalotermitidae (Isoptera) frass in conifer wood from a mid-Tertiary rainforest in central Queensland. *Lethaia* 24, 439-442.

149: Rozefelds, A.C.: 1991: Mid-Tertiary *Sarcopetalum* (Menispermaceae) from Glencoe, mid-eastern Queensland. *Alcheringa* 15, 145-149.

150: Basinger, J.F. & Christophel, D.C.: 1985: Fossil flowers and leaves of the Ebenaceae from the Eocene of southern Australia. *Canada J. Bot.* 63,10; 1825-1843.

151: Martin, H.A.: 1981: The Tertiary Flora. *in* A. Keast ed. *Ecological Biogeography of Australia*. Junk. The Hague.

152: Martin, H.A.: 1977: The history of *Ilex* (Aquifoliaceae) with special reference to Australia: evidence from pollen. *Aust. J. Bot.* 25, 655-673.

153: Martin, H.A.: 1982: Changing Cenozoic barriers and the Australian palaeobotanical record. *Ann. Missouri Bot. Gard.* 69, 625-667.

154: Martin, H.A.: 1978: Evolution of the Australian flora and vegetation through the Tertiary: evidence from pollen. *Alcheringa* 2, 181-202.

155: Campbell, J.D. & Holden, A.M.: 1984: Miocene Casuarinaceaean fossils from Southland and Central Otago, New Zealand. *N.Z. J. Bot.* 22, 159-167.

156: Pole, M.S., Holden, A.M. & Campbell, J.D.: 1989: Fossil legumes from the Manuherikia Group (Miocene) Central Otago, New Zealand. *J. Roy. Soc. N.Z.* 19,3; 225-228.

157: Campbell, J.D.: 1985: Casuarinaceae, Fagaceae and other plant megafossils from Kaikorai leaf beds (Miocene), Kaikorai Valley, Dunedin, New Zealand. *N.Z. J. Bot.* 23, 311-320.

158: Kovar, J.B., Campbell, J.D. & Hill, R.S.: 1987: *Nothofagus ninnisiana* (Unger) Oliver, from Waikato Coal Measures (Eocene-Oligocene) at Drury, Auckland, New Zealand. *N.Z. J. Bot.* 25, 79-85.

159: Vincent, W.F.: Microbial ecosystems of Antarctica. *in Geological Evolution of Antarctica.* Camb. Univ. Pr.

160: Holden, A.M.: 1981: Fossil *Nothofagus* from the Longford Formation, Murchison, New Zealand. *J. Roy. Soc. N.Z.* 12,1; 65-77.

161: Holden, A.M.: 1981: Fossil Lauraceae and Proteaceae from the Longford Formation, Murchison, New Zealand. *J. Roy. Soc. N.Z.* 12,1; 79-90.

162: Colhoun, E.A. ed.: 1988: Cainozoic vegetation of Tasmania. Handbook. *7th Int. Palynol. Conf.* Brisbane.

163: Hill, R.S. & Gibson, N.: 1986: Macrofossil evidence for the evolution of the alpine and subalpine vegetation of Tasmania. *in Flora and Fauna of Alpine Australia.* CSIRO, Melbourne.

164: Selkirk, D.R.: 1972: Fossil *Manginula*-like fungi and their classification. *Proc. Linn. Soc. NSW* 97, 141-149.

165: Pigram, C.J. & Davies, H.L.: 1987: Terranes and the accretion history of the New Guinea orogen. *B.M.R. J. Aust. Geol. Geophys.* 10, 193-211.

166: Brain, C.K.: 1985: Temperature-induced environmental changes in Africa as evolutionary stimuli. *in* E.S. Vrba ed. *Species and Speciation. Transvaal Mus. Monog.* 4, 45-52. Pretoria.

167: Brain, C.K.: 1981: The evolution of Man in Africa: was it a consequence of Cainozoic cooling? Alex. L. duToit Memorial Lecture. *J. Geol. Soc. Sth. Afr.* 84, 1-19.

168: Brain, C.K.: 1981: Hominid evolution and climatic change. *Sth. Afr. J. Sci.* 77, 104-105.

169: Hill, R.S., Harwood, D.M. & Webb, P.N.: 1991: Last remnants of Antarctica's Cenozoic flora: *Nothofagus* of the Sirius Group, Transantarctic Mountains. *Proc. Gondwana 8 Conf.*, Hobart.

170: Barrett, P.J., Adams, C.J., McIntosh, W.C., Swisher, C.C. & Wilson, G.S.: 1992: Radiometric ages support Antarctic deglaciation around 3 million years ago. *Nature* in press.

171: Holmes, W.B.K., Holmes, F.M. & Martin, H.A.: 1983: Fossil *Eucalyptus remains from the Middle Miocene Chalk Mountain Formation, Warrumbungle Mountains, New South Wales. Proc. Linn. Soc. N.S.W.* 106, 299-310.

172: Bowler, J.M.: 1982: Aridity in the Late Tertiary and Quaternary of Australia. *in* W.R. Barker & P.J.M. Greenslade eds. *Evolution of the Flora and Fauna of Arid Australia.* Peacock Publ.

173: Truswell, E.M. & Harris, W.K.: 1982: The Cainozoic palaeobotanical record in arid Australia: fossil evidence for the origins of an arid-adapted flora. *in* W.R. Barker & P.J.M. Greenslade eds. *Evolution of the Flora and Fauna of Arid Australia.* Peacock Publ.

174: Mildenhall, D.C.: 1980: New Zealand Late Cretaceous and Cenozoic plant biogeography: a contribution. *Palaeogeog. Palaeoclim. Palaeoecol.* 31, 197-233.

175: Holden, A.M.: 1983: Studies in New Zealand Oligocene and Miocene plant macrofossils. Ph.D. Thesis, Geology Dept., Victoria University, Wellington, N.Z., unpublished.

176: Ewart, A., Mateen, A. & Ross, J.A.: 1976: Review of mineralogy and chemistry of Tertiary central volcanic complexes in southeast Queensland and northeast New South Wales. *in* R.W. Johnson ed. *Volcanism in Australasia.* Elsevier. Amsterdam.

177: Archer, M., Hand, S.J. & Godthelp, H.: 1991: *Riversleigh. The story of animals in ancient rainforests in inland Australia.* Reed Books, Sydney.

178: Hope, J.H.: 1982: Late Cainozoic vertebrate faunas and the development of aridity in Australia. *in* W.R. Barker & P.J.M. Greenslade eds. *Evolution of the Flora and Fauna of Arid Australia.* Peacock Publ.

179: Murray, P.: 1984: *Australia's Prehistoric Animals.* Methuen.

180: Lamont, B.B., Hopkins, A.J.M. & Hnatiuk, R.J.: 1982: The flora — composition, diversity and origin. *in* J.S. Pate & J.S. Beard eds. *Kwongan, Plant Life of the Sandplain.* Univ. W.A. Press.

181: Benbow, M.C.: 1989: History of recognition and delineation of coastal aeolian landforms of the Eucla Basin, South Australia. *Q. geol. notes, Geol. Soc. S.A.* 109, 9-14.

182: Benbow, M.C.: 1990: Tertiary coastal dunes of the Eucla Basin, Australia. *Geomorphology* 3, 9-29.

183: Kiernan, K.: 1990: The extent of Late Cenozoic glaciation in the central highlands of Tasmania, Australia. *Arctic & Alpine Research* 22,4; 341-354.

184: Shafik, S.: 1990: The Maastrichtian and Early Tertiary record of the Great Australian Bight Basin and its onshore equivalents on the Australian southern margin. *B.M.R. J. Aust. Geol. Geophys.* 11, 473-497.

185: Bowler, J.M.: 1976: Aridity in Australia: age, origins and expression in aeolian landforms and sediments. *Earth Sci. Rev.* 12, 279-310.

186: Tapley, I.: 1990: Night thermal imagery reveals ancient drainage system of the Canning Basin. *in Exploration and Water.* W.A. Govt. Resource Review.

187: Ambrose, G.J. & Flint, R.B.: 1981: A regressive Miocene lake system and silicified strandlines in northern South Australia: implications for regional stratigraphy and silcrete genesis. *J. Geol. Soc. Aust.* 28, 81-94.

188: Aldam, R. & Kuang, K.S.: 1989: An investigation of the structures controlling natural discharge of artesian waters in the southwestern Great Artesian Basin. *Q. geol. notes. Geol. Surv. S.A.* 109, 1-8.

189: Kershaw, A.P.: 1990: Late Cainozoic vegetation of Australasia. *in* B. Huntley & T. Webb eds. *Vegetation History*. Junk, The Hague.

190: White, M.E.: 1974: Tertiary plant fossils from Melville Island, N.T. *Bur. Miner. Res. Record* 1974/5.

191: Prentice, M.L. & Denton, G.H.: 1988: The deep-sea oxygen isotope record, the global ice sheet system and hominid evolution. *in* F.E. Grine ed. *Evolutionary History of the 'Robust' Australopithicines*. Aldino de Gruyter.

192: Vrba, E.S., Denton, G.H. & Prentice, M.L.: 1989: Climatic influence on early hominid behaviour. *Ossa* 14, 127-156.

193: An Zhisheng, Bowler, J.M., Opdyke, N.D., Macumber, P.G. & Firman, J.B.: 1986: Palaeomagnetic stratigraphy of Lake Bungunnia: Plio-Pleistocene precursor of aridity in the Murray Basin, southeastern Australia. *Palaeogeog. Palaeoclim. Paleaoecol.* 54, 219-239.

194: Martin, H.A.: 1991: Tertiary stratigraphic palynology and palaeoclimate in the inland river systems in New South Wales. *in* M.A.J. Williams, P. DeDeckker & A.P. Kershaw eds. *The Cainozoic in Australia. Geol. Soc. Aust. spec. publ.* 18, 181-194.

195: Moore, R.F. & Simpson, C.J.: 1983: Image analysis — a new aid in morphotectonic studies. *Proc. 17th Int. Symp. Remote Sensing,* Ann Arbor.

196: Simpson, C.J.: 1990: Deep weathering, vegetation and fireburn: significant obstacles for geoscience remote sensing in Australia. *Int. J. Remote Sensing* 11,11; 2019-2034.

197: Wasson, R.J.: 1982: Landform development in Australia. *in* W.R. Barker & P.J.M. Greenslade eds. *Evolution of the Flora and Fauna of Arid Australia.* Peacock Publ.

198: Nix, H.: 1982: Environmental determinants of biogeography and evolution in Terra Australis. *in* W.R. Barker & P.J.M. Greenslade eds. *Evolution of the Flora and Fauna of Arid Australia.* Peacock Publ.

199: Noble, J.C.: 1982: The significance of fire in the biology and evolutionary ecology of mallee *Eucalyptus* populations. *in* W.R. Barker & P.J.M. Greenslade eds. *Evolution of the Flora and Fauna of Arid Australia.* Peacock Publ.

200: Kershaw, A.P., Band, J.G., D'Costa, D.M., Edney, P.A., Peterson, J.A. & Strickland, K.M.: 1991: A comparison of long Quaternary pollen records from the Atherton and Western Plains volcanic provinces, Australia. *in* M.A.J. Williams, P. DeDeckker & A.P. Kershaw eds. *The Cainozoic in Australia. Geol. Soc. Aust. spec. publ.* 18, 288-301.

201: Hill, R.S. & Macphail, M.K.: 1985: A fossil flora from rafted Plio-Pleistocene mudstones at Regatta Point, Tasmania. *Aust. J. Bot.* 33, 497-517.

202: Wasson, R.J.: 1989: Landforms. *in* J.C. Noble & R.A. Bradstock eds. *Mediterranean Landscapes in Australia. Mallee ecosystems and their management.* CSIRO, Melbourne.

203: Hope, G.: 1991: Quaternary Vegetation. *in* R.S. Hill ed. *Origins of the Australian Vegetation.* Camb. Univ. Pr.

204: Hope, G.: 1978: The Late Pleistocene to Holocene vegetational history of Hunter Island, northwestern Tasmania. *Aust. J. Bot.* 26, 493-514.

205: Hope, G. & Kirkpatrick, J.: 1988: The ecological history of Australian forests. *in* K.J. Frawley & N.M. Semple eds. *Australia's Ever-changing Forests. Dept. of Geography & Oceanography, Aust. Defence Force Acad. Spec. Publ. 1.*

206: McEwan Mason, J.R.C.: 1991: The Late Cainozoic magnetostratigraphy and preliminary palynology of Lake George, New South Wales. *in* M.A.J. Williams & P. DeDeckker eds. *The Cainozoic in Australia. Geol. Soc. Aust. spec. publ.* 18, 195-209.

207: Bowler, J.: 1990: The last 500,000 years. *in* N. McKay & D. Eastburn eds. *The Murray.* Murray-Darling Basin Commission, Canberra.

208: Chivas, A.R., DeDeckker, P., Nind, M., Thiriet, D. & Watson, G.: 1990: The Pleistocene palaeoenvironmental record of Lake Buchanan, an atypical Australian playa. SLEADS workshop.

209: McNamara, K.: 1983: *Pinnacles. W.A. Mus. Publ.* Perth.

210: Mabbutt, J.A.: 1988: Australian desert landscapes. *Geo J.* 16,4; 355-369.

211: Wasson, R.J.: 1983: The Cainozoic history of the Strzelecki and Simpson dunefields (Australia), and the origin of the desert dunes. *Z. Geomorph. N.F.* 45, 85-115.

212: Wasson, R.J.: 1983: Dune sediment types, sand colour, sediment provenance and hydrology in the Strzelecki-Simpson dunefield, Australia. *in* M.E. Brookfield & T.S. Ahlbrandt eds. *Eolian Sediments and Processes.* Elsevier, Amsterdam.

213: Wasson, R.J.: 1986: Geomorphology and Quaternary history of the Australian continental dunefields. *Geogr. Rev. Japan* 59,B1; 55-67.

214: Wasson, R.J.: 1983: Late Quaternary palaeoenvironments in the desert dunefields of Australia. *Proc. SASQUA Int. Symp.* Swaziland.

215: Wopfner, H. & Twidale, C.R.: 1988: Formation and age of desert dunes in the Lake Eyre depocentres in Central Australia. *Geolog. Rundschau* 77,3; 815-834.

216: Twidale, C.R. & Wopfner, H.: 1988: Dunefields. *in* M.J. Tyler, C.R. Twidale, M. Davies & C.B. Wells eds. *Natural History of the North East Deserts. Roy. Soc. S.A.* Adelaide.

217: Chen, X.Y. & Barton, C.E.: 1991: Onset of aridity and dune-building in Central Australia: sedimentological and magnetostratigraphic evidence from Lake Amadeus. *Palaeogeog. Palaeoclim. Palaeoecol.* 84, 55-73.

218: Baxter, J.L.: 1990: Heavy mineral sand deposits of Western Australia. *in* F.E. Hughes ed. *Geology of the Mineral Deposits of Australia and Papua New Guinea.* AIMM Melbourne.

219: Gardner, G.J., Mortlock, A.J., Price, D.M., Readhead, M.L. & Wasson, R.J.: 1987: Thermoluminescence and radiocarbon dating of Australian desert dunes. *Aust. J. Earth Sci.* 34, 343-358.

220: Pickett, J.W., Thompson, C.H., Kelley, R.A. & Roman, D.: 1985: Evidence of high sealevel during isotope stage 5C in Queensland, Australia. *Quat. Res.* 24, 103-114.

221: McKenzie, K.G. & Pickett, J.W.: 1984: Environmental interpretation of Late Pleistocene Ostracode assemblages from the Richmond River valley, New South Wales. *Proc. Roy. Soc. Vic.* 96,4; 227-242.

222: Galloway, R.W. & Kemp, E.M.: 1981: Late Cainozoic environments of Australia. *in* A. Keast ed. *Ecological Biogeography of Australia.* Junk, The Hague.

223: Fudali, R.F. & Ford, R.J.: 1979: Darwin glass and Darwin crater: A progress report. *Meteorites* 14,3; 283-296.

224: Ford, R.J.: 1988: An empirical model for the Australasian tektite field. *Aust. J. Earth Sci.* 35, 483-490.

225: Nanson, G.C., Young, R.W. & Price, D.M.: 1988: Stratigraphy, sedimentology and Late Quaternary chronology of the channel country of Western Queensland. *in* R.F. Warner ed. *Essays in Australian Fluvial Geomorphology*. Academic Press, Sydney.

226: DeDeckker, P.: 1986: What happened to the Australian aquatic biota 18,000 years ago? *in* P. DeDeckker & W.D. Williams eds. *Limnology in Australia*. Junk, The Hague.

227: Pickett, J.W.: 1981: A Late Pleistocene coral from Evans Head, New South Wales. *Alcheringa* 5, 71-83.

228: DeDeckker, P.: 1982: Holocene ostracods, other invertebrates and fish remains from cores of four maar lakes in southeastern Australia. *Proc. Roy. Soc. Vic.* 94,4; 183-220.

229: Lambeck, K.: 1990: Sealevel changes: Past and Present. *Proc. PACRIM 90 Congress*. AIMM.

230: DeDeckker, P, Cornege, T. & Head, J.: 1991: Late Pleistocene record of cyclic aeolian activity from tropical Australia suggesting the Younger Dryas is not an unusual climatic event. *Geology* 19, 602-605.

231: Currey, D.T.: 1964: The former extent of Lake Corangamite. *Proc. Roy. Soc. Vic.* 77,2; 377-387.

232: A-I *in Proc. CLIMANZ 1 and 3 Conferences, Canberra and Melbourne 1981 and 1987*. (Published 1983, 1989.)

232A: Hope, G.: 1989: Climatic implications of timberline changes in Australasia from 30,000 years B.P. to Present.

232B: Colhoun, E.A.: 1983: The climate of Tasmania 32 $\pm$ 5 KA.

232C: Lampert, R.J.: 1983: Kangaroo Island 18 $\pm$ 2 KA.

232D: Hays, J.D.: 1983: Evolution of the Antarctic Ocean.

232E: Hope, J.: 1983: The vertebrate record 15-10 KA in western New South Wales.

232F: Macumber, P: 1983: Hydrological evidence for Kow Swamp, northern Victoria 15-10,000 B.P.

232G: Ross, A.: 1983: An albedo map for Australia at 9000 B.P.

232H: Pittock, A,B. & Salinger, M.J.: 1983: The climatic optimum and a carbon dioxide-warmed Earth: The Australasian region.

232I: Chappell, J.: 1983: Sealevel changes: 0-40 K.A.

233: Colhoun, E.A.: 1985: Glaciation of the West Coast Range, Tasmania. *Quat. Res.* 24, 39-59.

234: Augustinus, P. & Colhoun, E.A.: 1986: Glacial history of the Upper Pieman and Boco Valleys, Western Tasmania. *Aust. J. Earth Sci.* 33, 181-191.

235: Colhoun, E.A. & van de Geer, G.: 1986: Holocene to middle last glacial vegetation history at Tullabardine Dam, Western Tasmania. *Proc. Roy. Soc. Lond.* B229, 177-207.

236: Shulmeister, J.: 1992: A Holocene pollen record from lowland tropical Australia. *in* E. Arnold ed. *The Holocene* 2,2: 107-116.

237: Arakel, A.V.: 1991: Evolution of Quaternary duricrust in the Karinga Creek drainage system, central Australian groundwater discharge zone. *Aust. J. Earth Sci.* 38, 333-347.

238: Tedford, R.H. & Wells, R.T.: 1990: Pleistocene deposits and vertebrate fossils from the 'dead heart of Australia'. *Mem. Qld. Mus.* 28,1; 263-284.

239: Schleiger, N.W.: 1982: Palaeowinds and the orientation of calcified tree remains and dune forests in coastal calcarenite, Bellarine Peninsula, Central Victoria. *Archaeometry*. A.N.U. Canberra. 137-143.

240: Jacobson, G., Arakel, A.V. & Chen, Y.: 1988: The central Australian groundwater discharge zone: Evolution of associated calcrete and gypcrete zones. *Aust. J. Earth Sci.* 35; 549-566.

241: Arakel, A.V. & Wakelin-King, G.A.: 1990: Chemical sedimentation and facies relationships in the Karinga Creek palaeodrainage system, Amadeus Basin, Northern Territory. *Palaeogeog. Palaeoclim. Palaeoecol.* 581-589.

242: Colhoun, E.A. & van de Geer, G.: 1987: Pleistocene macro- and micro-plant fossils from Rosebery, Western Tasmania. *Pap. Proc. Roy. Soc. Tas.* 121, 89-92.

243: Colhoun, E.A. & van de Geer, G.: 1987: Vegetation history and climate before the maximum of the last glaciation at Crotty, Western Tasmania. *Pap. Proc. Roy. Soc. Tas.* 121, 69-74.

244: Hutton, J.T., Prescott, J.R. & Twidale, C.R.: 1984: Thermoluminescent dating of coastal dune sand related to a higher stand of Lake Woods, Northern Territory. *Aust. J. Soil Res.* 22, 15-21.

245: Chappell, J.: 1991: Late Quaternary environmental changes in eastern and central Australia, and their climatic interpretations. *Quat. Sci. Rev.* 10, 377-390.

246: Torgersen, T, Luly, J., DeDeckker, P., Jomes, M.R., Searle, D.E., Chivas, A.R. & Ullman, W.J.: 1988: Late Quaternary environments of the Carpentaria Basin, Australia. *Palaeogeog. Palaeoclim. Palaeoecol.* 67, 445-461.

247: Wyroll, K.H.: 1979: Late Quaternary climates of Western Australia: evidence and mechanisms. *J. Roy. Soc. W.A.* 62,1-4; 129-142.

248: Nanson, G.C. & Young, R.W.: 1987; Comparison of thermoluminescence and radiocarbon age determination from Late Pleistocene alluvial deposits near Sydney, Australia. *Quat. Res.* 27, 263-269.

249: Pickett, J.W.: 1983: An annotated bibliography and review of Australian fossil sponges. *Mem. Ass. Australas. Palaeontols.* 1, 93-120.

250: Pickett, J.W.: 1982: *Vaceletia progenitor*, the first Tertiary Sphinctozoan (Porifera). *Alcheringa* 6, 241-247.

252: Barbetti, M.: 1980: Geomagnetic strength over the last 50,000 years and changes in atmospheric carbon-14 concentration: emerging trends. *Radiocarbon* 22,2; 192-199.

253: Kershaw, A.P.: 1981: Quaternary vegetation and environments. *in* A. Keast ed. *Ecological Biogeography of Australia*. Junk, The Hague.

254: Humphreys, W.F., Brooks, R.D. & Vine, B.: 1990: Rediscovery of the palm *Livistona alfredii* on the North West Cape Peninsula. *Rec. W.A. Mus.* 14,4; 647-650.

255: Humphreys, W.F. pers. comm.

256: Bowler, J.M.: 1981: Australian salt lakes. *Hydrobiologia* 82, 431-444.

257: DeDeckker, P., Kershaw, A.P. & Williams, M.A.J.: Past environmental analogues. *in* G.I. Pearman ed. *Greenhouse. Planning for climate change. CSIRO Div. Atmosph. Res.* Brill, Leiden.

258: Nicholls, N.: 1989: How old is ENSO? *Climatic Change* 14, 111-115. Kluwer.

259: Nicholls, N.: 1991: The El Nino/Southern Oscillation and Australian vegetation. *Vegetation* 91, 23-36.

260: Nicholls, N.: in press: Historical ENSO variability in the Australasian region.

261: Nicholls, N. & Wong, K.K.: 1990: Dependence of rainfall variability on mean rainfall, latitude and the Southern Oscillation.

262: Nicholls, N.: 1988: El Nino-Southern Oscillation and rainfall variability. *J. Climate* 1, 418-421.

263: Norman, F.I. & Nicholls, N.: 1991: The Southern Oscillation and variations in waterfowl abundance in southeastern Australia. *Aust. J. Ecol.* 16, 485-490.

264: Churchill, D.M.: 1973: The ecological significance of tropical mangroves in the Early Tertiary flora of southern Australia. *Geol. Soc. Aust. spec. publ.* 4, 79-86.

265: Cosgrove, R.: 1989: Thirty thousand years of human colonisation in Tasmania: New Pleistocene dates. *Science* 243, 1706-1708.

266: Orchiston, D.W.: 1981: Prehistoric man in the Bass Strait region. *SEE Australia* 2,3; 6-11.

267: Nanson, G.C., Young, R.W. & Stockton, E.D.: 1987: Chronology and palaeo-environment of the Cranebrook Terrace (near Sydney) containing artefacts more than 40,000 years old. *Archaeology in Oceania* 22, 72-86.

268: Anderson, I.:1991: Global warming rings true. *New Scientist* Sept. p.13.

269: Francey, R. & Cook, E.: 1990: Tasmanian tree rings: Prospects for palaeo-environment reconstructions. In press.

270: Cook, E., Bird, T., Peterson,M., Barbetti, M., Buckley, B., D'Arrigo, R., Francey, R. & Tans, P.: 1991: Climatic change in Tasmania inferred from a 1089-year tree-ring chronology of Huon Pine. *Science* 253, 1266-1268.

271: Strahan, R. ed.: 1983: *The Australian Museum Complete Book of Australian Mammals.* Angus and Robertson, Sydney.

272: Noble, J.C.: 1975: The effects of emus (*Dromaius novaehollandiae* Latham) on the distribution of Nitre Bush (*Nitraria billardieri* DC). *J. Ecol.* 63, 979-984.

273: Noble, J.C. & Whalley, R.D.B.: 1978: The biology and autecology of *Nitraria* L. in Australia. 1. Distribution, morphology and potential utilisation. 2. Seed germination, seedling establishment and response to salinity. *Aust. J. Ecol.* 3, 141-177.

274: Noble, J.C.: 1985: Fires and emus: The population ecology of some woody plants in arid Australia. *in Studies in Plant Demography* Academic Pr. London.

275: Symon, D.E.: 1984: A checklist of plants of Dalhousie Springs and their immediate environs. *J. Adelaide Bot. Gard.* 7,1; 127-134.

276: Barbetti, M.F. & McElhinney, M.W.: 1976: The Lake Mungo geomagnetic excursion. *Phil. Trans. Roy. Soc. Lond.* 281,1305; 515-542.

277: Hodgkinson, K.C.: 1990: The ecological basis of prescribed burning for shrub control in the semi-arid pastoral zone of Australia. *Proc. Int. Symp. Grassland Vegetation.* Science Press, Beijing, China.

278: Greene, R.S.B., Chartree, C.J. & Hodgkinson, K.C.: 1990: The effects of fire on the soil in the degraded semi-arid woodland. 1. Cryptogam cover and physical and micromorphological properties. *Aust. J. Soil Res.* 28, 60-70.

279: Hodgkinson, K.C.: 1991: Shrub recruitment response to intensity and season of fire in semi-arid woodland. *J. Applied Ecol.* 28, 60-70.

280: Hodgkinson, K.C., Gordon, W.T. & Smith, D.L.: 1982: A field device for integrating temperatures sensed in grass or forest fires. *Aust. J. Ecol.* 7, 403-409.

281: Hodgkinson, K.C.: 1986: Responses of rangeland plants to fire in water-limited environments. *in Rangelands; A Resource under Siege.* 437-441. *Proc. 2nd Int. Rangelands Conf.* Academy of Science, Canberra.

282: Noble, J.C., Harrington, G.N. & Hodgkinson, K.C.: 1986: The ecological significance of irregular fire in Australian rangelands. *in Rangelands: A Resource under Siege.* 577-580. *2nd Int. Rangelands Conf.* Academy of Science, Canberra.

283: Hodgkinson, K.C. & Oxley, R.E.: 1990: Influence of fire and edaphic factors on germination of the arid zone species *Acacia aneura, Cassia nemophila* and *Dodonea viscosa. Aust. J. Bot.* 38, 269-279.

284: Griffin, G.F. & Hodgkinson, K.C.: 1985: The use of fire for management of the Mulgaland vegetation in Australia. *in Proc. Mulga Lands Symp.* Brisbane. 93-97. *Roy Soc. Qld.*

285: Hodgkinson, K.C., Harrington, G.N., Griffin, G.F., Noble, J.C. & Young, M.D.: 1984: Management of vegetation with fire. *in Management of Australia's Rangelands.* 141-156. CSIRO, Melbourne.

287: Hodgkinson, K.C. & Harrington, G.N.: 1985: The case for prescribed burning to control shrubs in eastern semi-arid woodlands. *Aust. Rangeland J.* 7,2; 64-74.

288: Saxon, E.C. ed.: 1990: *Anticipating the Inevitable. A patch-burn strategy for fire management at Uluru (Ayers Rock-Mt Olga) National Park.* CSIRO.

289: Nicholls, N.: 1991: ENSO, drought and flooding rain in Southeast Asia. *in Proc. Conf. 'Towards a sustainable environmental future for the Southeast Asian region'.* Indonesia.

290: Braithwaite, R.: 1992: *Kapalga Research Station: Program for the ecology and conservation management of tropical forests and savannahs.* CSIRO, Darwin.

291: Brain, C.K. & Sillen, A.: 1991: Fire in the Swartkrans Caves. *Nature.* In press.

292: Hodgkinson, K.C. & Griffin, G.F.: 1982: Adaptation of shrub species to fires in the arid zone. *in* W.R. Barker & P.J.M. Greenslade eds. *Evolution of the Flora and Fauna of Arid Australia,* 145-152. Peacock Publ.

293-301 *in Fire and the Australian Biota* A.M. Gill, R.H. Groves & I.R. Noble eds., Academy of Science, Canberra.

293: Kemp, E.M.: 1981: Pre-Quaternary fire in Australia. 3-21.

294: Singh, G., Kershaw, A.P. & Clark, R.: 1981: Quaternary vegetation and fire history in Australia. 23-54.

295: Gill, A.M.: 1981: Post-settlement fire history in Victorian landscapes. 77-98.

296: Gill, A.M.: 1981: Adaptive responses of Australian vascular plant species to fires. 243-271.

297: Ashton, D.H.: 1981: Fire in the tall open forests (wet sclerophyll forests). 339-366.

298: Christensen, P., Recher, H. & Hoare, J.: 1981: Responses of open forest (dry sclerophyll) to fire regimes. 367-393.

299: Specht, R.L.: 1981: Responses to fires in heathlands and related shrublands. 395-415.

300: Suijdendorp, H.: 1981: Responses of the hummock grasslands of northwestern Australia to fire. 417-424.

301: Stocker, G.C. & Mott, J.J.: 1981: Fire in the tropical forests and woodlands of northern Australia. 425-439.

302: Blainey, G.: 1975: *Triumph of the Nomads*. Macmillan, Melbourne.

303: Kershaw, A.P.: 1986: Climatic change and Aboriginal burning in north-east Australia during the last two glacial/interglacial cycles. *Nature* 322, 47-49.

304: Giles, E.: 1889: *Australia Twice Traversed; five exploring expeditions into and through central South Australia and Western Australia, from 1872 to 1876*. Sampson Low, London.

305: Braithwaite, R.W.: 1991: Aboriginal fire regimes of monsoonal Australia in the 19th century. *Search* 22,7; 247-249.

306: Mitchell, T.L.: 1848: *Journal of an expedition into the interior of tropical Australia in search of a route from Sydney to the Gulf of Carpentaria*. London.

307: Sattler, P.S. ed.: 1986: *The Mulga Lands. Proc. Symp. Brisbane 1985. Roy. Soc. Qld. Publ.*

308: Noble, J.C.: 1989: Fire regimes and their influence on herbage and mallee coppice dynamics. *in* J.C. Noble & R.A. Bradstock eds. *Mediterranean Landscapes in Australia: Mallee ecosystems and their management*. CSIRO, Melbourne.

309: Noble, J.C.: 1984: Mallee. *in* G.N. Harrington, A.D. Wilson & M.D. Young eds. *Management of Australia's Rangelands*. CSIRO, Melbourne.

310: Noble, J.C.: 1989: Fire studies in mallee (*Eucalyptus* spp.) communities of western New South Wales: The effects of fires applied in different seasons on herbage productivity and their implications for management. *Aust. J. Ecol.* 14, 169-187.

311: Noble, J.C.: 1982: The significance of fire in the biology and evolutionary ecology of mallee *Eucalyptus* populations. *in* W.R. Barker & P.J.M. Greenslade eds. *Evolution of the Flora and Fauna of Arid Australia*. Peacock Publ.

312: Stafford Smith, D.M. & Morton, R.S.: 1990: A framework for the ecology of arid Australia. *J. Arid Environments*. 18, 255-278.

313: Smith, G.D., Lynch, R.M., Jacobson, G. & Barnes, C.J.: 1990: Cyanobacterial nitrogen fixation in arid soils of Central Australia. *FEMS Microbiol. Ecol.* Elsevier, Amsterdam.

314: Pate, J.S.: 1990: Seedling growth and storage characteristics of seeder and resprouter species of Mediterranean-type ecosystems of southwestern Australia. *Ann. Bot.* 65, 585-601.

315: Davidson, N.J., True, K.C. & Pate, J.S.: 1989: Water relations of the parasite:host relationship between the mistletoe *Amyema linophyllum* and *Casuarina obesa*. *Oecologia* 80, 321-330.

316: Pate, J.S.: 1989: Australian micro stilt plants. *Tree* 4,2; 45-49.

317: Lamont, B.: 1982: Mechanisms for enhancing nutrient uptake in plants, with particular reference to Mediterranean South Africa and Western Australia. *Bot. Rev.* 48,3; 597-689.

318: Van der Moezel, P.G. & Bell, D.T.: 1990: Saltland reclamation: selection of superior Australian tree genotypes for discharge sites. *Proc. Ecol. Soc. Aust.* 16, 545-549.

319: Griffin, G.F.: 1984: Hummock grasslands. *in* G.N. Harringtom, A.D. Wilson & M.D. Young eds. *Management of Australia's Rangelands*. 271-284. CSIRO.

320: Griffin, G.F.: 1990: Characteristics of three spinifex alliances in Central Australia. *J. Veg. Sci.* 1, 435-444.

321: Griffin, G.F. & Friedel, M.H.: 1985: Discontinuous change in central Australia: some implications of major ecological events for land management. *J. Arid Environments* 9, 63-80.

322: Morton, S.R.: 1990: The impact of European settlement on the vertebrate animals of arid Australia: a conceptual model. *Proc. Ecol. Soc. Aust.* 16, 201-213.

323: Nicholson, P.H.: 1981: Fire and the Australian Aborigine — an enigma. *in Fire and the Australian Biota*. Academy of Science, Canberra.

324: AUSLIG: 1990: *Atlas of Australian Vegetation*. Govt. Printer, Canberra.

325: Beale, B. & Fray, P.: 1990: *The Vanishing Continent*. Hodder & Stoughton.

326: Nance, C. & Speight, D.L. eds.: 1986: *A Land Transformed: Environmental change in South Australia*. Longman Cheshire.

327: Jacobs, S.W.L.: 1982: Relationships, distribution and evolution of *Triodia* and *Plectrachne* (Gramineae). *in* W.R. Barker & P.J.M. Greenslade eds. *Evolution of the Flora and Fauna of Arid Australia*. Peacock Publ.

328: Griffin, G.F.: 1991: Will it burn — should it burn?: management of the spinifex grasslands of inland Australia. *in* S. Chapman ed. *Desertified Grasslands: Their biology and management*. Linn. Soc. London.

329: Griffin, G.F., Price, N.F. & Portlock, H.F.: 1983: Wildfires in central Australian rangelands. *J. Environmental Management* 17, 311-323.

330: Stebbins, G.L.: 1950: *Variations and Evolution in Plants*. New York.

331: van Dijk, D.C.: 1980: Salt and soil reaction patterns in the Tara Brigalow lands, southeast Queensland. *Div. Soils. Rep.* 47. CSIRO, Brisbane.

332: Rogers, R.W.: 1982: Lichens of arid Australia. *J. Hattori Bot.* 53, 351-355.

333: Rogers, R.W.: 1972: Soil surface lichens in arid and subarid southeastern Australia: the relationship between distribution and environment. *Aust. J. Bot.* 20, 301-316.

334: Rogers, R.W.: 1974: Lichens from the T.G. Osborn vegetation reserve at Koonamore in arid South Australia. *Trans. Roy. Soc. S.A.* 98,3; 113-124.

335: Rogers, R.W.: 1971: Distribution of the lichen *Chondropsis semiviridis* in relation to heat and drought resistance. *New Phytologist* 70, 1069-1077.

336: Rogers, R.W.: 1971: Soil surface lichens in arid and subarid southeastern Australia: phytosociology and geographic zonation. *Aust. J. Bot.* 20, 215-227.

337: Rogers, R.W.: 1990: Soil surface biotic crusts on semi-arid soils. *in Conf. pap. Degradation of Vegetation in Semi-arid Regions: climate impact and implications.* Macquarie University.

338: Rogers, R.W.: 1989: Blue-green algae in southern Australian rangeland soils. *Aust. Rangeland J.* 11,2; 67-73.

339: Parr-Smith, G.A.: 1982: Biogeography and evolution in the shrubby Australian species of *Atriplex* (Chenopodiaceae). *in* W.R. Barker & P.J.M. Greenslade eds. *Evolution of the Flora and Fauna of Arid Australia.* Peacock Publ.

340: Anderson, D.J.: 1982: Environmentally adaptive traits in arid zone plants. *in* W.R. Barker & P.J.M. Greenslade eds. *Evolution of the Flora and Fauna of Arid Australia.* Peacock Publ.

341: Braithwaite, R.W.: 1988: Burning for conservation in the north. *N.T. Rural News. March 1988.* 31-32.

342: Lonsdale, W.M. & Braithwaite, R.W.: 1991: Assessing the effects of fire on vegetation in tropical savannas. *Aust. J. Ecol.* 16, 363-374.

343: Russel-Smith, J. & Dunlop, C.: 1987: The status of monsoon vine forests in the Northern Territory: a perspective. *in The Rainforest Legacy. Aust. Nat. Rainforest Study.* Govt. Printer.

344: Braithwaite, R. & Estbergs, J.: 1988: Fire-birds of the Top End. *Aust. Nat. Hist.* 22,7; 299-302.

345: Russell-Smith, J. & Bowman, D.M.J.S.: 1992: Conservation of monsoon rainforest isolates in the Northern Territory, Australia. *Biol. Conserv.* 59, 5-63.

346: Braithwaite, R.: 1987: Effects of fire regimes on lizards in the wet-dry tropics of Australia. *J. Trop. Ecol.* 3, 265-275.

347: Woinarski, J.C.Z.: 1990: Effects of fire on bird communities of tropical woodlands in northern Australia. *Aust. J. Ecol.* 15, 1-22.

348: Lovelock, J.: 1979: *GAIA: A New Look at Life on Earth.* Oxford Univ. Pr.

349: Lovelock, J.: 1988: *The Ages of GAIA: A biography of the living Earth.* Oxford Univ. Pr.

350: Rolls, E.C.: 1969: *They All Ran Wild.* Angus & Robertson, Sydney.

351: Weston, G., Dexter, N., O'Brien, P. & Bomford, M.: 1992: *Pest Animals in Australia.* Bur. Rural Resourc. & Kangaroo Press.

352: van Oosterzee, P.: 1987: Rabbits. The Desert Makers. *Aust. Geographic* 7, 90-101.

353: van Oosterzee, P.: 1991: *The Centre. The natural history of Australia's desert regions.* Reed Books, Sydney.

354: Erickson, R., George, A.S., Marchant, N.G. & Morcombe, M.K.: 1973: *Flowers and Plants of Western Australia.* Reed Books, Sydney.

355: Beard, J.S.: 1990: *Plant Life of Western Australia.* Kangaroo Press.

356: Bettenay, E.: 1986: Salt affected soils in Australia. *Reclam. Reveg. Res.* 5, 167-179.

357: Bell, A.: 1986: Evaporation from a 'dry' salt lake. *Ecos* 47, 16-18.

358: Arakel, A.V.: 1991: Land degradation — A geological perspective. *Australas. Sci. Mag.* 3, 52-57.

359: Evans, R., Brown, C. & Kellett, J.: 1990: Geology and Groundwater. *in* N. MacKay & D. Eastburn eds. *The Murray.* Murray-Darling Basin Comm.

360: Jacobson, G., Calf, G.E., Jankowski, J. & McDonald, P.S.: 1989: Groundwater chemistry and palaeo-recharge in the Amadeus Basin, Central Australia. *J. Hydrol.* 109, 237-266.

361: Kellett, J.R.: 1989: The Ivanhoe Block — its structure, hydrology and effect on groundwaters of the Riverine Plain of New South Wales. *B.M.R. J. Aust. Geol. Geophys.* 11, 333-353.

362: Evans, R.S.: 1989: Saline water disposal options in the Murray Basin. *B.M.R. J. Aust. Geol. Geophys.* 11, 167-185.

363: Lau, J.E. & Jacobson, G.: 1990: Aquifer characteristics and groundwater resources of the Amadeus Basin. *Bur. Miner. Res. Bull.* 236, 563-579.

364: Jacobson, G.: 1988: Hydrology of Lake Amadeus, a groundwater discharge playa in central Australia. *B.M.R. J. Aust. Geol. Geophys.* 10, 301-308.

365: Calf, G.E.: McDonald, P.S. & Jacobson, G.: 1991: Recharge mechanism and groundwater age in the Ti Tree Basin, Northern Territory. *Aust. J. Earth Sci.* 38, 299-306.

366: Macumber, P.: 1990: The salinity problem. *in The Murray.* CSIRO.

367: Tapley, I.J.: 1990: Relation between landform and soil parameters and NOAA-AVHRR derived temperatures across palaeochannels in the Great Sandy Desert region of Western Australia. *Proc. Remote Sensing Symp.* Perth.

368: McFarlane, M.J.: 1987: The key role of micro-organisms in the process of bauxitisation. *Mod. Geol.* 11, 325-344.

369: Ollier, C.D.: 1988: Morphotectonics of Papua New Guinea. *Z. Geomorph. N.F.* 69, 1-16.

370: McCulloch, M.T., Mortimer, G.E. & Xianhua Li: Application of U/TH dating using thermal ionisation mass spectrometry. in press.

371: deWit, M., Jeffrey, M., Bergh, H. & Nocolaysen, L.: 1988: *Geological Map of sectors of Gondwana.* Am. Ass. Petrol. Geol. & Univ. Witwatersrand.

372: Griffith, J.R.: 1971: Reconstruction of the South-West Pacific margin of Gondwanaland. *Nature* 234, 203-205.

373: Habermehl, M.A.: 1980: The Great Artesian Basin, Australia. *B.M.R. J. Aust. Geol. Geophys.* 5, 9-38.

INDEX:

Maps, Diagrams, Charts, Graphs

Maps

Diagrams, Charts, Graphs

INDEX

A

B

C

D

E

F

G

H

I

N

O

P

Q

R

S

T

U

V

W

X

Y

Z